Photoaffinity Labeling for Structural Probing Within Protein

Yasumaru Hatanaka • Makoto Hashimoto
Editors

Photoaffinity Labeling for Structural Probing Within Protein

Springer

Editors
Yasumaru Hatanaka
University of Toyama
Toyama
Japan

Makoto Hashimoto
Graduate School of Agriculture
Hokkaido University
Sapporo
Hokkaido
Japan

ISBN 978-4-431-56568-0 ISBN 978-4-431-56569-7 (eBook)
DOI 10.1007/978-4-431-56569-7

Library of Congress Control Number: 2017954302

Printed on acid-free paper

This Springer imprint is published by Springer Nature
The registered company is Springer Japan KK
The registered company address is: Chiyoda First Bldg. East, 3-8-1 Nishi-Kanda, Chiyoda-ku, Tokyo 101-0065, Japan

Foreword

Harnessing the Protein

The highly ordered molecular assembly of living systems represents an endless frontier for the application of chemical biology. A useful chemical approach to define specific molecular recognition is covalent cross-linking of a ligand and its receptor in an affinity-based manner. The chemical basis for affinity cross-linking is classified into two categories: ground-state reactions secure the residue-selective nature of affinity labeling and excited-state reactions generate highly reactive species to characterize the powerful feature of photoaffinity labeling. The major advantage for photoaffinity labeling is that the probe is inert before irradiation to prevent ground-state side reactions with surrounding molecules in a non-specific manner. Ideally, the irradiation immediately generates an extremely reactive species to complete specific and stable cross-links which comes in contact with any one of the residues located close by. Since Frank H. Westheimer originally introduced this unique idea of the method in 1962, photoaffinity labeling remains and should continue to be a principal chemical method for the identification of a particular target among the complex mixture of biomolecules.

Photoaffinity labeling has become increasingly important in association with the development of rationally designed powerful probes. Simply stated, probes are prepared by installing a photoactivatable functional group (photophore) on the framework of biological ligands. For target-specific labeling, probes are always needed to satisfy conflicting requirements. The probe structure should be close to the original ligand whereas the introduction of the photophore and detection tag cause an increase in the size of probes resulting in the decrease of affinity. For designing probes, the generated reactive species should react rapidly within the functional domain depending on the affinity between the ligand and the target but should not indiscriminately react with co-existing molecules.

This book is composed of 12 chapters that describe recent topics of photoaffinity labeling mainly by taking advantage of the use of diazirine photophore. Leading experts have written their chapters by focusing on the photophore's application for illuminating the interesting protein world. The first three chapters describe the rational design of efficient probes, followed by two chapters that consider the rapid and specific approach for the analysis of trace photo-labeled products. Chapters introducing synthetic, chemoselective, and genetic approaches for installing the diazirine photophore suggest new aspects of the methodology for the elucidation of bio-molecular assembly. Successful applications for glucose transporter and peroxisome represent the power of the methodology to analyze their structure and function. The last two chapters introduce the recent topic of photoaffinity labeling in the process of drug discovery and development.

Since the first report of photoaffinity labeling, the probe has been continuously improved through several important innovations to unleash the full power of the methodology. Now, photoaffinity labeling is well defined to ensure its specificity on labeled sites, which reveals the method to be an important approach of chemical biology for entering important areas of life science. An interesting application should be the structural probing of proteins that are difficult to crystallize. Photoaffinity labeling also could be a reliable strategy to reveal the molecular target during the early stage of drug discovery and development.

April 2017

Yuichi Kanaoka, Ph.D.
Toyama International Institution
Toyama, Japan

Hokkaido University
Hokkaido, Japan

Pharmaceutical Society of Japan
Tokyo, Japan

Contents

Chapter 1
Multifunctional Photoprobes for Identification of Ligand Sites Within Biomolecules

Makoto Hashimoto

Abstract The technique of photoaffinity labeling has become increasingly appreciated as a powerful methodology for post-genome field because it is one of the attractive methods to elucidate the interactions between bioactive ligand and biomolecule. The combinations with detection and isolation methods are essential to identify photolabeled components. There are several methods to introduce detection and isolation methods, which are so-called "tag," for photolabeled components from the photoaffinity label mixture. High detection limits of tag enable us to identify the photolabeled components. The introduction of detection and isolation tags in the ligand skeleton is one of the ways to archive identification of photolabeled components because the labeled components have been only introduced the tag. On the other hands, the specific biological interactions for target biomolecules are also utilized to identify photolabeled components. The chapter summarized that the several combinations of photoaffinity labeling and "tag" to study labeled components effectively.

Keywords Photoaffinity labeling • Chemical biology • Ligand binding sites • Radioisotope • Immunochemistry • Avidin–biotin

1.1 Introduction

Elucidation of protein functions based on the structure–activity relationship is one of the greatest interests for scientists to reveal the mechanisms of homeostasis functions in life. In the human body, many proteins are activated and/or inactivated by various ligands to maintain homeostasis. Understanding the mechanism

---

M. Hashimoto
Division of Applied Science, Graduate School of Agriculture, Hokkaido University, Kita 9, Nishi 9, Kita-ku, Sapporo 060-8589, Japan
e-mail: hasimoto@abs.agr.hokudai.ac.jp

Y. Hatanaka, M. Hashimoto (eds.), *Photoaffinity Labeling for Structural Probing Within Protein*, DOI 10.1007/978-4-431-56569-7_1

of molecular interactions between small bioactive ligands and proteins is an important step in rational drug design and discovery. For these purposes, a genetic approach provides an efficient and indirect route for pinpointing functional amino acids within proteins by deleting or mutating the native amino acid alignment of proteins. When the target protein can be expressed in a large quantity, 3D structural determinations which include NMR spectroscopy and homology modeling are a powerful and direct approach for analyzing the three-dimensional structure of proteins at the atomic level. Chemical methods, which are fundamental in chemical biology, provide an alternative route for the direct identification of target proteins in biomolecule mixtures as well as their ligand binding site structure because these analyses are based on the affinity between the ligand and target protein. Affinity-based chemical modification introduces a useful tag for analyzing the target protein. For example, single molecular imaging with fluorophore (Wallrabe and Periasamy 2005) visualizes target biomolecules in complex systems for imaging the localization of biomolecules and processing the flow of bioactive compounds in cell compartments. For specifically attaching the desired tag on the target protein, affinity labeling (Wold 1977) has a limited role because the method requires the presence of nucleophilic residues near to a ligand binding site for preventing nonspecific introduction of the tag to a site different from the binding site or on other coexisting biomolecules. But recent progress of molecular biology prompted site-directed labeling using cysteine mutants to overcome this problem (Foucaud et al. 2001).

Photoaffinity labeling (PAL) (Brunner 1993; Kotzyba-Hibert et al. 1995; Hatanaka et al. 1996a, b; Dormán and Prestwich 2000; Hashimoto and Hatanaka 2008a) largely increased the capability of specific tagging. Photochemically generated highly reactive species (Chap. 2) introduce covalent bonds between ligand and protein in a nonselective manner, and any amino acid in the binding site can be tagged.

One of the successful applications of photoaffinity labeling is the identification of target biomolecules in crude extracts using radioisotope-labeled probes as highly sensitive detection tags. The covalent bond fixes the tag to the contact point even though affinity was destroyed by the denaturation condition, which allows further sophistication in detailed structural analysis, and the detection of tagged components may reflect the status of conformational changes of target molecules.

The other future is to identify the labeled regions of target biomolecules, and combinations with specific purification methods for labeled components are needed (Hatanaka et al. 1996a, b; Hashimoto and Hatanaka 2008a, b; Tomohiro and Hatanaka 2014; Hatanaka 2015).

Here we briefly describe the fundamentals of several concepts.

1. PAL with radioisotope tag
2. PAL with immunological methods
3. PAL with avidin–biotin system

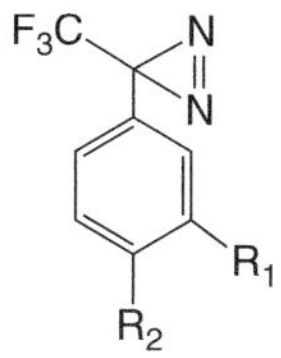

1 R_1 = H, R_2 = CT_2OH
2 R_1 = T, R_2 = COOH
3 R_1 = OCHTCOOH, R_2 = H
4 R_1 = OCT_3, R_2 = $COOCH_3$
5 R_1 = H, R_2 = ^{14}COOH
6 R_1 = NO_2, R_2 = $OCH_2{}^{14}COOH$
7 R_1 = $O^{14}CH_3$, R_2 = COOSu
8 R_1 = ^{125}I, R_2 = H
9 R_1 = ^{125}I, R_2 = CH_2OH
10 R_1 = NCS, R_2 = ^{125}I

Fig. 1.1 Radioisotope contained phenyldiazirines. **1** (Latli et al. 1998), **2** (Ambroise et al. 2001), **3** (Rimoldi et al. 1993), **4** (Hatanaka et al. 1994a, b), **5** (Bender et al. 2007), **6** (Hatanaka et al. 1992, 1994a, b), **7** (Hatanaka et al. 1995), **8** (Brunner and Semenza 1981), **9** (Blanton et al. 1998, 2000), **10** (Dolder et al. 1990)

1.2 PAL with Radioisotope Tag

Radioisotope is one of the highest detection methods for low amounts photolabeled component (Filer 2009). Tritium (Latli et al. 1998; Ambroise et al. 2001; Rimoldi et al. 1993; Hatanaka et al. 1994a, b), ^{14}C (Bender et al. 2007; Hatanaka et al. 1992, 1994a, b, 1995), and ^{125}I (Brunner and Semenza 1981; Blanton et al. 1998, 2000; Dolder et al. 1990) were well selected for these purposes. Several radioisotope-labeled phenyldiazirine derivatives are summarized in Fig. 1.1. The introduction of radioisotope elements must be considered in the late synthetic stage. Researchers have to perform pre-experiment with unlabeled reagents to well-handled hole of experiments. Major withdraws to utilize radioisotope are limited amount of radiolabeled reagents for derivatizations. It was well observed that chemical synthetic yields in radioisotope are less than twice than in unlabeled synthetic methods because less amount reagents are available in radioisotope methods. For example, methylation of diazirinyl phenolic alcohol with methyl iodide was very effective in unlabeled reaction but was less than 30% with radiolabeled methyl iodide to afford 4 (Hatanaka et al. 1994a) or 7 (Hatanaka et al. 1995). Furthermore, purification steps (Preparative TLC and HPLC) for less amount compounds sometimes drawback to utilize radioisotope methods. Radioisotope was very useful as detection tag, but was less effective for isolation tag for photolabeled components due to overlap with unlabeled components in the purification steps. The details on applications are mentioned in a later chapter (Chaps. 4 and 5).

The recent progress of mass spectrometry for biomolecules promoted researchers to apply stable isotope for photoaffinity labeling. Detailed synthesis (Chap. 6) and applications (Chap. 2) are summarized in the later chapters.

1.3 PAL with Immunological Systems

Immunological methods are very useful to purify the photolabeled components with antibody, which consisted with target molecules. Immunoprecipitates were applied for photolabeled components and the antibody recognized the peptides derived

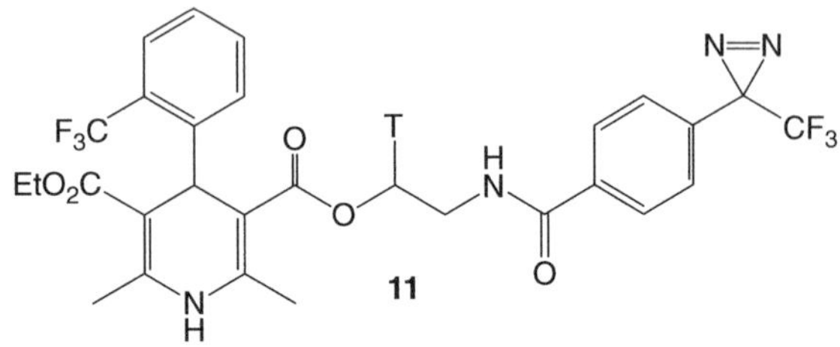

Fig. 1.2 [^{3}H]-Diazepam **11** for photoaffinity labeling of calcium channel

from both unlabeled and photolabeled biomolecules. Furthermore radioisotope introduced photolabeled ligand to detect low limit amounts.

These strategies were applied to elucidation of binding site of the ligands to ion channels. Antibody was prepared for predicted peptide sequences, and utilized photolabeled mixture for target ion channel with radiolabel. The correct predictions were caused immunoprecipitates, which contained radioisotopes. These experiments revealed binding site of [^{3}H]-diazepam **11** to calcium channels (Fig. 1.2, Nakayama et al. 1991, 1993).

It is very effective and the predictions of ligand binding site to biomolecules were easy to estimate because of its primary sequence. The details on applications are mentioned in later chapters (Chap. 5).

But sometimes it is difficult to apply these methods for unknown sequences for target biomolecules.

An antibody for the ligand skeleton is also available for the detection of the labeled components. Anti-galactosylceramide and anti-ceramide antibody was utilized to detect diazirinyl containing ceramide derivatives on TLC or PVDF membrane (Hashimoto and Hatanaka 2004).

1.4 PAL with Avidin–Biotin System

It is essential for predicted sequence to detect photolabeled components with immunological methods. It is very difficult to apply this method to unknown biomolecules. Specific interaction between avidin and biotin has been well studied. Avidin, which consists of tetrameric subunits, recognizes small molecule biotin with very strong affinity ($K_d = 10^{-15}$ M) (Bayer and Wilchek 1990). Other biotin-specific recognized biomolecules are streptavidin, anti-biotin antibody, and monomeric avidin. Streptavidin (Bayer et al. 1990) is purified from *Streptomyces avidinii* and has the same affinity for biotin as avidin, but the pI of the protein is between 5 and 6, whereas the pI of avidin is almost basic (pI 10).

Furthermore, streptavidin is non-glycosylated. These differences suppress the unspecific binding of biotinylated compounds to streptavidin rather than avidin. The strong interaction between (strept)avidin and biotin was very useful for detection of less amounts (10^{-13} mol) of biotinylated biomolecules with high detection methods (i.e., chemiluminescence, fluorescence, etc.) to the same degree as radioisotope detection. Researchers also utilized anti-biotin antibody

(Dakshinamurti and Rector 1990), which is less sensitive than (strept)avidin ($K_d = 10^{-9}$ M), to detect biotinylated components when it is too hard to make (strept)avidin–biotin complex.

Monomeric avidin, which is only prepared on a solid support, has less affinity than avidin ($K_d = 10^{-9}$ M) (Green and Toms 1973), and dissociation of the avidin–biotin complex is easier than the tetrameric form (2 mM biotin or glycine buffer pH 2.0).

The combination was first developed to a peptide ligand, because photophores and biotin can be introduced separately at lysine α- and ε-amino groups (Gilbert and Rando 1995), but not all ligands have many modification sites in the ligand skeleton. One resolution is the synthesis of biotinylated photophore.

We described these methodologies to apply identifications of acceptor binding site of bovine β-1,4-galactosyltransferase (Gal T).

Glycosyltransferases, which had not been cleared functional analysis, were suitable examples to analysis with photoaffinity labeling. The asparagine moiety in *N*-acetyl glucosamine-asparagine skeleton, which is a acceptor substrate for β-1,4-galactosyltransferase (EC 2.4.1.38), was modified with biotinylated diazirinyl photophore, and the binding site in the protein was revealed for the first time (Fig. 1.3 and Table 1.1) (Hatanaka et al. 1996a, b; Hatanaka and Kanaoka 1998).

Our previous study on β-1,4-galactosyltransferase demonstrated that photoaffinity labeling can reveal conformational change during the enzymatic reaction. The enzyme has two substrates, UDP-galactose as a donor and *N*-acetyl-D-glucosamine (GlcNAc) as an acceptor to generate *N*-acetyllactosamine. The biotinylated photoreactive GlcNAc derivative **12** was prepared for this purpose. Specific photoincorporation of photoreactive GlcNAc derivatives was achieved only in the presence of UMP, which is a UDP-Gal analogue at 37 °C. The results indicated that the two substrates should be orientated near position in the active site to promote conformational change of the enzyme structure (Hashimoto and Hatanaka 1999; Hatanaka

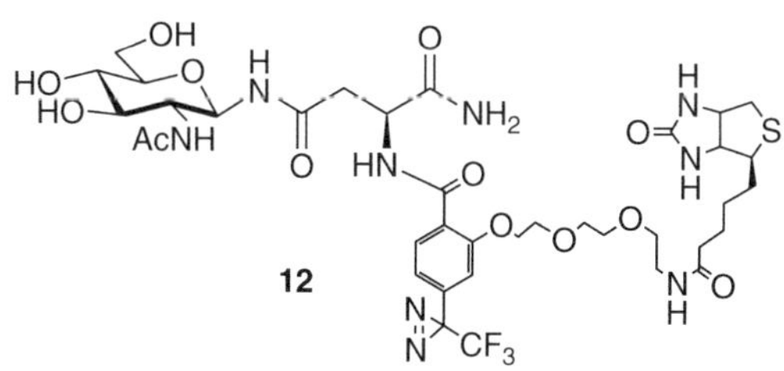

12

Fig. 1.3 Biotinylated photoreactive GlcNAc derivative for photoaffinity labeling of bovine β-1,4-galactosyltransferase

Table 1.1 Photoaffinity labeling yield of bovine GalT with compound **12** in various conditions

Compound **12**	+	+	+	+
UMP	–	+	+	+
Temperature (°C)	37	37	23	4
Specific labeling (%)	1.7	4.0	2.5	0.6

et al. 1998, 2001). It was also reported that dynamic structural investigations on the torpedo nicotinic acetylcholine receptor were performed by time-resolved photoaffinity labeling (Mourot et al. 2006).

The enzyme also utilized donor substrate UDP-Gal, which is common donor substrate for other galactosyltransferases. The kinetic study revealed that the photoreactive acceptor substrate analogue had enough affinity for target enzymes (observed Km were 0.17 and 0.06 mM for GlcNAc and 12, respectively). The photoaffinity labeling of GalT with compound **12** was set up, at ten times higher substrate concentration against Km value. The enzymatic reaction mixture was incubated at optimal temperature (37 °C) for 10 min to make complex with substrate, followed by irradiation at 4 °C. The sample was subjected to detect biotinylated proteins by avidin–biotin specific interaction with chemiluminescence. The chemiluminescence signals were quantified with external biotinylated proteins. The results show that both incubation at optimal temperature and the presence of donor substrate, UMP as inhibitor for UDP-Gal, were important to detect specific labeling. Photoaffinity labeling enables us to judge molecular dynamics (conformational change) under setting up labeling conditions.

Preparative scale of photoaffinity labeling for GalT was setting up from preliminary results, including UMP, which was one of the substrate mimics for GalT. Streptavidin–peroxidase conjugate was applied for photoaffinity label experiments, and immobilized streptavidin was utilized for isolation of photoaffinity-labeled components. The cell lysate overexpressed GalT in *E. coli* cell was applied to photolabeling with compound **12**. The results revealed the biotin components were introduced to GalT specifically in the cell lysate, even though the photolabeled efficiency of compound **12** and GalT was calculated less 4%.

Photoaffinity-labeled mixture was subjected to immobilized avidin to isolate photolabeled GalT specifically. The unlabeled GalT was passed through the affinity column and the labeled Galt was recovered with SDS sample buffer at 100 °C (for streptavidin) or excess biotin (for monomeric_avidin). The purified photolabeled GalT was subjected to enzymatic digestion, followed by HPLC analysis. The several peaks were detected and each fraction was submitted blotting on PVDF membrane to detect biotin components. But the small peptides did not retain PVDF membrane due to increase hydrophobicity. *p*-Nitrodiazirine was modified on PVDF membrane followed by irradiation with black light followed by reduction of nitro group to amine to form amide bond with carboxylic acid of peptides in the presence of carbodiimide. The biotin-based chemiluminescence was detected in one of the major peptides, and the sequence analysis revealed it as Y197-R208 peptide. Combination of molecular docking study and photoaffinity-labeled results strongly suggests that the carboxylate group of Asp318 could be involved in the activation of acceptor sugar 4-OH for the efficient galactosyltransfer (Fig. 1.4).

Although avidin–biotin interaction is very strong, and formation of the complex is very fast and rigid, it is too difficult to dissociate the biotin molecule from the complex as a native form. Very harsh conditions are needed for the biomolecules

0 MKFREPLLGG SAAMPGASLQ RACRLLVAVC ALHLGVTLVY YLAGRDLRRL PQLVGVHPPL

60 QGSSHGAAAI GQPSGELRLR GVAPPPPLQN SSKPRSRAPS NLDAYSHPGP GPGPGSNLTS

120 APVPSTTTRS LTACPEESPL LVGPMLIEFN IPVDLKLVEQ QNPKVKLGGR YTPMDCISPH

180 KVAIIIPFRN RQEHLK**YWLY YLHPILQR**QQ LDYGIYVINQ AGESMFNRAK LLNVGFKEAL

240 KDYDYNCFVF SDVDLIPMND HNTYRCFSQP RHISVAMDKF GFSLPYVQYF GGVSALSKQQ

300 FLSINGFPNN YWGWGGE**D**DD IYNRLAFRGM SVSRPNAVIG KCRMIRHSRD KKNEPNPQRF

360 DRIAHTKETM LSDGLNSLTY MVLEVQRYPL YTKITVDIGT PS

Fig. 1.4 β-1,4-Galactosyltransferase (bovine) sequence. *Bold underlined* sequence (Y197-R208) was photolabeled sequence with compound **12**. *Bold* at ^{318}D could play important roles for the activation of acceptor with docking study

(70% HCOOH at r.t., autoclave, or 2% SDS—8 M urea with boiling), and the recovery yield of the biotinylated biomolecule is not quantitative.

Since biotin–(strept)avidin binding is essentially irreversible, several approaches have been investigated to achieve the efficient recovery of biotinylated products from (strept)avidin-immobilized matrix.

Several approaches were examined for this problem. In one, modified avidin, which is described above, was used. Another introduced the chemical cleavable tags between the photophore and biotin to isolate photolabeled components from the avidin–biotin complex. Diol (Hashimoto et al. 2001) and disulfide moieties (Hashimoto et al. 2001) are well used for this purpose with isolation with thiol and periodate, respectively; however these moieties are also present in biomolecules (disulfide bond and sugar moiety), so it is not a specific method for labeled biomolecules. Periodate-treated diol, which generates aldehyde, can react with hydrazine as post-functional modification, and the detection limit of aldehyde and biotin hydrazide was estimated over 0.2 pmol (Hashimoto and Hatanaka 2005b)

It was reported that nitrophenyl (Olejnik et al. 1995; Fang et al. 1998; Fang and Bergstrom 2003) and sulfonamide moiety (Park et al. 2005) were suitable for the purpose. These moieties are less abundant in biomolecules and the cleavable condition is very mild in almost all biomolecules.

The final possibility is to utilize site-specific digestion enzymes for the ligand skeleton (Hashimoto et al. 2004). The protected amino acid residue was inserted between diazirine and biotin and was easily deprotected and then acted as a substrate for digestion enzymes. Folding biomolecules were not digested under this condition, although the introduced amino acid also became a substrate when the protecting groups were removed. Digestion was regulated by a methyl ester, which is a basic protecting group in organic chemistry. This combination was utilized in not only protein mixtures but also digested mixtures. The combination of the glutamic acid γ-methyl ester and V8 protease revealed that effective retrieval of labeled components is possible. For labeled biocomponents, it is first enzymatically digested and then deprotected, followed by redigestion to cleave the avidin–biotin complex to isolate the labeled peptides. Furthermore, the combination of linker amino acid and digested enzyme had many patterns to facilitate handling of the labeled components.

1.5 PAL with Other Specific Interactions

Oligo acidic or basic amino acids could shift the isoelectric point of target proteins and shield with protecting groups, which are widely used in amino acid chemistry. The e-protected penta-Lys ($[Lys(Boc)]_5$) was neutralized, but acid treatment, which deprotected Boc group, caused the compound to have high basicity. The properties combined with photoaffinity labeling. The diazirine derivative was condensed with C-terminal of biotin-$[Lys(Boc)]_5$, and the reagents were subjected to chromatofocusing. Acid treatment promoted, which deprotected Boc group, the isoelectric point which changed the pH from 5 to 7 because all e-amino group was deprotected. It is very convenience that the protecting group can be easily distinguished from amino acid moiety from protein. The properties can apply to the enzymatic digestion. Lys in the peptides is recognized by endopeptidase Lys-C or trypsin. The synthetic compound without any treatment was resisted to the trypsin digestion, but was easily digested after deprotection with acid followed by trypsin treatment. The cleavable property caused the isoelectric point to return to the original value. The properties are also utilized in the photolabeled component that can change the isoelectric point to isolate the original protein and return the original isoelectric point after acid treatment followed by digestion to release penta-Lys moiety (Hashimoto and Hatanaka 2006).

1.6 Limitation

Although photoaffinity biotinylation was very useful to identify the labeled region, the large modification with the ligand skeleton sometimes causes decreased affinity of the synthetic compounds. For example, fatty acid derivatives, which contained diazirine photophore and biotin (Fig. 1.5a) (Hashimoto and Hatanaka 2005a, b), cannot be utilized for photoaffinity labeling of sphingolipid ceramide *N*-deacylase (SCDase). Biotin and photophore have to be introduced at different part of ceramide to recognized by the enzyme (Fig. 1.5b) (Hashimoto and Hatanaka 2008b). The limitation promoted the development of post-modification of labeled components, which is summarized in the next chapter.

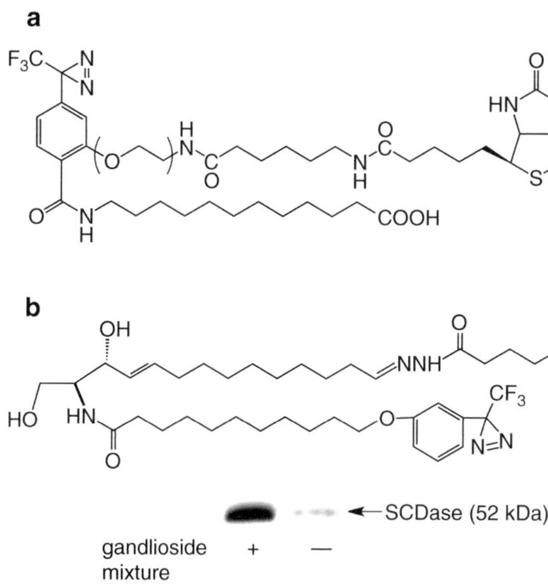

Fig. 1.5 Photoaffinity biotinylation probes for sphingolipid ceramide *N*-deacylase (SCDase). (**a**) Photoaffinity biotinylation fatty acid derivative, which was not recognized by SCDase. (**b**) Photoaffinity biotinylation ceramide derivative, which was recognized by SCDase

1.7 Conclusion

The combinations of photoaffinity labeling and avidin–biotin systems will be developed not only to identify and to detect photolabeled biomolecules but also to identify photolabeled regions. The low detection limits for avidin–biotin systems are very useful, but irreversible dissociations make some troubles to isolation of labeled components. Many efforts to overcome these problems, which are based on chemical structure of ligand derivatives and modified avidin to dissociate effectively, have been made over the last two decades.

Acknowledgments M.H. is very grateful to Prof. Y. Hatanaka (University of Toyama), G.D. Holman (University of Bath), and Prof. Y. Kanaoka (Toyama College) for valuable advice throughout the manuscript.

References

Ambroise Y, Pillon F, Mioskowski C, Valleix A, Rousseau B (2001) Synthesis and tritium labeling of new aromatic diazirine building blocks for photoaffinity labeling and cross-linking. Eur J Org Chem 2001:3961–3964

Bayer EA, Wilchek M (1990) Application of avidin–biotin technology to affinity-based separations. J Chromatogr 510:3–11

Bayer EA, Ben-Hur H, Wilchek M (1990) Isolation and properties of streptavidin. Methods Enzymol 184:80–89

Bender T, Huss M, Wieczorek H, Grond S, von Zezschwitz P (2007) Convenient synthesis of a [1-^{14}C]diazirinylbenzoic acid as a photoaffinity label for binding studies of a V-ATPase inhibitors. Eur J Org Chem 2007:3870–3878

Blanton MP, McCardy EA, Huggins A, Parikh D (1998) Probing the structure of the nicotinic acetylcholine receptor with the hydrophobic photoreactive probes [^{125}I]TID-BE and [^{125}I]TIDPC/16. Biochemistry 37:14545–14555

Blanton MP, McCardy EA, Gallagher MJ (2000) Examining the noncompetitive antagonist-binding site in the ion channel of the nicotinic acetylcholine receptor in the resting state. J Biol Chem 275:3469–3478

Brunner J (1993) New photolabeling and crosslinking methods. Annu Rev Biochem 62:483–514

Brunner J, Semenza G (1981) Selective labeling of the hydrophobic core of membranes with 3-(trifluoromethyl)-3-(*m*-[^{125}I]iodophenyl)diazirine, a carbene-generating reagent. Biochemistry 20:7174–7182

Dakshinamurti K, Rector ES (1990) Monoclonal antibody to biotin. Methods Enzymol 184:111–119

Dolder M, Michel H, Sigrist H (1990) 3-(Trifluoromethyl)-3-(*m*-isothiocyanophenyl)diazirine: synthesis and chemical characterization of a heterobifunctional carbene-generating crosslinking reagent. J Protein Chem 9:407–415

Dormán G, Prestwich GD (2000) Using photolabile ligands in drug discovery and development. Trends Biotechnol 18:64–77

Fang S, Bergstrom DE (2003) Fluoride-cleavable biotinylation phosphoramidite for 5′-end-labeling and affinity purification of synthetic oligonucleotides. Nucleic Acids Res 31:708–715

Fang K, Hashimoto M, Jockusch S, Turro NJ, Nakanishi K (1998) A bifunctional photoaffinity probe for ligand/receptor interaction studies. J Am Chem Soc 120:8543–8544

Filer CN (2009) Tritium labelled photoaffinity agents. J Radioanal Nucl Chem 281:521–530

Foucaud B, Perret P, Grutter T, Goeldner M (2001) Cysteine mutants as chemical sensors for ligand-receptor interactions. Trends Pharmacol Sci 22:170–173

Gilbert BA, Rando RR (1995) Modular design of biotinylated photoaffinity probes—synthesis and utilization of a biotinylated pepstatin photoprobe. J Am Chem Soc 117:8061–8066

Green NM, Toms EJ (1973) The properties of subunits of avidin coupled to Sepharose. Biochem J 133:687–698

Hashimoto M, Hatanaka Y (1999) Identification of photolabeled peptides for the acceptor substrate binding domain of β1,4-galactosyltransferase. Chem Pharm Bull 47:667–671

Hashimoto M, Hatanaka Y (2004) Simple synthesis of deuterium and ^{13}C labeled trifluoromethyl phenyldiazirine derivatives as stable isotope tags for mass spectrometry. Chem Pharm Bull 52:1385–1386

Hashimoto M, Hatanaka Y (2005a) Post-biotinylation of photocrosslinking by Staudinger–Bertozzi ligation of preinstalled alkylazide tag. Chem Pharm Bull 53:1510–1512

Hashimoto M, Hatanaka Y (2005b) Diol derivative of (3-trifluoromethyl)phenyldiazirine for post-labeling of photocrosslink. Heterocycles 66:531–534

Hashimoto M, Hatanaka Y (2006) Positively coded photoaffinity label for altering isoelectric points of proteins. Bioorg Med Chem Lett 16:5998–6000

Hashimoto M, Hatanaka Y (2008a) Recent progress in diazirine-based photoaffinity labeling. Eur J Org Chem 2008:2513–2523

Hashimoto M, Hatanaka Y (2008b) A novel biotinyl and diazirinyl ceramide analogue for photoaffinity labeling. Bioorg Med Chem Lett 18:650–652

Hashimoto M, Hatanaka Y, Yang J, Dhesi J, Holman GD (2001) Synthesis of biotinylated bis(D-glucose) derivatives for glucose transporter photoaffinity labeling. Carbohydr Res 331:119–127

Hashimoto M, Okamoto S, Nabeta K, Hatanaka Y (2004) Enzyme cleavable and biotinylated photoaffinity ligand with diazirine. Bioorg Med Chem Lett 14:2447–2450

Hatanaka Y (2015) Development and leading-edge application of innovative photoaffinity labeling. Chem Pharm Bull 63:1–12

Hatanaka Y, Kanaoka Y (1998) Biotinyl diazirine photophore—an approach to high-resolution photoaffinity-labeling for probing receptor-ligand interface. Heterocycles 47:625–632

Hatanaka Y, Nakayama H, Kanaoka Y (1992) Photoaffinity labeling of the electroplax sodium channel with a photoreactive μ-conotoxin carrying a radioactive and chromogenic diazirine. Chem Pharm Bull 40:2537–2539

Hatanaka Y, Hashimoto M, Kurihara H, Nakayama H, Kanaoka Y (1994a) A novel family of aromatic diazirines for photoaffinity labeling. J Org Chem 59:383–387

Hatanaka Y, Hashimoto M, Nakayama H, Kanaoka Y (1994b) Syntheses of nitro-substituted aryl diazirines. An entry to chromogenic carbene precursors for photoaffinity labeling. Chem Pharm Bull 42:826–831

Hatanaka Y, Hashimoto M, Hidari KI-PJ, Sanai Y, Nagai Y, Kanaoka Y (1995) A carbene-generating biotinylated lactosylceramide analog as novel photoreactive substrate for GM3 synthase. Bioorg Med Chem Lett 5:2859–2862

Hatanaka Y, Hashimoto M, Nishihara S, Narimatsu H, Kanaoka Y (1996a) Synthesis and characterization of a carbene-generating biotinylated N-acetylglucosamine for photoaffinity labeling of β-1,4-galactosyltransferase. Carbohydr Res 294:95–108

Hatanaka Y, Nakayama H, Kanaoka Y (1996b) Diazirine-based photoaffinity labeling: chemical approach to biological interfaces. Rev Heteroatom Chem 14:213–243

Hatanaka Y, Hashimoto M, Kanaoka Y (1998) A rapid and efficient method for identifying photoaffinity biotinylated sites within proteins. J Am Chem Soc 120:453–454

Hatanaka Y, Ishiguro M, Hashimoto M, Gastinel LN, Nakagomi K (2001) A model of photoprobe docking with β1,4-galactosyltransferase identifies a possible carboxylate involved in glycosylation steps. Bioorg Med Chem Lett 11:411–413

Kotzyba-Hibert F, Kapfer I, Goeldner M (1995) Recent trends in photoaffinity labeling. Angew Chem Int Ed Engl 34:1296–1312

Latli B, Morimoto H, Williams PG, Casida JE (1998) Photoaffinity radioligand for NADH:ubiquinone oxidoreductase: [S-C_3H_2](trifluoromethyl)diazirinyl-pyridaben. J Label Compd Radiopharm 41:191–199

Mourot A, Grutter T, Goeldner M, Kotzyba-Hibert F (2006) Dynamic structural investigations on the torpedo nicotinic acetylcholine receptor by time-resolved photoaffinity labeling. Chembiochem 7:570–583

Nakayama H, Taki M, Striessnig J, Glossmann H, Catterall WA, Kanaoka Y (1991) Identification of 1,4-dihydropyridine binding regions within the α1 subunit of skeletal muscle calcium channels by photoaffinity labeling with diazipine. Proc Natl Acad Sci U S A 88:9203–9207

Nakayama H, Hatanaka Y, Taki M, Yoshida E, Kanaoka Y (1993) Identification of ligand-binding sites that form external mouth of ion pore in calcium and sodium channels. Ann N Y Acad Sci 707:349–351

Olejnik J, Sonar S, Krzymanska-Olejnik E, Rothschild KJ (1995) Photocleavable biotin derivatives: a versatile approach for the isolation of biomolecules. Proc Natl Acad Sci U S A 92:7590–7594

Park JJ, Sadakane Y, Masuda K, Tomohiro T, Nakano T, Hatanaka Y (2005) Synthesis of diazirinyl photoprobe carrying a novel cleavable biotin. Chembiochem 6:814–818

Rimoldi JM, Kingston DGI, Chaudhary AG, Samaranayake G, Grover S, Hamel E (1993) Modified taxols, 9. Synthesis and biological evaluation of 7-substituted photoaffinity analogs of taxol. J Nat Prod 56:1313–1330

Tomohiro T, Hatanaka Y (2014) Diazirine-based multifunctional photo-probes for affinity-based elucidation of protein-ligand interaction. Heterocycles 89:2697–2727

Wallrabe H, Periasamy A (2005) Imaging protein molecules using FRET and FLIM microscopy. Curr Opin Biotechnol 16:19–27

Wold F (1977) Affinity labeling—an overview. Methods Enzymol 46:3–14

Chapter 2
Tag-Creation Approaches for Highly Efficient Profiling of Interacting Proteins and Domains

Takenori Tomohiro

Abstract Diazirine-based photoaffinity labeling is recognized as one of the most reliable methods for identification of biomolecular interactions because of its excellent chemical and physical properties. To avoid time-consuming steps in the analysis of a tiny amount of labeled product, functionalization of photoprobe should be an essential subject in this method. However, addition of functions often affects affinity of bioactive molecule. In this chapter, multifunctional diazirine-based photocross-linkers and their strategies for rapid target protein profiling of bioactive molecules are described, especially tagging methods after cross-linking including post-labeling using cleavable function, tandem labeling using clickable function, and fluorogenic labeling. Further, a unique target-visualization strategy is presented for facile identification of labeled site within protein using isotope-coded fluorescent tag, which can easily distinguish the target from the enormous range of biomolecules in analytical process using LC-MS/MS. Without any chemical treatments, a coumarin tag is photochemically generated on ligand-interacting surface of protein through structural change from nonfluorescent photocross-linker unit with accompanying cleavage of ligand molecule.

Keywords Photoaffinity labeling • Diazirine • Target identification • Tandem labeling • Post-labeling • Multifunction • Fluorogenic

2.1 Introduction

To expand capability and to gain efficiency in photoaffinity labeling (PAL)-based protein identification from the enormous range of biomolecules, sorting functions such as detection and isolation tags will be necessary for efficient

T. Tomohiro
Graduate School of Medicine and Pharmaceutical Sciences, University of Toyama, 2630 Sugitani, Toyama 930-0194, Japan
e-mail: ttomo@pha.u-toyama.ac.jp

Y. Hatanaka, M. Hashimoto (eds.), *Photoaffinity Labeling for Structural Probing Within Protein*, DOI 10.1007/978-4-431-56569-7_2

discrimination and purification of unknown PAL-captured proteins (Tomohiro and Hatanaka 2014; Hatanaka 2015). Each tag should have high sensitivity or a high association constant together with high specificity, in order to provide ultra-high resolution, because only a tiny amount of the labeled target molecules is present in the proteome. In order to ascertain and facilitate the analysis of photo-cross-linked biomolecules without the necessity of any post-chemical treatments, a photoactivatable group needs to be introduced into the structure of the ligand (inhibitor) to prepare the PAL probe, concurrent with the attachment of a chemical reporter. For example, biotin has shown excellent properties with respect to detection and isolation because of its almost irreversible binding to avidin protein. Consequently, biotinylated proteins can be efficiently concentrated from the lysate using an avidin-immobilized support. The eluted proteins are digested directly or after further purification using SDS-PAGE wherein these proteins can be sensitively detected with chemiluminescence using the avidin–HRP conjugate. The candidates for target proteins are determined using peptide mass fingerprinting (PMF) analysis with amino acid sequences of the proteolytic product.

Despite the excellent properties of biotin in purification/detection, biotin-incorporated photoprobes often show low solubility in both water and organic solvents. This often results in severe difficulty in probe synthesis and thus usage of biotin. To improve solubility and the binding ability with avidin, a PEG-like spacer unit has often been incorporated between biotin and the photocross-linker (Hatanaka et al. 1994; Tomohiro et al. 2005). Further, the biotin moiety possibly affects the membrane permeability of the probe (Taldone et al. 2013; Luo et al. 2003), which may give rise to limitations in its use in experiments using living cells. In addition, intrinsically biotinylated proteins, such as carboxylases, in cells often cause high background noise in detection and purification of the labeled target proteins, because the quantity of the photochemically biotinylated product should be fairly low in contrast to those of the proteins. Alternatively, as a nonradioactive detection tag, fluorophores allow effective detection and isolation of labeled products (Maurel et al. 2010) such as the following: aplyronine A (**1**, Fig. 2.1) (Kuroda et al. 2006), which is an actin-depolymerizing substance of marine origin; bisphenol A (**2**) (Chan et al. 2004), which is a common environmental endocrine disruptor; and peptidyl hydroxamates (**3**) (Maezawa et al. 2007). However, common fluorophores are relatively large in size, characteristic in physical properties such as charge, and unstable in conditions of probe synthesis and UV irradiation. These features may affect the specificity of the biological activity of the ligand as it may cause steric disruption of its key interactions with the target molecule. To address these disadvantages, these chemical tags have been often introduced after photocross-linking.

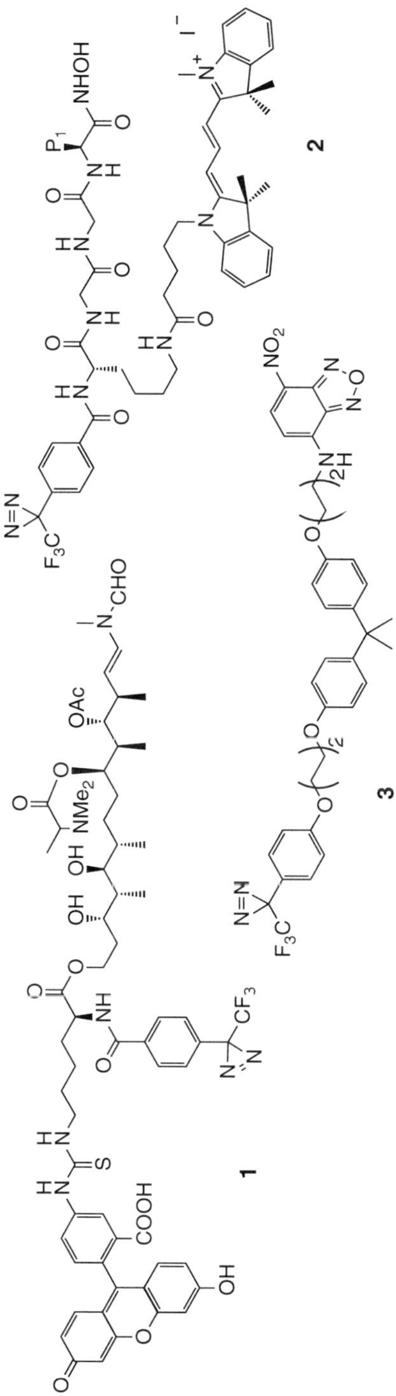

Fig. 2.1 Some examples of fluorophore-tethering photoprobes

2.2 Post-Labeling Using Cleavable Function

Some methodologies for post-tagging of functions have been developed as one of the strategies to decrease the change in the characteristics of the active molecule, keeping the key interactions as native as possible. Namely, the target protein is covalently cross-linked by UV irradiation at the first step with the photoprobe bearing a chemical reporter, where an analytical tag can be selectively introduced in the second step.

The insertion of a cleavable linker between a cross-linker and a ligand of interest is one of the effective ways to add a function to a PAL probe without increasing its size. A disulfide bond has often been used as the simplest linker that can be scissile under reductive conditions, and the resultant thiol group can be easily modified with a thiol-selective tagging agent. This approach has been applied to a fluorescence-based saccharide-sensing system that uses a photoactivatable saccharide and has a disulfide bond between a phenyldiazirine and the saccharide (**4**, Fig. 2.2) (Nagase et al. 2003). After cross-linking to concanavalin A (Con A), a thiol group was produced by the reductive cleavage of the disulfide. A dansyl group, as an environment-sensitive fluorophore, was subsequently introduced by post-labeling into the region close to the saccharide-binding site. The dansyl group emitted strong fluorescence in the pocket and weak fluorescence outside on binding with a saccharide. Although various fluorophores were successfully installed to the target protein in this method, the labeling yield of diazirine photolysis was generally low and it required prior protection of all intrinsic SH groups on the protein surface. Furthermore, the disulfide bond is generally labile with reducing agents, which makes it difficult to use under biological conditions. Another generation of thiol group-containing molecules for post-labeling, a photoactivatable GTP modified with a phenyldiazirine at the terminal phosphate via a phosphorothioate linkage (**5**) (Kaneda et al. 2007), was synthesized to label H-Ras, a family of small GTPases. The P–S bond was easily hydrolyzed under basic conditions to produce a thiol group that was post-labeled with a biotin group for high-sensitive chemiluminescence detection of labeled H-Ras. However, these kinds of post-labeling methods can be hardly applied to proteomic identification of unknown targets.

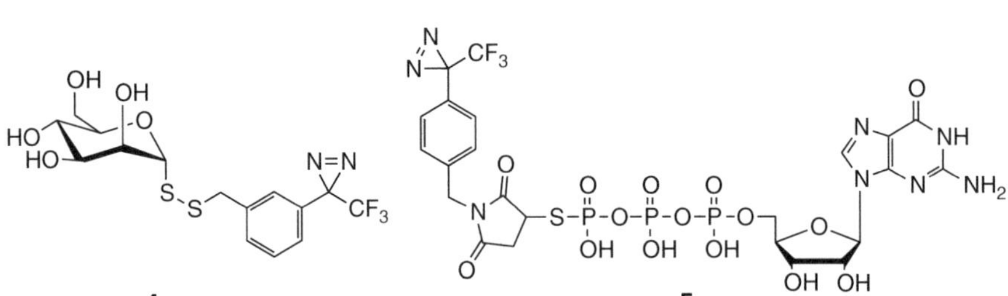

4 **5**

Fig. 2.2 Photoprobes having a cleavable function

2.3 Tandem Labeling Using Two Different Cross-Linkers

2.3.1 Essential Features for Post-Labeling in PAL

To be considered as a chemical reporter in a post-tagging strategy with application in processing of PAL-based target identification in the proteome, the chemical reporter must satisfy a number of requirements: First, because the cross-linked protein will be present in a very small quantity in a complex solution, the reaction must be highly selective between functional groups of the reactive partners to avoid side reactions with biological compounds, without the need for any treatment for protection of functional groups. Secondly, kinetics must be another significant factor to be considered (Patterson et al. 2014); the reaction must be rapid enough to complete in a short period with the cross-linked target at very low concentration and must proceed in biological, aqueous conditions. These features are necessary in order to accurately track a small signal from the target surrounded by nonspecific contaminating materials. Furthermore, the reporter tag should be better off without antigenicity and not affect membrane permeability in cases using cells. In addition, the covalent bond-linking tag should be stable during purification, digestion, and analytical processes.

2.3.2 Clickable Diazirine Photophore

Bioorthogonal chemistry has been widely applied in chemical biology for in vivo functionalization of biomolecules such as monitoring cell proliferation and studying glycan processing, without interfering with native biochemical processes. In particular, copper-mediated 1,3-dipolar cycloaddition of small azide and ethynyl groups, the so-called click reaction, shows excellent chemoselectivity and reactivity (Boutureira and Bernardes 2015). These features allow highly specific and efficient enrichment of trace amounts of target proteins, which resulted in giving advantages for activity-based protein profiling (ABPP), identifying protein targets, and probing the structure of proteins. In fact, clickable chemical cross-linkers combined with mass spectrometry have been employed in the analysis of three-dimensional protein structures and protein–protein complexes (Sohn et al. 2012). Additional properties, such as aliphatic azide and terminal alkyne exhibiting stability under UV irradiation and physiological conditions, give them potential to be embedded, clickable tags incorporated into photoprobes as chemical reporters. Namely, the bio-probes including two controllable cross-linkers responding to different external stimuli, light and a copper catalyst, should be able to trap the target by light and then isolate/detect it sensitively via a post-tagging reaction. For example, fluorophores can be stably and selectively introduced to the target protein using Staudinger–Bertozzi ligation or the Huisgen reaction (Ballell et al. 2006). Therefore, the use of tandem target labeling has rapidly increased for detection of the target in a broad range of biomolecules such as dasatinib (**6**, Fig. 2.3) (Shi et al. 2012) and fungal

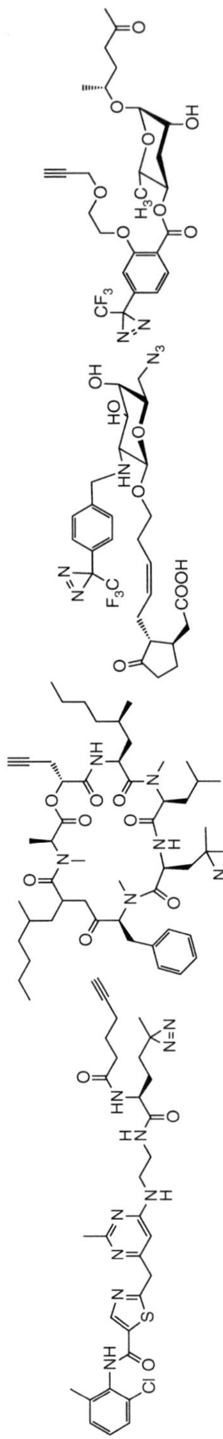

6 7 8 9

Fig. 2.3 Some photoprobes having a clickable function

cyclodepsipeptide (**7**) (MacKinnon et al. 2007) with rhodamine and β-D-glucopyranosyl 12-hydroxyjasmonic acid (**8**) (Nakamura et al. 2011) with biotin. Park et al. successfully identified two G-protein-coupled receptors, which cooperatively mediate ascaroside perception using a photobleaching, fluorescent energy transfer assay system (**9**) (Park et al. 2012). Recently, other efficient bioorthogonal coupling reagents have been developed for copper-free clickable ligation such as the *trans*-cyclooctene-tetrazine system used for intracellular live-cell labeling (Devaraj et al. 2010; Blizzard et al. 2015).

2.3.3 Strategies for Introduction of Multifunctionality

Photo-probing of a ligand of interest is generally performed by utilizing a three-directional scaffold to connect the functions of a ligand, a photophore (diazirine), and a reporter tag through multiple amide bonds (Fig. 2.4) (Ziegler et al. 2013). Commonly, trifunctional amino acids such as lysine and glutamic acid have been used for this purpose. However, modification of probes has resulted in a significant decrease of the solubility as well as an increase of the probe size. As a solution, multifunctional, "all-in-one," cross-linker units have been designed and synthesized. For example, various photoactivatable amino acids have been developed through the direct attachment of a photophore to part of the side chain, such as *p*Bpa (Chin et al. 2002a, b; Hino et al. 2005), *p*Azpa (Chin et al. 2002a, b), TmdPhe (Tanaka et al. 2008; Chin and Schultz 2002), L-Photo-Trp (Murai et al. 2012; Wartmann and Lindel 2013), L-Photo-Met, L-Photo-Leu (Suchanek et al. 2005), L-Photo-Ile, and L-Photo-Pro (Srinivas et al. 2010) (Fig. 2.5). These multifunctional photo-amino acids have exhibited good solubility in aqueous media and excellent advantages in the preparation of photoactivatable peptides of interest with an automatic synthesizer, for example, the calmodulin binding peptide (Nakashima et al. 2006), a representative cell-penetrating peptide (Kawaguchi et al. 2013), as well as in the preparation of photoactivatable proteins through photo-amino acid incorporation using site-directed mutagenesis in living cell systems (Chin et al. 2002a, b). Another synthetic approach to avoid using multiple amide bonds in probe structure has been focused on the modification of the benzene ring of the 3-trifluoromethyl-3-phenyldiazirine photophore, which has shown good stability under

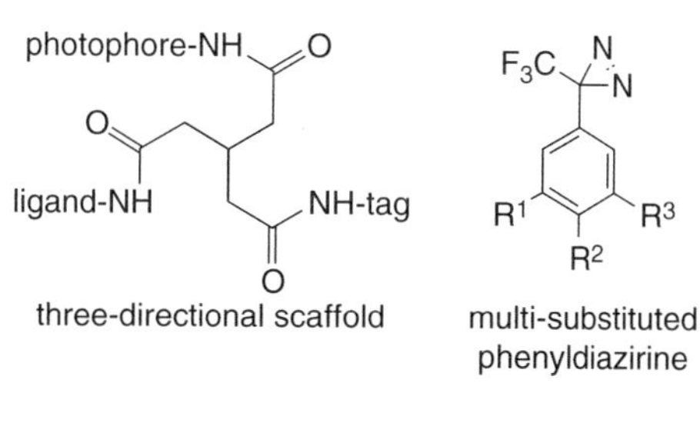

Fig. 2.4 Multifunctional scaffold for probing

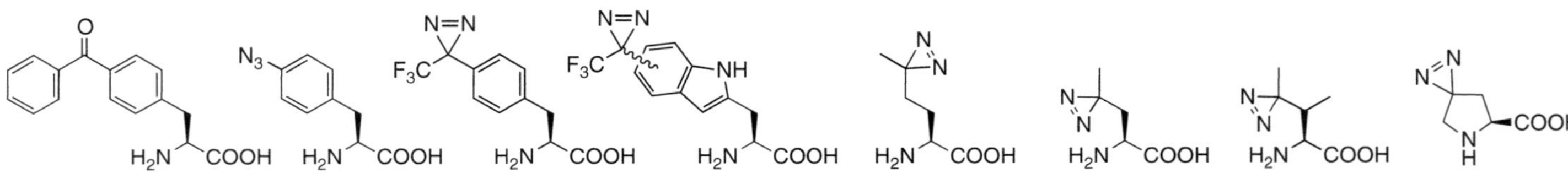

pBpa, pAzpa, TmdPhe, L-Photo-Trp, L-Photo-Met, L-Photo-Leu, L-Photo-Ile, L-Photo-Pro

Fig. 2.5 Structure of photo-amino acids

general synthetic and biological conditions. Furthermore, in this diazirine derivative, unfavorable side reactions, such as rearrangement reactions induced from carbene intermediates and alkylation reactions by diazo intermediates, can be avoided by substitution of the benzene ring and the CF_3 group. A trifunctional unit equipped with a biotin moiety on the benzene ring, via an ethylene glycol-type linkage, provided high performance with multifunctionality in terms of a relatively small size, with better solubility and increased stability. As a tandem labeling agent, a bis-azido compound bearing both aliphatic and aromatic azides (**10**, Fig. 2.6) (Hosoya et al. 2004) was initially developed to selectively detect phenyl azide-based PAL-captured proteins with the attachment of a fluorescent tag after post-labeling of the alkyl azide. Phenyldiazirines tethering an aliphatic azido (**11** Hosoya et al. 2004, **12** Hashimoto and Hatanaka 2005) or ethynyl (**13** Hosoya et al. 2005, **14** Mayer and Maier 2007) group were then synthesized and used for post-labeling of PAL products by the Staudinger–Bertozzi ligation or Huisgen reaction.

A unique synthetic approach for novel, clickable photocross-linkers has been reported; this involves the introduction of an azide (**15**) (Kumar and Young 2009) or an ethynyl (**16**) (Hiramatsu et al. 2007) group at a trifluoromethyl group (Fig. 2.7). The former derivative was stable and coupled with an azido-biotin reagent in order to facilitate purification of the labeled target. The position of derivatization is essentially important for designing all-in-one photophores. Except for some examples (**17** (Lee et al. 2013), **18** (Bell et al. 2013), Fig. 2.8), trifluoromethylphenyldiazirine, a relatively large unit, is not generally installed as a part of a ligand's structure but as an additional function to the ligand. Therefore, the unit should be attached to the ligand at the position where the affinity should not be disturbed, and the diazirine moiety is usually located around the lip of the binding pocket during interactions. In those cases, the substituent on the trifluoromethyl group should be directed to the outside of the protein, which might not largely affect the affinity. In fact, as another development, phenyldiazirine compounds bearing a perfluoroalkyl chain instead of the trifluoromethyl group were applied to target identification (**19**, **20**, Fig. 2.9) (Son and Zhang 2009; Burkard et al. 2010). A perfluoroalkyl chain shows differing hydrophobicity from general alkyl chains, which resulted in its application for highly selective purification using fluorous solid phase extraction instead of the biotin–avidin system. The fluorous tags can avoid the problems of conventional purification on a solid phase, such as nonspecific adsorption on the surface and incomplete elution.

Alternatively, clickable, aliphatic diazirine derivatives, alkyne-DA (**21** (Dubinsky et al. 2009), **22**, **23** (Li et al. 2013a, b)), have been chosen for proteomics owing to their smaller size (Fig. 2.10). Although aliphatic diazirines have potential for the generation of side products arising from rearrangement reactions, the smaller photophore unit can be incorporated as a part of the ligand structure. This allows the unit to be located inside the binding domain, which provides a high labeling yield, and to be used to delineate three-dimensional structures of the ligand-bound state in the cognate site. For example, two photoactivatable, n-octan-1-ol geometric isomers were used to delineate an anesthetic site, in which a diazirine group was modified on either the third or seventh carbon, 3- and 7-azioctanol (**24**, **25**) (Addona et al.

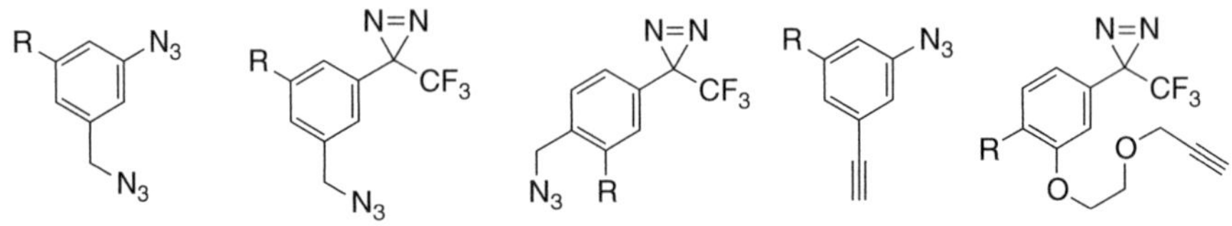

10 **11** **12** **13** **14**

Fig. 2.6 Clickable aromatic photophores

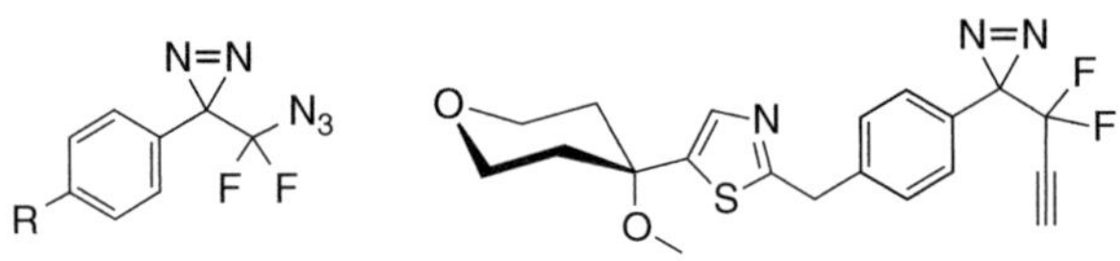

15 **16**

Fig. 2.7 Other clickable phenyldiazirines

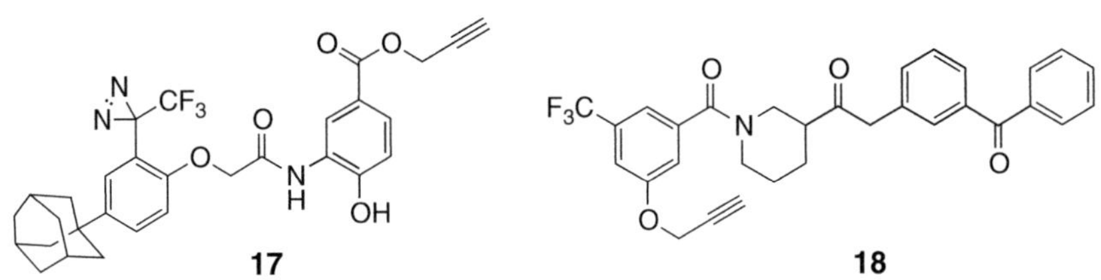

17 **18**

Fig. 2.8 Photoprobes having a clickable function

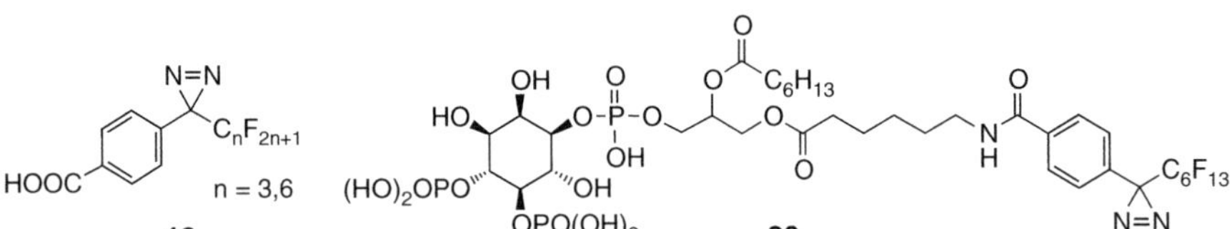

19 **20**

Fig. 2.9 Perfluoroalkylphenyldiaizirines

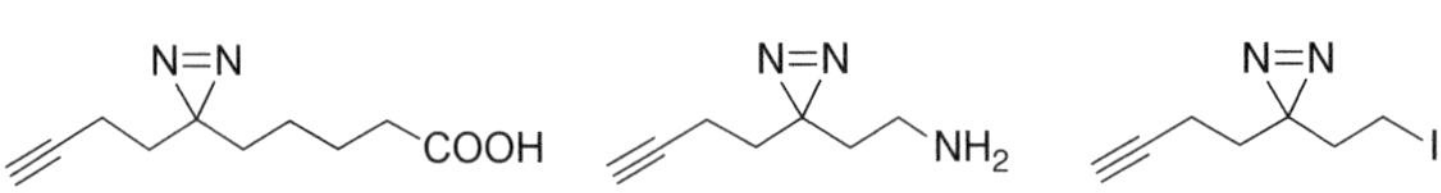

21 **22** **23**

Fig. 2.10 Clickable aliphatic diazirines

2002), respectively (Fig. 2.11). The ligand-binding position was estimated at an atomic level from the labeled residues. In addition, intracellular labeling reagents should possess high membrane permeability in addition to fulfill the stability, toxicity, and chemoselectivity requirements of "click" chemistry. Some photoactivatable biomolecules such as kinase inhibitors (**26**) (Li et al. 2013a, b) and steroids (**27**) (Hulce et al. 2013) have been prepared and successfully applied in protein profiling (Fig. 2.12). Similarly, cycloaddition of cyclopropene (**28**, Fig. 2.13) (Li et al. 2014a, b) and tetrazine into the structure of the photocross-linker was finally performed for imaging of the intracellular target protein.

Parallel synthesis using combinatorial chemistry has been applied to the preparation of a library of small-molecule probes with multifunctionality, including a photocross-linker and a clickable function, which resulted in a ~30 member probe library based on two scaffolds by fusing to the benzophenone backbone, 5-benzoyl indole and 7-benzoyl-benzo-1,4-diazepin-2,5-dione (**29**, Fig. 2.14) (Cisar and Cravatt 2012), and a ~60-member probe library based on 1,5-disubstituted tetrazole

Fig. 2.11 Azioctanols

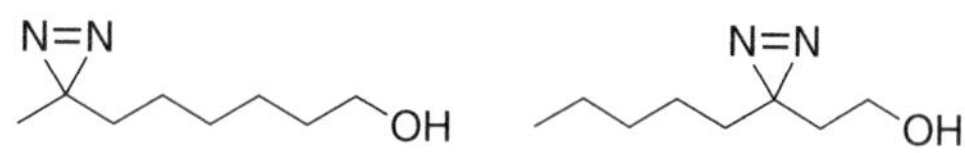

24 **25**

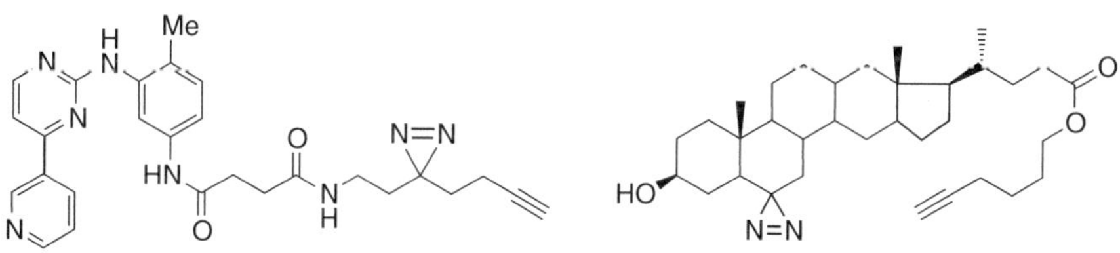

26 **27**

Fig. 2.12 Photoprobes tethering an aliphatic diazirine

Fig. 2.13 Aliphatic diazirine having a cyclopropene

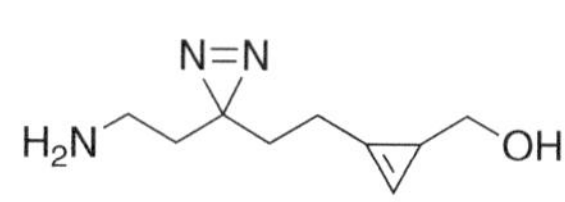

28

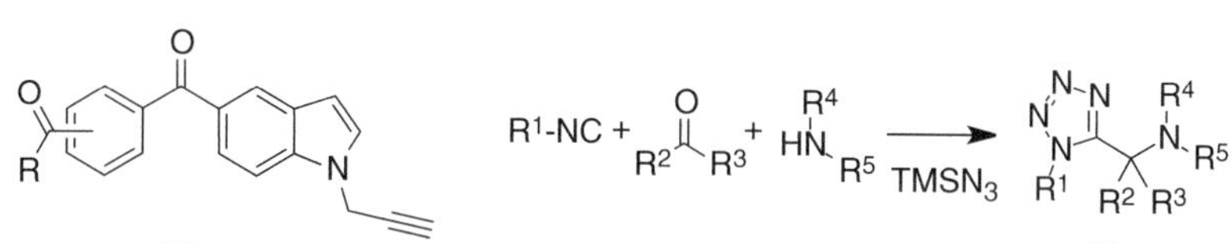

29 **30**

Fig. 2.14 Combinatorial probe synthesis having multifunctions

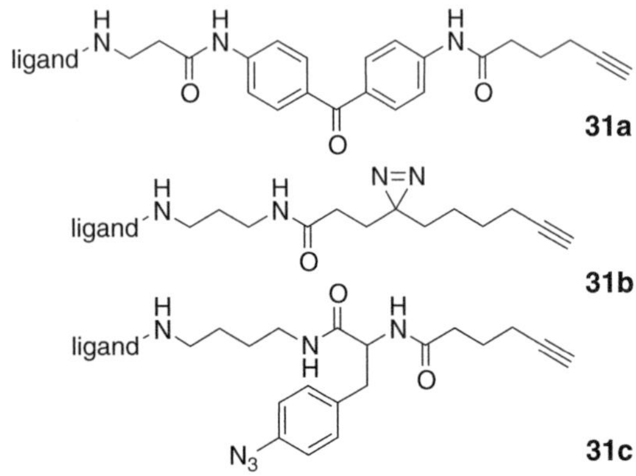

Fig. 2.15 Multifunctional bioconjugates with different photophores

containing a diazirine group prepared from the Ugi–azide multicomponent condensation reaction (**30**) (Kambe et al. 2014). Corresponding targets of these compounds were further identified by gel-based profiling followed by the quantitative MS analysis using SILAC (stable-isotope labeling by amino acids in cell culture (Ong et al. 2003)) strategy. It successfully resulted in giving the profiling of probe–protein interactions in vivo. As another example of PAL-based phenotypic screening of the small molecules in living cells, a combination of photochemical cross-linking and click-based visualization of each target protein with fluorescence in two-dimensional gel electrophoresis could aid in clear differentiation of target proteins from nonspecific binding proteins, by comparison of labeling patterns using three kinds of photophores (**31**, Fig. 2.15) (Park et al. 2016).

Another unique probing strategy is DNA-programmed PAL (Li et al. 2013a, b). The ligand probe is conjugated with an oligonucleotide. After incubation with the probe, the oligonucleotide can formed a duplex with its complementary oligonucleotide tethering a photophore. The target protein can be then photocross-linked with the photophore. This is a modular system in that the functions of target binding, covalent capture, and isolation are separated into different probes. In this method, the same complementary probe allows to be used in any cases.

2.3.4 Tandem Labeling for Profiling of Cognate Proteins

One of the strategies of using tandem cross-linking is for the profiling of cognate proteins. In this case, a photoreactive group can be used as a "nonselective" cross-linker in the second step to trap any cognate proteins. For example, a label transfer reagent—a heterobifunctional cross-linker consisting of an active ester, a photoreactive group, and a cleavable moiety—is often used for investigations of protein–protein interactions in signal transduction pathways. In a typical reaction, probing of a protein of interest is first performed by a chemical coupling reaction of the reagent with a functional group, such as the amino group of a surface Lys residue,

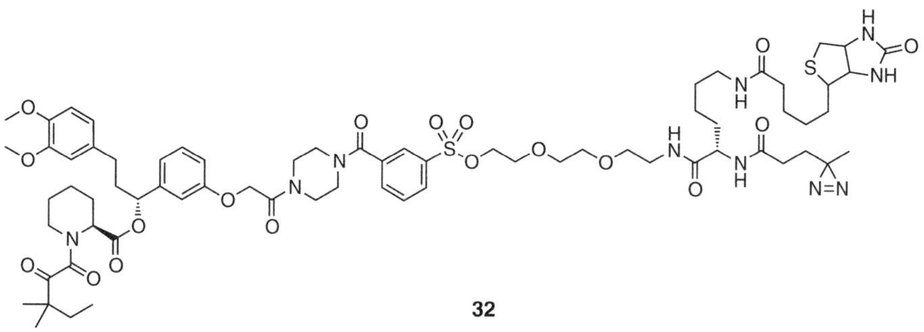

32

Fig. 2.16 LDA probe for tandem labeling

and then photochemically cross-linking proteins that interact with the bait protein, forming a covalent bond that is tethered to a radioactive (RI), fluorescent, or biotin label. Recently, a unique, all-in-one fluorophore-transfer reagent with scissile function was developed and applied to visualize cells through fluorophore labeling of antibodies expressed on the living cell membrane (Tomohiro et al. 2011). Alternatively, ABPP and ligand-directed affinity labeling (LDA) methods have been developed, which provide increase in target specificity in the chemical labeling step. The former method has been mainly used in screenings based on enzymatic activity (Evans and Cravatt 2006; Yang and Liu 2015), where the reagent cross-links to the active residue acting as an irreversible inhibitor. This means that labeling often occurs inside the active domain. The probe used for the latter method consists of a ligand molecule, a detection tag, and a displaceable bonding group such as tosylate between them instead of using an active ester for the label transfer method (**32**, Fig. 2.16). The tosylate could react with the nucleophile located close to the ligand-binding domain (outside of the domain) within the bait protein, successfully reducing nonspecific chemical modifications in vivo (Tsukiji et al. 2009). When the tag is a photoreactive group, the photoreactive group installed on the bait protein can cross-link to its cognate protein by UV irradiation. This methodology was demonstrated for identification of interacting proteins of FK506 binding protein 12 labeled by an LDA-mediated reaction in vitro and in living cells (Tamura et al. 2012).

2.4 In Situ Generation of Fluorophores for Microanalysis of Cross-Linked Sites Within Proteins

2.4.1 Fluorogenic Labeling for Target Imaging/Identification

The combined technique of photocross-linking with fluorescence was mainly used for the profiling of binding proteins with high sensitivity, which is accordingly performed where the fluorescent tag is introduced covalently around the interacting site of the protein, in the manner of pre- or post-tagging. After irradiation, the cross-linked protein was generally purified by SDS-PAGE and detected through specific fluorescence emission. The fluorophore must possess the ability of producing a strong emission at a longer wavelength, such as fluorescein, rhodamine, Cy3, and 7-aminocoumarin, to distinguish the target protein, even at exceptionally low levels, from the noise due to the inevitable background emissions from the gel, unlabeled probes, and Raman scattering. Since these compounds are generally unstable under synthetic conditions and easily breached in the cell by irradiation, post-fluorophore labeling was developed to overcome these problems as previously described. Alternatively, cross-linkers having a fluorogenic function were constructed and gained a high signal-to-noise ratio of emission by coupling with a quenching system (Li et al. 2014a, b). Fluorogenic compounds have been extensively investigated for studies of protein functions in living cells as low-background imaging agents because unreacted fluorophores need not have to be washed out from the cells (Kanaoka 1977). In particular, click reactions have been focused on target-specific fluorescent imaging in vivo, since ethynyl and azide groups have been found useful for the efficient fluorescence quenching of compounds (Le Droumaguet et al. 2010) such as coumarins (Sivakumar et al. 2004; Zhou and Fahrni 2004; Hsu et al. 2007), naphthalimides (Sawa et al. 2006), boron-dipyrromethene (Beatty et al. 2011), and fluorescein (Shieh et al. 2012). These compounds were nonfluorescent until the click reaction proceeded. This resulted in target-specific fluorescent imaging in vivo without washing or the exclusion of unreacted probes.

A pioneering study on fluorogenic PAL was performed with 8-azido-1,N^6-ethenoadenosine 3′,5′-cyclic monophosphate (8-N_3-εcAMP, **33**) as a fluorescent ligand molecule probe, with the direct attachment of an azide group (Keeler and Cambell 1976) (Fig. 2.17) that mimics cyclic AMP in its ability to stimulate the phosphotransferase activity of protein kinases. After PAL, the observed fluorescence was blue-shifted and highly polarized, suggesting that 8-N_3-εcAMP labeled the hydrophobic cAMP receptor site of the beef heart protein kinase. The study suggested that the fluorogenic PAL method could be more amenable to interpretation due to the better knowledge of the location of the ligand within the protein of interest. To date, fluorogenic PAL experiments have been performed using cross-linkers with azide groups installed directly into the aromatic rings of fluorophores such as 1,N^6-ethenoadenosine (Dreyfuss et al. 1978), coumarin, naphthalene, pyrene (Kanaoka et al. 1984), nitrobenzoxadiazole (Lord et al. 2010), and dicyanomethylenedihydrofuran (Lee et al. 2010). The cross-linker units were maintained to a size small, without having a large branching structure separately composed of a photocross-linker, a ligand, and a fluorophore. For example, photoactivatable uridine (**34**) prepared from 7-azido-4-(bromomethyl)coumarin was incorporated into various oligonucleotides (Kellner et al. 2011). Upon UV irradiation, the cognate protein such as hnRNP A1 protein was labeled with a coumarin fluorophore. Compound **35**, as a HaloTag substrate, was incorporated into HaloEnzyme-fused proteins in an enzymatic manner. By decomposition of the azide group with irradiation, the protein localization patterns in living *Caulobacter crescentus* bacteria were successfully highlighted, which helped elucidated the mechanisms of asymmetric

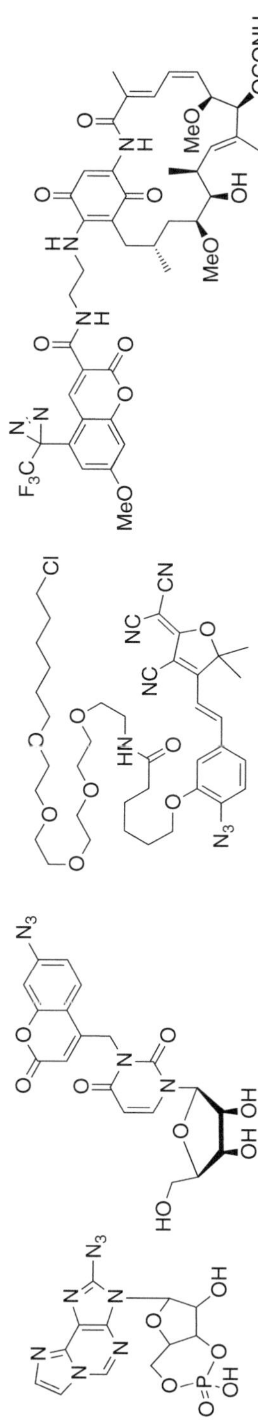

33 34 35 36

Fig. 2.17 Fluorogenic photoprobes

cell division (Lee et al. 2010). We have revealed that the diazirine moiety has the ability to quench the fluorescence of coumarin. The 3-trifluoromethyldiazirinyl group was attached at the 5- or 7-position of the coumarin skeleton to afford the smallest coumarin-based aromatic diazirine compounds, a new class of fluorogenic cross-linker, by the sharing of the benzene ring of the coumarin fluorophore (Tomohiro et al. 2013a, b). The cross-linker units were easily synthesized from 2-hydroxy-6-methoxy-4-[(3-trifluoromethyl)-3*H*–diazirin-3-yl]benzaldehyde via a Knoevenagel condensation reaction. The 5-position-substituted derivatives showed strong emission that was initially quenched efficiently and was then extensively recovered in response to its decomposition by irradiation with 365 nm light. Interestingly, ^{1}H and ^{19}F NMR studies of the photolysis did not indicate the existence of a photochemical pathway to produce the diazo intermediate that the common diazirines have partly converted, indicating that unfavorable, nonspecific labeling via the diazo intermediate should be reduced. The 5-position-substituted derivatives were installed at position C-17 of geldanamycin (GA), a potent, therapeutic cancer inhibitor, to yield photoactivatable GA probes (**36**) that were photochemically cross-linked to heat shock protein 90, accompanied with a fluorescent tag.

2.4.2 Fluorogenic PAL Coupled with a Cleavage Function for Target Identification

When the labeled protein can be detected on the gel by utilizing a detection tag, the protein band is usually cut out and digested with protease followed by MS-based identification such as PMF analysis. Hence, the digestion products contained significant amounts of concomitants in addition to non-labeled target proteins, which often resulted in listing many candidates as results of the MS Mascot search. Other procedures for the purification of the labeled protein, such as the biotin–avidin system, often face the same situation. To provide the guarantee of accurate identification, the repeated enrichment/purification, in order to obtain highly purified labeled protein, as well as the additional amino acid sequencing of the peptide fragments, especially of the labeled peptide, is required. Using conventional PAL methods, however, the identification of the labeled peptide has often resulted in failure. The handling of unknown peptides of small quantity is generally difficult because individual peptide fragments possess different physical properties. The target peptide has been significantly lost during purification due to nonspecific adsorption on the support or vessel. The small difference of the target signal in comparison with the control experiments could be hardly distinguished due to poor reproducibility. A typical example was an early PAL study on the tetrodotoxin (TTX)-binding in the eel sodium channel protein in 1992 (Nakayama et al. 1992). Using a diazirine-tethering TTX probe with a trithium, a minute amount of radioactive fragment was clearly observed by HPLC of the digested sample. However, this signal disappeared in the second HPLC performed to purify the target peptide. Through years of enormous efforts, the target peptide was finally identified via enrichment by immunoprecipitation using many kinds of antibodies. The inherent high speed and sensitivity of mass spectrometry makes it well suited for high-throughput analysis of minute sample amounts, and the current development of LC-MS-based analysis has accelerated a comprehensive analysis of biological processes. Even though high-performance MS equipment and analysis software can be used for exhaustive analysis of all fragments in proteomics, specification of the labeled peptide by LC-MS has often been complicated.

The highly sensitive fluorescence property given into the labeled peptide should be quite useful for selecting it among non-labeled peptides in HPLC, and the fluorescent peak can be directly identified by following mass analysis. However, the determination of the target peptide could be very hampered in a solution containing numerous contaminants from proteome. We developed a unique PAL strategy for rapid target identification using a diazirine-based fluorogenic cross-linker, which consists of an *ortho*-hydroxycinnamide moiety and a trifluoromethyldiazirinyl group directly substituted on the benzene ring. The size of this unit was similar to that of conventional trifluoromethylphenyldiazirine cross-linkers, in which a double bond was inserted between a benzene ring and a carbonyl group. This compound's particular features enable it to undergo two photoreactions, including photolysis of the diazirine ring and photoisomerization, from the *E* to *Z* form, of the double bond. When the initial *E* form was changed to the *Z* form under UV irradiation, the hydroxy group at the *ortho* position became spatially close to the carbonyl group and then an easily induced intramolecular substitution reaction followed, leading to the formation of a coumarin ring (Fig. 2.18). After photoreactions, the coumarin fluorophore should be installed at the interacting site within the target protein without any chemical treatments. The initial study was applied to fluorescent visualization of living B cells using a dihydroxylcinnamate-type cross-linker (**37**, Fig. 2.19) (Tomohiro et al. 2011). Photoactivatable lysozyme (photo-Lyz) was prepared by the attachment of the cross-linker unit to lysine residues on the protein surface. Upon brief photo-irradiation after addition of photo-Lyz into a cell solution, the blue emission was detected from B cells expressing anti Lyz on cell membrane. The coumarin fluorophore was successfully transferred between interacting proteins, from photo-Lyz, the bait protein, to anti-Lyz, the prey protein, via two sequential

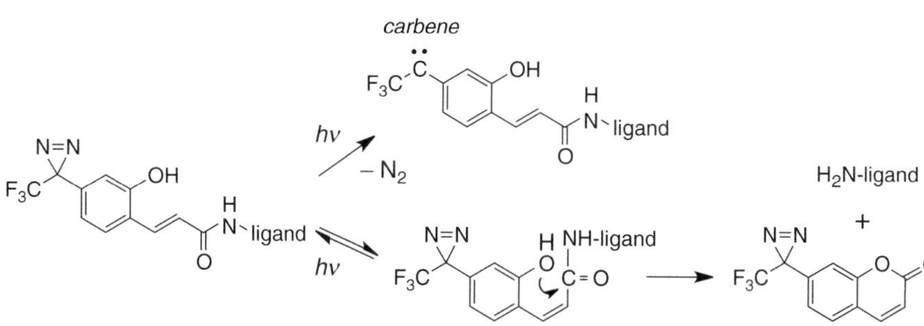

Fig. 2.18 Two kinds of photoreactions of *o*-hydroxycinnamate diazirine derivative

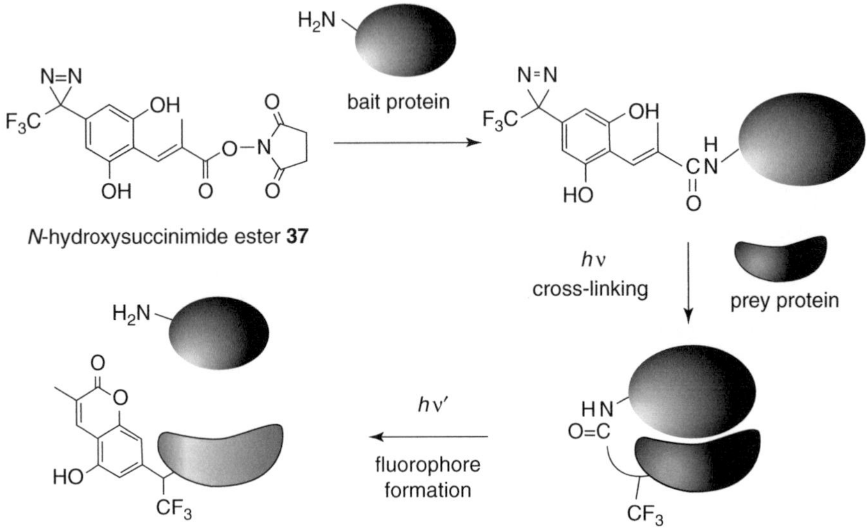

Fig. 2.19 Strategy of fluorogenic label transfer to cognate protein

photoreactions, including the specific cross-linking of each of the proteins and coumarin formation on the prey protein, accompanied with the leaving of the bait protein.

The photolysis of diazirine group of compound **40** was efficiently achieved under 360 nm light at 0 °C, which was verified from the absorption decay at 360 nm due to the n-π* transition (Fig. 2.20a). It has been also supported by the ^{19}F NMR studies, since signals from the CF_3 group in ^{19}F NMR appeared at different chemical shifts: for the diazirine compounds at around −65 ppm, for methanol (solvent) adducts at −76 ppm, and for the corresponding diazo compounds at −57 ppm (Tomohiro et al. 2011). The photolysis of the diOH derivative (**40**) proceeded quite slowly compared to that of the OH-free derivative (**38**). This indicated that diazirine photolysis largely depends on the hydroxy substituent at the ortho position of the benzene ring, even though the π-conjugated system of benzene does not interact with that of diazirine. The *E*/*Z* ratio finally reached equilibrium at ca. 1/2 for compound **38** and ca. 0.5 for compound **39** (mono-OH derivative). From these reasons mono-OH derivative has been chosen for the following studies. In contrast, Fig. 2.20b illustrated the coumarin formation of ester (**40**) and amide (**41**) derivatives, which indicated the *E*-*Z* photoisomerization efficiently occurred under irradiation of light at both 313 and 360 nm. ^{1}H NMR studies also supported this phenomenon. However, the intramolecular substitution of amide derivative **41** became slow at 0 °C. The initial reaction rate of compound **40** was ca. 19 times faster than that of compound **41** under the light at around 315 nm. To label the coumarin onto the target protein, the intramolecular cyclization (coumarin formation/cleavage reaction) must occur after cross-linking by diazirine photolysis. The intramolecular cyclization (coumarin formation) could be regulated with the reaction

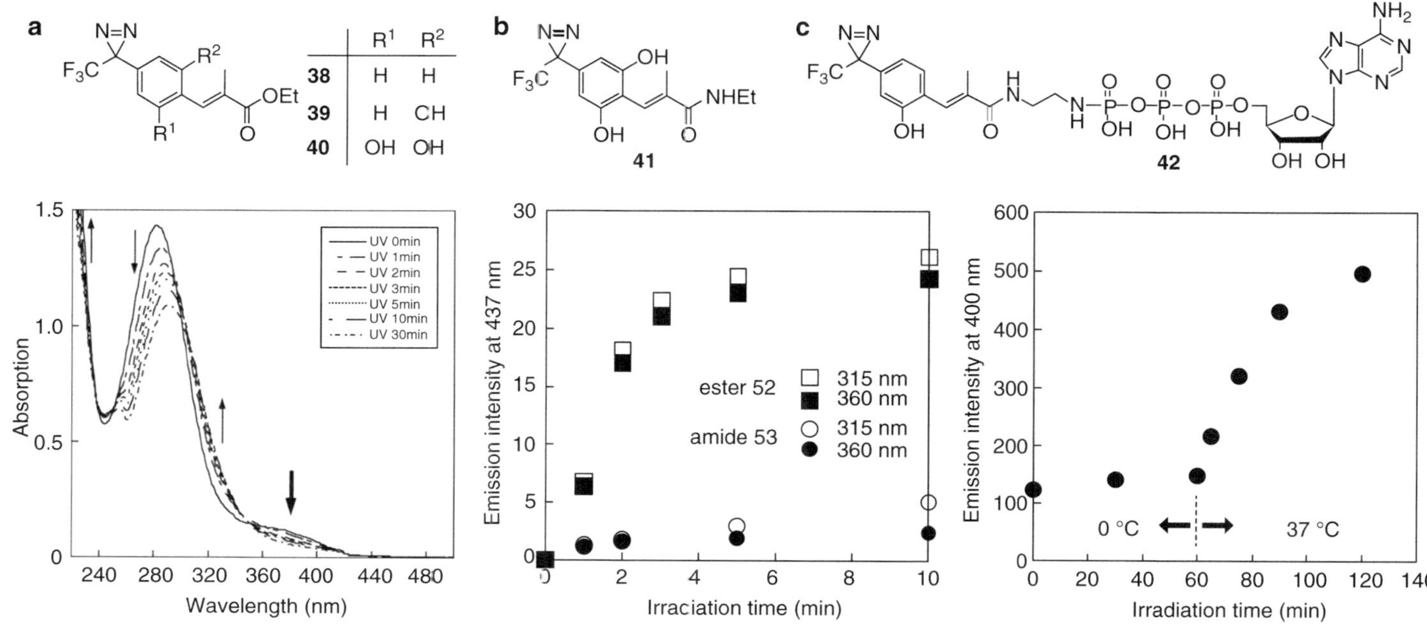

Fig. 2.20 (**a**) Absorption decay spectra of compound **40** under UV irradiation at 360 nm at 0 °C. (**b**) Coumarin formation of ester (**40**, *square*) and amide (**41**, *circle*) derivatives under UV irradiation at 315 (*open*) and 360 nm (*filled*) at 0 °C. (**c**) Temperature dependence of coumarin formation of ATP probe **42** under UV irradiation at 360 nm

temperature by connecting with the ligand via a stable amide bond instead of an ester bond. In fact, the intramolecular substitution reaction of ATP probe **42** could be resisted under irradiation at 0 °C for 60 min but efficiently took place at 37 °C, indicated by an increase of fluorescence at 400 nm (λ_{ex} = 320 nm, Fig. 2.20c) (Tomohiro et al. 2013a, b).

In this method, the cross-linker itself is changed to a fluorophore. This means that a fluorescent tag need not have to be pre-installed into a probe structure, which will be greatly advantageous in probe design and synthesis in terms of stability, solubility, simplicity, size, and ease of preparation. In addition, the intramolecular cyclization (coumarin formation) can be considered as a scissile function, which can be utilized for selective isolation (enrichment) of cross-linked proteins from a solid support for purification. Despite the simple structure, this unit exhibited multifunctional properties, including photocross-linking, fluorogenicity, and cleavability, in addition to possessing a carboxy group for ligand introduction. In fact, this unit was used for analysis of the interaction between a signal peptide and its transporter protein in soybean, and the photoreactions were evaluated in terms of efficiency in target identification (Morimoto et al. 2013). The photoactivatable unit was conjugated to the N-terminus of the signal peptide (ILRAFY, the vacuolar sorting determinant (VSD) peptide) via biocytin containing a biotin moiety, resulting in the production of probe **43** (Fig. 2.21). Upon photolysis, the vacuolar sorting receptor (VSR) of soybean was biotinylated by cross-linking with the probe, and after SDS-PAGE, the labeled VSR was detected by a chemiluminescent method using the avidin–HRP conjugate. The emission intensity of the band was not changed over long irradiation, which suggested that the second photoreaction did not efficiently proceed at 0 °C. The cleavage was subsequently carried out with irradiation at room temperature, and the intensity decreased with the irradiation time in response to the release of the biotin group.

The biotinylated VSR was trapped on the avidin-immobilized support and was then selectively eluted by the second irradiation after thoroughly washing the support. The labeled VSR was recovered from the support at about 60% yield, and it showed a fluorescent emission due to the corresponding coumarin. Tryptic digest of the eluted protein was subjected to HPLC, and the coumarin-labeled peptide frag-

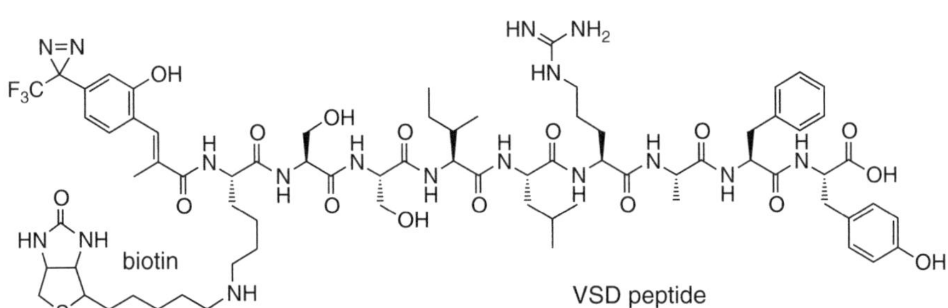

43

Fig. 2.21 Structure of photoactivatable signal peptide tethering a biotin

ments were selectively detected. The following ESI-MS of two major fluorescent peaks successfully indicated the coumarin-modified peptides as FVVEK and VWNAQK, which were shown to be true by the amino acid sequence determined by MS/MS analysis. Assignment of the labeled amino acid residues was completed at the same time in a short period without repeating the problematic purification procedure.

2.4.3 IsoFT-PAL for Rapid Target Identification

The use of stable-isotope labeling and the selective detection of the distinctive isotopic patterns for labeled peptides using mass spectrometry have given an efficient and powerful method for quantitative proteomics. Stable-isotope labeling techniques include the use of mixed isotope tags and isobaric tags, for example, isotope-coded affinity tag (ICAT) and stable-isotope labeling with amino acids in cell culture (SILAC) for the former technique, and isobaric tags such as isobaric tags for relative and absolute quantitation (iTRAQ) and deuterium isobaric amine reactive tags (DiART) for the latter one. Although incorporation of isotopic label splits single MS signals, resulting in reducing the MS signal intensity, further combination with biotin techniques has made the minute, labeled peptides easy to distinguish and able to be clearly determined. PAL reagents in combination with stable isotopes have been practically useful for target identification, such as phenyldiazirine incorporating deuterium and carbon-13 (**44**, Fig. 2.22) (Hashimoto and Hatanaka 2004; Murai et al. 2010) and deuterated benzophenone (**45**) with a biotin tag for identifying the interface region of the immunosuppressive drug cyclosporin A within cyclophilin A (Lamos et al. 2006). In addition, an isotope-coded PAL tag (**46**) was also developed with the attachment of a perfluoroalkyl chain as a purification tag (Song et al. 2012).

Recently, we established an efficient PAL methodology, a stable-isotope-based strategy, that has been integrated with the coumarin-tagging technique as described above, which uses an isotope-coded fluorescent tag (IsoFT) as a fully functionalized cross-linker (Tomohiro et al. 2014). A deuterated ethyl group was introduced at the α-position of the cinnamate unit to be transferred into the target protein as a part of the coumarin tag. The structure of the cross-linker unit maintains the size to be typical of a cross-linker, which thus has minimal effects on the affinity of the ligand for

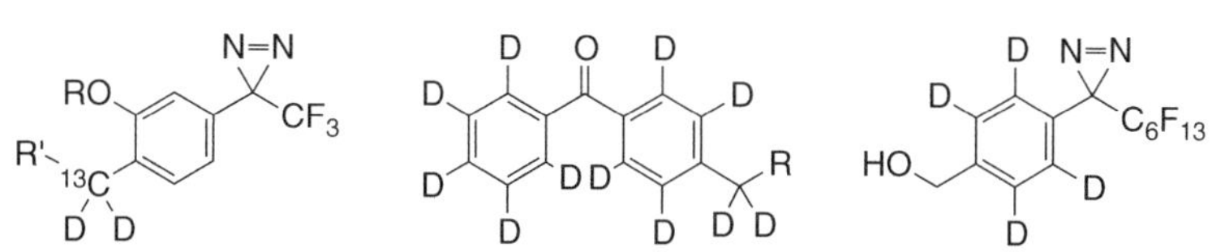

44 **45** **46**

Fig. 2.22 Isotope-coded photophores

the target protein. After UV irradiation, the stable-isotopic coumarin fluorophore should be created on the interacting interface of the target protein. In LC-MS analysis of the proteolytic digest, the labeled peptides can be selectively detected by the measurement of coumarin fluorescence in HPLC and distinguished as doublet peaks with a mass difference of 5.03 u ($z = 1$) in MS by the use of mixed isotope tags (light and heavy ethyl groups) in equal amounts (Fig. 2.23).

The detailed procedure is as follows: in PAL experiment of avidin with a biotin probe (**47**, Fig. 2.24) among a four-protein mixture, identification of the labeled peptide and the labeled amino acid residue has been carried out to establish the methodology and to evaluate the efficacy in microlevel analysis to get structural information on the binding state. After incubation of a protein mixture of avidin, bovine serum albumin, transferrin, and carbonic anhydrase (50 ng of each) with the probe **47**, the solution was irradiated with 365 nm light using a 250 W high-pressure mercury lamp with a bandpass filter (fwhm = 10 nm) at 0 °C. After reductive alkylation and the removal of small compounds by ultrafiltration, the sample solution was irradiated with 365 nm light at 37 °C and then digested with trypsin/Lys-C. Although the HPLC profile of the digested products showed many peptide fragments through the detection of absorption at 215 nm, only two peaks appeared through the detection

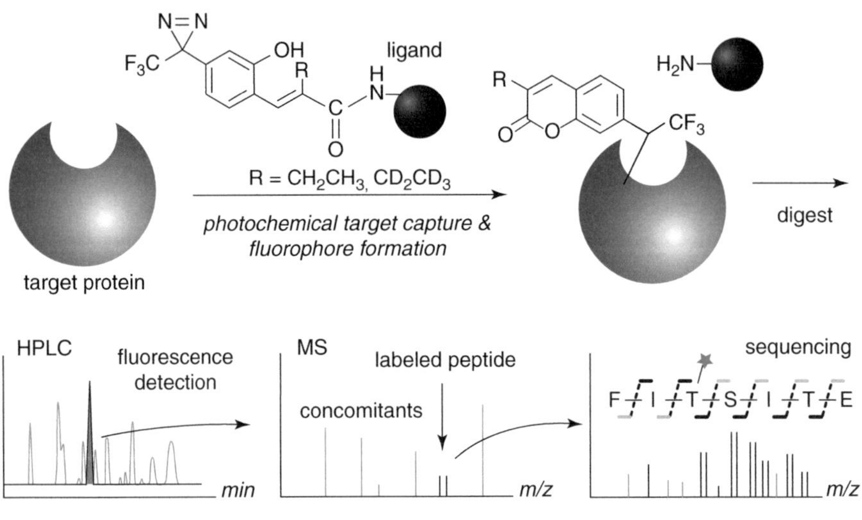

Fig. 2.23 Strategy of IsoFT-PAL

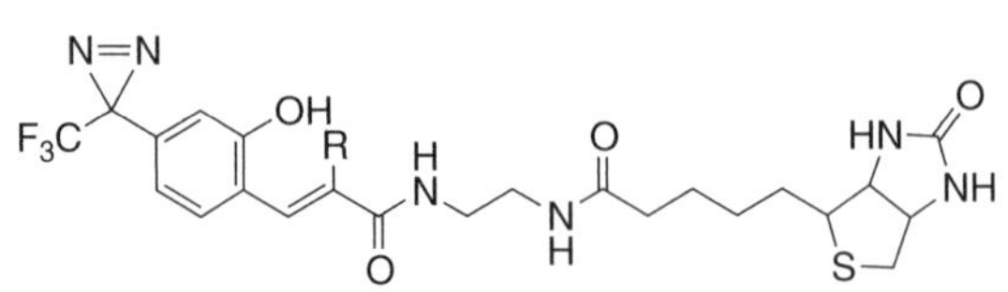

47

	R
a	CH_2CH_3
b	CD_2CD_3

Fig. 2.24 Structure of photoactivatable biotin probes

of fluorescence at 410 nm, due to coumarin moiety. The decrease of these peak intensities in the presence of biotin as a competitor indicated the specificity of the probe to avidin. A doublet peak at *m/z* 744.8 and 747.3 ($z = 2$) could be simply chosen as the target while these peaks were small compared with other MS peaks of the first fluorescent peak in HPLC, which was expected to be the peptide fragment from residues 101–111 of avidin. Another labeled peptide at *m/z* 1122.5 and 1124.9 ($z = 2$) in the second fluorescence peak was easily selected and expected to be the peptide fragment from residues 27–45. The assignment was determined by the MS/MS amino acid sequence analysis. The mass difference, $\Delta m = 5$ u, may be suitable for MS/MS-based sequencing, as the number of MS peaks can be limited by reducing the mass range around the target peak for isolation. The MS/MS spectrum of peptide 2 displayed a series of b_n ions and y_n ions where the fragments of b_{15}–b_{18} and y_5–y_{18} with a coumarin modification appeared as doublet peaks, whereas unmodified fragments appeared as singlet peaks. The labeled sites were identified as Asn104 (indicated by an asterisk) in the 11-amino-acid peptide (SSVN*DIGDDWK) and Ser41 of the 19-amino-acid peptide (GEFTGTYTTAVTATS*NEIK). Thus, just by looking at doublet peaks, the sequencing with a high depth of coverage as well as the assignment of the labeled residue can be achieved without difficulty. The small coumarin tag is very advantageous for analysis of all MS/MS fragments in a relatively small mass range without decreasing the MS resolution. Furthermore, this tag is stable under denaturation conditions, digestion procedures, and collision-induced dissociation at typical energy levels. These features can succeed in curbing complex fragmentation and the reduction of signal intensity, in addition to avoiding the fragmentation derived from the ligand molecule upon releasing.

Two labeled amino acid residues are located on the lip of the biotin-binding pocket (Fig. 2.25) which are reasonable positions considering the probe structure. The labeling yield of avidin was calculated as about 8% from the chemiluminescent method after SDS-PAGE, which indicated about 230 fmol as the total of the two labeled peptide fragments. However, it should be noted that these peaks were not observed in the HPLC of the PAL products using only avidin through the

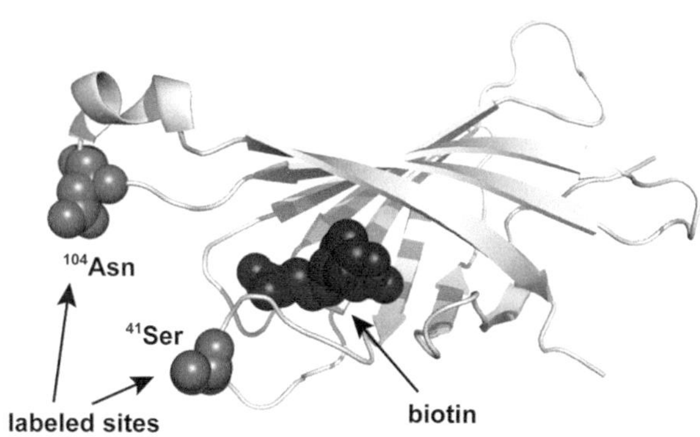

Fig. 2.25 Labeled residues within avidin with the biotin probe **47** (crystal data from PDB ID: 1AVD (Pugliese et al. 1993))

measurement of absorption at 215 nm, while these were clearly detected through the measurement of emission at 410 nm (λ_{ex} = 340 nm). These results suggested that some of these peptides were expected to be lost during the procedure. Identification of the labeled peptides from the four-protein mixture was finally accomplished in several repeated experiments, including an inhibition assay, without any further purification procedures. It should be also important to note that the assignment of multiple labeled sites has been completed in a few days after irradiation without using MS analysis software, whereas by conventional methods, and with great effort, only major labeled residues can be identified. This indicates that this method can raise the reliability of the data and can give more structural information on the ligand-binding interface (Masuda et al. 2015).

PAL-based target identification was then performed using HeLa cell lysate. In proteomics, however, it is important that a tiny amount of labeled proteins should be concentrated for the microlevel analysis. Biotin is also an excellent tag for enrichment of some enzymes, such as biotin-dependent carboxylases, as well as being a ligand. Therefore, these labeled proteins can be enriched and selectively recovered from the avidin-immobilized support by the second round of photo-irradiation. The lysate (cytoplasm, 2 mg of proteins in total) containing the probe (100 μM) was irradiated at 0 °C and then treated with reductive alkylation reagents. The biotinylated proteins were enriched by trapping on avidin-immobilized agarose gel, thoroughly washed with a 0.2% SDS-containing buffer solution, and then eluted with the second irradiation at 37 °C. This enrichment process was also useful to wash out the small fluorescent compounds in the cell lysate that are disturbing factors in the following LC-MS analysis and to avoid co-elution of intrinsically biotinylated proteins. HPLC profiles of the tryptic digest showed some fluorescent peaks with intensities reduced in the presence of biotin, acting as a competitor. A doublet peak at *m/z* 473.7 and 476.2 ($z = 2$) was clearly distinguished in the ESI-MS profile of fluorescent peak 1. The following MS/MS analysis revealed that the sequence was DT*QAMK (with coumarin) and that it was a part of pyruvate carboxylase (PC, 1178 a.a.), an enzyme that catalyzes the biotin-dependent production of oxaloacetate. In fact, TagIdent analysis of the amino acid sequence, DTQAMK, identified three candidate human proteins: (1) mitochondrial PC, (2) cleft lip and palate transmembrane protein 1-like protein, and (3) isoform 2 of cleft lip and palate transmembrane protein 1-like protein. Then, we easily confirmed PC as the target. Alternatively, PMF analysis of the digests of eluted proteins, which was the same sample as that used in the LC-MS analysis, gave hundreds of candidates including abundant proteins in HeLa cells according to the Mascot database. PC was listed as a minor candidate in the Mascot analysis, since the peptide coverage of PC was about 13%. The result indicated the difficulty in purification of minute protein even if using this selective elution method of the biotin–avidin system. PC is a mitochondrial protein that contains the biotin carboxylase (BC), carboxyltransferase (CT), and biotin-carboxyl carrier protein (BCCP) domains. The labeled amino acid residue, Thr1092, is located in the end region of PC tetramerization domain (PT), between the CT and BCCP domains, where it exists in the central area of the tetrameric structure (X-ray crystallography data, PBD ID 3BG3 (Xiang and Tong 2008)). The residue, Thr1092, is oriented toward the catalytic pocket (Fig. 2.26). Due to the chemical property of carbene as a reactive species, Thr1092 should be spatially close to the diazirine moiety of the probe. The result supports the fact that the C-terminal BCCP domain allosterically moves in the proximity of the catalytic site when a biotin molecule binds in the pocket (see left panel of Fig. 2.26) (Lietzan and St Maurice 2013). Another labeled peptide in the fluorescent peak 2 (22.2 min) was successfully selected among numerous contaminants and was assigned as IT*IGNK. TagIdent analysis suggested five candidates including four isoforms of acetyl-CoA carboxylase (ACC)-1 and an ACC-2. Although human ACCs have not been crystallized in their entirety, the corresponding labeled threonine in each ACC is located adjacent to the BCCP domain where it is in a similar position to that observed with PC. These ACCs were not listed in the PMF analysis. These results indicated that sequencing of the labeled peptide in addition to the PMF analysis will be important in providing more accurate information regarding the target protein and the ligand-binding site.

The IsoFT strategy using an all-in-one, multifunctional cross-linker demonstrates excellent features for LC-MS-base target identification. Despite the progress of MS apparatus and software, the identification of the labeled site at microlevel analysis for structural elucidation has still often failed using conventional PAL probes. Highly efficient focusing to the protein target from the proteome has been achieved by three different, sequential, detection steps: photochemical elution of proteins enriched using a biotin–avidin interaction, fluorescent detection in HPLC, and isotopic signal detection in MS. This enables the identification processes to continue forward and to be completed even if the sample was contaminated. The high clarity of the target signal, which can be distinguished easily, can drastically reduce the time and quantity required for the identification from years to weeks and from mg to μg, respectively. Simple manipulations such as photo-irradiation, washing, digestion, and operation of LC-MS without any special techniques will give general versatility in use. In addition, the simple structure of all-in-one cross-linkers,

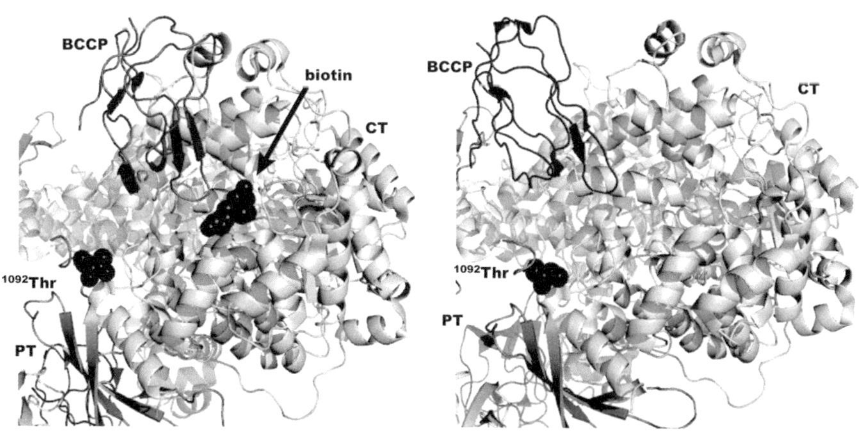

Fig. 2.26 X-ray crystal structures of human pyruvate carboxylase (crystal data from PDB ID: 3BG3 (Xiang and Tong 2008))

which can be converted themselves into fluorophores, will be beneficial in minimizing the damage to ligand specificity, and the small and stable tag can avoid complex fragmentation of the labeled peptide and prevent the decrease of the high MS resolution of MS/MS analysis.

2.5 Future Aspects

Solid systems have been used for simplification of the washing procedure, to exclude reactants, and for parallel analysis of multiple biomolecules. In PAL-based protein profiling, solid supports such as glass, plastics, and PDVF membranes have been used for immobilization of small molecules to trap cognate proteins, peptides to pick up the labeled peptide from digestion products, or proteins to make protein chip in conjugation with photoactivatable polymers (Tomohiro et al. 2009). Among them, a unique immobilizing method of inert bioactive molecules on the solid surface has been developed using diazirine-fabricated glass slides or particles without prior chemical modification (Kanoh et al. 2003, 2005). Upon irradiation, diazirine forms a covalent bond with the bioactive molecules in a nonselective manner (Suzuki et al. 2015), which results in various kinds of surface presentation—the ligand molecule has been immobilized in different ways. Some of ligands on the surface might maintain their activity, which allows the investigation of the interaction against various proteins. Despite these methods show excellent potency for primary screening of bioactive compounds, unfortunately, these cannot be applied to the approaches using living cells.

Another cumbersome step in PAL is the preparation of proper photoprobes, especially for complicated natural products as well as simple, small compounds without having any functional groups that may be derivatized. The nonselectivity of diazirine-inducing cross-linking allows compound pools to be prepared in a random manner without an enormous effort, expense, and time period for the proper probing of the ligand to determine if it is in a bioactive form. Parallel phenotype screening using some compounds in the compound pool will be achieved with the efficient target selection techniques using multifunctional cross-linkers, which will be promising for the determination of biologically relevant targets of small molecules. In association with the development of analysis software and hardware, further development of the efficient strategies for PAL-based identification will accelerate in situ protein profiling in cells and/or organisms in the near future.

Acknowledgments The author is delighted to recognize the many contributions of my collaborators, whose names appear in the references, and especially grateful to Professor Yasumaru Hatanaka for his pioneering works on multifunctional diazirines. His far-sighted ideas and superior insights have been impressed many researchers in this field. The author is grateful to JSPS KAKENHI (the Grants-in-Aids for Scientific Research from the Japan Society for the Promotion of Sciences) for financial support.

Dr. Takenori Tomohiro is associate professor of the Graduate School of Medicine and Pharmaceutical Sciences, University of Toyama. He received both B.Sc. (1986) and Ph.D. degrees (1992) from the University of Tsukuba. He had worked at the National Chemical Laboratory for Industry (National Institute of Advanced Industrial Science and Technology at present) since 1986 and was a postdoctoral research fellow in the University of Oxford for 2 years from 1992. He joined the faculty at the Toyama Medical and Pharmaceutical University in 2002 (University of Toyama at present). His research interests are in the areas of chemical biology related to biomolecular interaction and medicinal chemistry.

References

Addona GH, Husain SS, Stehle T, Miller KW (2002) Geometric isomers of a photoactivable general anesthetic delineate a binding site on adenylate kinase. J Biol Chem 277:25685–25691

Ballell L, van Scherpenzeel M, Buchalova K, Liskamp RM, Pieters RJ (2006) A new chemical probe for the detection of the cancer-linked galectin-3. Org Biomol Chem 4:4387–4394

Beatty KE, Szychowski J, Fisk JD, Tirrell DA (2011) A BODIPY-cyclooctyne for protein imaging in live cells. ChembioChem 12:2137–2139

Bell JL, Haak AJ, Wade SM, Sun Y, Neubig RR, Larsen SD (2013) Design and synthesis of tag-free photoprobes for the identification of the molecular target for CCG-1423, a novel inhibitor of the Rho/MKL1/SRF signaling pathway. Beilstein J Org Chem 9:966–973

Blizzard RJ, Backus DR, Brown W, Bazewicz CG, Li Y, Mehl RA (2015) Ideal bioorthogonal reactions using a site-specifically encoded tetrazine amino acid. J Am Chem Soc 137:10044–10047

Boutureira O, Bernardes GJ (2015) Advances in chemical protein modification. Chem Rev 115:2174–2195

Burkard N, Bender T, Westmeier J, Nordmann C, Huss M, Wieczorek H, Grond S, von Zezschwitz P (2010) New fluorous photoaffinity labels (F-PAL) and their application in V-ATPase inhibition studies. Eur J Org Chem 2010:2176–2181

Chan EW, Chattopadhaya S, Panicker RC, Huang X, Yao SQ (2004) Developing photoactive affinity probes for proteomic profiling: hydroxamate-based probes for metalloproteases. J Am Chem Soc 126:14435–14446

Chin JW, Schultz PG (2002) In vivo photocrosslinking with unnatural amino acid mutagenesis. ChembioChem 3(11):1135–1137

Chin JW, Martin AB, King DS, Wang L, Schultz PG (2002a) Addition of a photocrosslinking amino acid to the genetic code of Escherichiacoli. Proc Natl Acad Sci U S A 99:11020

Chin JW, Santoro SW, Martin AB, King DS, Wang L, Schultz PG (2002b) Addition of p-azido-L-phenylalanine to the genetic code of Escherichia coli. J Am Chem Soc 124:9026–9027

Cisar JS, Cravatt BF (2012) Fully functionalized small-molecule probes for integrated phenotypic screening and target identification. J Am Chem Soc 134:10385–10388

Devaraj NK, Hilderbrand S, Upadhyay R, Mazitschek R, Weissleder R (2010) Bioorthogonal turn-on probes for imaging small molecules inside living cells. Angew Chem Int Ed 49:2869–2872

Dreyfuss G, Schwartz K, Blout ER, Barrio JR, Liu FT, Leonard NJ (1978) Fluorescent photoaffinity labeling: Adenosine 3′,5′-cyclic monophosphate receptor sites. Proc Natl Acad Sci U S A 75:1199–1203

Dubinsky L, Jarosz LM, Amara N, Krief P, Kravchenko VV, Krom BP, Meijler MM (2009) Synthesis and validation of a probe to identify quorum sensing receptors. Chem Commun (47):7378–7380

Evans MJ, Cravatt BF (2006) Mechanism-based profiling of enzyme families. Chem Rev 106:3279–3301

Hashimoto M, Hatanaka Y (2004) Simple synthesis of deuterium and ^{13}C labeled trifluoromethyl phenyldiazirine derivatives as stable isotope tags for mass spectrometry. Chem Pharm Bull 52:1385–1386

Hashimoto M, Hatanaka Y (2005) Post-biotinylation of photocrosslinking by Staudinger–Bertozzi ligation of preinstalled alkylazide tag. Chem Pharm Bull 53:1510–1512
Hatanaka Y (2015) Development and leading-edge application of innovative photoaffinity labeling. Chem Pharm Bull 63:1–12
Hatanaka Y, Hashimoto M, Kanaoka Y (1994) A novel biotinylated heterobifunctional cross-linking reagent bearing an aromatic diazirine. Bioorg Med Chem 2:1367–1373
Hino N, Okazaki Y, Kobayashi T, Hayashi A, Sakamoto K, Yokoyama S (2005) Protein photo-cross-linking in mammalian cells by site-specific incorporation of a photoreactive amino acid. Nat Methods 2:201–206
Hiramatsu T, Guo Y, Hosoya T (2007) 3-Azidodifluoromethyl-3*H*-diazirin-3-yl group as an all-in-one functional group for radioisotope-free photoaffinity labeling. Org Biomol Chem 5:2916–2919
Hosoya T, Hiramatsu T, Ikemoto T, Nakanishi M, Aoyama H, Hosoya A, Iwata T, Maruyama K, Endo M, Suzuki M (2004) Novel bifunctional probe for radioisotope-free photoaffinity labeling: compact structure comprised of photospecific ligand ligation and detectable tag anchoring units. Org Biomol Chem 2:637–641
Hosoya T, Hiramatsu T, Ikemoto T, Aoyama H, Ohmae T, Endo M, Suzuki M (2005) Design of dantrolene-derived probes for radioisotope-free photoaffinity labeling of proteins involved in the physiological Ca^{2+} release from sarcoplasmic reticulum of skeletal muscle. Bioorg Med Chem Lett 15:1289–1294
Hsu TL, Hanson SR, Kishikawa K, Wang SK, Sawa M, Wong CH (2007) Alkynyl sugar analogs for the labeling and visualization of glycoconjugates in cells. Proc Natl Acad Sci U S A 104:2614–2619
Hulce JJ, Cognetta AB, Niphakis MJ, Tully SE, Cravatt BF (2013) Proteome-wide mapping of cholesterol-interacting proteins in mammalian cells. Nat Methods 10:259–264
Kambe T, Correia BE, Niphakis MJ, Cravatt BF (2014) Mapping the protein interaction landscape for fully functionalized small-molecule probes in human cells. J Am Chem Soc 136:10777–10782
Kanaoka Y (1977) Organic fluorescence reagents in the study of enzymes and proteins. Angew Chem Int Ed 16:137–147
Kanaoka Y, Kobayashi A, Sato E, Nakayama H, Ueno T, Muno D, Sekine T (1984) Multifunctional cross-linking reagents. I. Synthesis and properties of novel photoactivable, thiol-directed fluorescent reagents. Chem Pharm Bull 32:3926–3933
Kaneda M, Masuda S, Tomohiro T, Hatanaka Y (2007) A simple and efficient photoaffinity method for proteomics of GTP-binding proteins. ChembioChem 8:595–598
Kanoh N, Kumashiro S, Simizu S, Kondoh Y, Hatakeyama S, Tashiro H, Osada H (2003) Immobilization of natural products on glass slides by using a photoaffinity reaction and the detection of protein-small-molecule interactions. Angew Chem Int Ed 42:5584–5587
Kanoh N, Honda K, Simizu S, Muroi M, Osada H (2005) Photo-cross-linked small-molecule affinity matrix for facilitating forward and reverse chemical genetics. Angew Chem Int Ed 44:3559–3562
Kawaguchi Y, Tanaka G, Nakase I, Imanishi M, Chiba J, Hatanaka Y, Futaki S (2013) Identification of cellular proteins interacting with octaarginine (R8) cell-penetrating peptide by photo-crosslinking. Bioorg Med Chem Lett 23:3738–3740
Keeler EK, Campbell P (1976) A fluorescent photo-affinity label for cyclic AMP binding proteins. Biochem Biophys Res Commun 72:575–580
Kellner S, Seidu-Larry S, Burhenne J, Motorin Y, Helm M (2011) A multifunctional bioconjugate module for versatile photoaffinity labeling and click chemistry of RNA. Nucleic Acids Res 39:7348–7360
Kumar NS, Young RN (2009) Design and synthesis of an all-in-one 3-(1,1-difluoroprop-2-ynyl)-3*H*-diazirin-3-yl functional group for photo-affinity labeling. Bioorg Med Chem 17:5388–5395

Kuroda T, Suenaga K, Sakakura A, Handa T, Okamoto K, Kigoshi H (2006) Study of the interaction between actin and antitumor substance aplyronine A with a novel fluorescent photoaffinity probe. Bioconjug Chem 17:524–529

Lamos SM, Krusemark CJ, McGee CJ, Scalf M, Smith LM, Belshaw PJ (2006) Mixed isotope photoaffinity reagents for identification of small-molecule targets by mass spectrometry. Angew Chem Int Ed 45:4329–4333

Le Droumaguet C, Wang C, Wang Q (2010) Fluorogenic click reaction. Chem Soc Rev 39:1233–1239

Lee HL, Lord SJ, Iwanaga S, Zhan K, Xie H, Williams JC, Wang H, Bowman GR, Goley ED, Shapiro L, Twieg RJ, Rao J, Moerner WE (2010) Superresolution imaging of targeted proteins in fixed and living cells using photoactivatable organic fluorophores. J Am Chem Soc 132:15099–15101

Lee K, Ban HS, Naik R, Hong YS, Son S, Kim BK, Xia Y, Song KB, Lee HS, Won M (2013) Identification of malate dehydrogenase 2 as a target protein of the HIF-1 inhibitor LW6 using chemical probes. Angew Chem Int Ed 52:10286–10289

Li G, Liu Y, Liu Y, Chen L, Wu S, Liu Y, Li X (2013a) Photoaffinity labeling of small-molecule-binding proteins by DNA-templated chemistry. Angew Chem Int Ed 52:9544–9549

Li Z, Hao P, Li L, Tan CY, Cheng X, Chen GY, Sze SK, Shen HM, Yao SQ (2013b) Design and synthesis of minimalist terminal alkyne-containing diazirine photo-crosslinkers and their incorporation into kinase inhibitors for cell- and tissue-based proteome profiling. Angew Chem Int Ed Engl 52:8551–8556

Li X, Gao X, Shi W, Ma H (2014a) Design strategies for water-soluble small molecular chromogenic and fluorogenic probes. Chem Rev 114:590–659

Li Z, Wang D, Li L, Pan S, Na Z, Tan CY, Yao SQ (2014b) "Minimalist" cyclopropene-containing photo-cross-linkers suitable for live-cell imaging and affinity-based protein labeling. J Am Chem Soc 136:9990–9998

Lietzan AD, St Maurice M (2013) A substrate-induced biotin binding pocket in the carboxyltransferase domain of pyruvate carboxylase. J Biol Chem 288:19915–19925

Lord SJ, Lee HL, Samuel R, Weber R, Liu N, Conley NR, Thompson MA, Twieg RJ, Moerner WE (2010) Azido push-pull fluorogens photoactivate to produce bright fluorescent labels. J Phys Chem B 114:14157–14167

Luo Y, Vassilev PM, Li X, Kawanabe Y, Zhou J (2003) Native polycystin 2 functions as a plasma membrane Ca^{2+}-permeable cation channel in renal epithelia. Mol Cell Biol 23:2600–2607

MacKinnon AL, Garrison JL, Hegde RS, Taunton J (2007) Photo-leucine incorporation reveals the target of a cyclodepsipeptide inhibitor of cotranslational translocation. J Am Chem Soc 129:14560–14561

Maezawa N, Tsuchikawa H, Katsumura S, Kubo T, Imaoka S (2007) Synthesis of fluorescent and photoaffinity-labeled derivatives of bisphenol A and their inhibitory activity toward hypoxic expression of erythropoietin. Bioorg Med Chem Lett 17:5121–5124

Masuda S, Tomohiro T, Yamaguchi S, Morimoto S, Hatanaka Y (2015) Structure-assisted ligand-binding analysis using fluorogenic photoaffinity labeling. Bioorg Med Chem Lett 25:1675–1678

Maurel D, Banala S, Laroche T, Johnsson K (2010) Photoactivatable and photoconvertible fluorescent probes for protein labeling. ACS Chem Biol 5:507–516

Mayer T, Maier ME (2007) Design and synthesis of a tag-free chemical probe for photoaffinity labeling. Eur J Org Chem 2007:4711–4720

Morimoto S, Tomohiro T, Maruyama N, Hatanaka Y (2013) Photoaffinity casting of a coumarin flag for rapid identification of ligand-binding sites within protein. Chem Commun 49:1811–1183

Murai Y, Takahashi M, Muto Y, Hatanaka Y, Hashimoto M (2010) Simple deuterium introduction at α-position of carbonyl in diazirinyl derivatives for photoaffinity labeling. Heterocycles 82:909–915

Murai Y, Masuda K, Sakihama Y, Hashidoko Y, Hatanaka Y, Hashimoto M (2012) Comprehensive synthesis of photoreactive (3-trifluoromethyl)diazirinyl indole derivatives from 5- and 6- trifluoroacetylindoles for photoaffinity labeling. J Org Chem 77:8581–8587

Nagase T, Nakata E, Shinkai S, Hamachi I (2003) Construction of artificial signal transducers on a lectin surface by post-photoaffinity-labeling modification for fluorescent saccharide biosensors. Chem Eur J 9:3660–3669

Nakamura Y, Inomata S, Ebine M, Manabe Y, Iwakura I, Ueda M (2011) "Click-made" biaryl-linker improving efficiency in protein labelling for the membrane target protein of a bioactive compound. Org Biomol Chem 9:83–85

Nakashima H, Hashimoto M, Sadakane Y, Tomohiro T, Hatanaka Y (2006) Simple and versatile method for tagging phenyldiazirine photophores. J Am Chem Soc 128:15092–15093

Nakayama H, Hatanaka Y, Yoshida E, Oka K, Takanohashi M, Amano Y, Kanaoka Y (1992) Photolabeled sites with a tetrodotoxin derivative in the domain III and IV of the electroplax sodium channel. Biochem Biophys Res Commun 184:900–907

Ong SE, Foster LJ, Mann M (2003) Mass spectrometric-based approaches in quantitative proteomics. Methods 29:124–130

Park D, O'Doherty I, Somvanshi RK, Bethke A, Schroeder FC, Kumar U, Riddle DL (2012) Interaction of structure-specific and promiscuous G-protein-coupled receptors mediates small-molecule signaling in Caenorhabditis elegans. Proc Natl Acad Sci U S A 109:9917–9922

Park J, Koh M, Koo JY, Lee S, Park SB (2016) Investigation of specific binding proteins to photoaffinity linkers for efficient deconvolution of target protein. ACS Chem Biol 11:44–52

Patterson DM, Nazarova LA, Prescher JA (2014) Finding the right (bioorthogonal) chemistry. ACS Chem Biol 9:592–605

Pugliese L, Coda A, Malcovati M, Bolognesi M (1993) Three-dimensional structure of the tetragonal crystal form of egg-white avidin in its functional complex with biotin at 2.7 A resolution. J Mol Biol 231:698–710

Sawa M, Hsu TL, Itoh T, Sugiyama M, Hanson SR, Vogt PK, Wong CH (2006) Glycoproteomic probes for fluorescent imaging of fucosylated glycans in vivo. Proc Natl Acad Sci U S A 103:12371–12376

Shi H, Zhang CJ, Chen GY, Yao SQ (2012) Cell-based proteome profiling of potential dasatinib targets by use of affinity-based probes. J Am Chem Soc 134:3001–3014

Shieh P, Hangauer MJ, Bertozzi CR (2012) Fluorogenic azidofluoresceins for biological imaging. J Am Chem Soc 134:17428–17431

Sivakumar K, Xie F, Cash BM, Long S, Barnhill HN, Wang Q (2004) A fluorogenic 1,3-dipolar cycloaddition reaction of 3-azidocoumarins and acetylenes. Org Lett 6:4603–4606

Sohn CH, Agnew HD, Lee JE, Sweredoski MJ, Graham RL, Smith GT, Hess S, Czerwieniec G, Loo JA, Heath JR, Deshaies RJ, Beauchamp JL (2012) Designer reagents for mass spectrometry-based proteomics: clickable cross-linkers for elucidation of protein structures and interactions. Anal Chem 84:2662–2669

Song Z, Zhang Q (2009) Fluorous aryldiazirine photoaffinity labeling reagents. Org Lett 11:4882

Song Z, Huang W, Zhang Q (2012) Isotope-coded, fluorous photoaffinity labeling reagents. Chem Commun 48:3339–3341

Srinivas N, Jetter P, Ueberbacher BJ, Werneburg M, Zerbe K, Steinmann J, Van der Meijden B, Bernardini F, Lederer A, Dias RL, Misson PE, Henze H, Zumbrunn J, Gombert FO, Obrecht D, Hunziker P, Schauer S, Ziegler U, Käch A, Eberl L, Riedel K, DeMarco SJ, Robinson JA (2010) Peptidomimetic antibiotics target outer-membrane biogenesis in Pseudomonas aeruginosa. Science 327:1010–1013

Suchanek M, Radzikowska A, Thiele C (2005) Photo-leucine and photo-methionine allow identification of protein-protein interactions in living cells. Nat Methods 2:261–267

Suzuki T, Okamura T, Tomohiro T, Iwabuchi Y, Kanoh N (2015) Third generation photo-cross-linked small-molecule affinity matrix: a photoactivatable and photocleavable system enabling quantitative analysis of the photo-cross-linked small molecules and their target purification. Bioconjug Chem 26:389–395

Taldone T, Rodina A, Dagama Gomes EM, Riolo M, Patel HJ, Alonso-Sabadell R, Zatorska D, Patel MR, Kishinevsky S, Chiosis G (2013) Synthesis and evaluation of cell-permeable biotinylated PU-H71 derivatives as tumor Hsp90 probes. Beilstein J Org Chem 9:544–556

Tamura T, Tsukiji S, Hamachi I (2012) Native FKBP12 engineering by ligand-directed tosyl chemistry: labeling properties and application to photo-cross-linking of protein complexes in vitro and in living cells. J Am Chem Soc 134:2216–2226
Tanaka Y, Bond MR, Kohler JJ (2008) Photocrosslinkers illuminate interactions in living cells. Mol BioSyst 4:473–480
Tomohiro T, Hatanaka Y (2014) Diazirine-based multifunctional photo-probes for affinity-based elucidation of protein-ligand interaction. Heterocycles 89:2697–2727
Tomohiro T, Hashimoto M, Hatanaka Y (2005) Cross-linking chemistry and biology: development of multifunctional photoaffinity probes. Chem Rec 5:385–395
Tomohiro T, Tachi N, Azuma Y, Hatanaka Y (2009) Hydrophilic diazirine polymer for one-step photo-fabrication of proteins on polypropylene surface. Heterocycles 79:897–908
Tomohiro T, Kato K, Masuda S, Kishi H, Hatanaka Y (2011) Photochemical construction of coumarin fluorophore on affinity-anchored protein. Bioconjug Chem 22:315–318
Tomohiro T, Yamamoto A, Tatsumi Y, Hatanaka Y (2013a) [3-(Trifluoromethyl)-3*H*-diazirin-3-yl] coumarin as a carbene-generating photocross-linker with masked fluorogenic beacon. Chem Commun 49:11551–11553
Tomohiro T, Inoguchi H, Masuda S, Hatanaka Y (2013b) Affinity-based fluorogenic labeling of ATP-binding proteins with sequential photoactivatable cross-linkers. Bioorg Med Chem Lett 23:5605–5608
Tomohiro T, Morimoto S, Shima T, Chiba J, Hatanaka Y (2014) An isotope-coded fluorogenic cross-linker for high-performance target identification based on photoaffinity labeling. Angew Chem Int Ed 53:13502–13505
Tsukiji S, Miyagawa M, Takaoka Y, Tamura T, Hamachi I (2009) Ligand-directed tosyl chemistry for protein labeling in vivo. Nat Chem Biol 5:341–343
Wartmann T, Lindel T (2013) L-Phototryptophan. Eur J Org Chem 2013(9):1649–1652
Xiang S, Tong L (2008) Crystal structures of human and Staphylococcus aureus pyruvate carboxylase and molecular insights into the carboxyltransfer reaction. Nat Struct Mol Biol 15:295–302
Yang P, Liu K (2015) Activity-based protein profiling: recent advances in probe development and applications. ChembioChem 16:712–724
Zhou Z, Fahrni CJ (2004) A fluorogenic probe for the copper(I)-catalyzed azide-alkyne ligation reaction: modulation of the fluorescence emission via $^3(n, \pi)$-$^1(\pi,\pi)$ inversion. J Am Chem Soc 126:8862–8863
Ziegler S, Pries V, Hedberg C, Waldmann H (2013) Target identification for small bioactive molecules: finding the needle in the haystack. Angew Chem Int Ed 52:2744–2792

Chapter 3
Recent Developments in Designing Compact Biological Photoprobes

Amberlyn M. Wands and Jennifer J. Kohler

Abstract Identification and characterization of small molecule–protein interactions are common needs for both basic science and drug discovery programs. Photoaffinity probe molecules (photoprobes) have been used for more than 40 years to label the targets of small molecules, but identification of those targets has remained challenging. Recently, renewed interest in the use of photoprobes has been spurred by the advent of modern mass spectrometry methods that facilitate target identification. In addition, development of new chemoselective labeling reactions (i.e., "click chemistry" approaches) now enables facile purification of photocrosslinked complexes for analysis. Photoprobe technology is being applied for a variety of purposes, including identifying the direct binding partner(s) of a small molecule, obtaining information about the nature of the ligand binding site in the absence of a three-dimensional structure, determining whether ligand binding occurs through a primary or allosteric site, and investigating the specificity determinants of ligand binding. In this chapter, we discuss a selection of compact photoprobes that have been reported in the past 10 years. The chapter describes photoprobes containing each of the three common photoactivatable functional groups—aryl azide, benzophenone, and diazirine. We highlight the molecular design strategies that have yielded functional photoprobes, including compact construction, choice of photoactivatable functional groups, use of tags for chemoselective labeling, and linker design. We conclude by discussing remaining challenges that stand in the way of widespread adoption of photoprobe reagents.

Keywords Photocrosslinking • Diazirine • Benzophenone • Photoaffinity • Click chemistry • Chemoselective • Proteomics • Ligand binding

A.M. Wands • J.J. Kohler (✉)
Department of Biochemistry, University of Texas Southwestern Medical Center, Dallas, TX 75390, USA
e-mail: jennifer.kohler@utsouthwestern.edu

Y. Hatanaka, M. Hashimoto (eds.), *Photoaffinity Labeling for Structural Probing Within Protein*, DOI 10.1007/978-4-431-56569-7_3

3.1 Introduction

Development of a photoaffinity labeling reagent (or photoprobe) can be a useful biochemical approach for studying the proteinaceous binding partner(s) of a ligand, with such applications as target identification within a proteome or three-dimensional mapping of a ligand binding site. In general, the design of an effective photoprobe relies on the modification of a pharmacophore with a photoreactive group, which upon photoirradiation can form a covalent bond with its target before dissociation occurs. To achieve this, researchers typically utilize information from a structure–activity relationship (SAR) study on the parent compound to determine optimal substitution positions. Three main photoreactive groups are used to date, namely, the aryl azide, the benzophenone, and the diazirine, each of which has distinct advantages and disadvantages that need to be taken into consideration for the intended application. Additionally, a method of detection of photoaffinity-labeled proteins must also be considered. Traditionally, radioisotopes were incorporated into the photoprobe, but more recently researchers have moved toward use of a bioorthogonal functional group that can be further functionalized after crosslinking through the use of a "click chemistry" reaction. For example, photoprobe reagents may incorporate an alkyne, which can be further reacted with an azide-containing reagent to add a fluorescent tag for in-gel fluorescence visualization or a biotin tag as a purification handle for enrichment. Furthermore, the use of a "clickable" handle rather than direct tethering of a fluorophore or biotin to the ligand helps to minimize the overall size of the photoprobe, thereby facilitating photocrosslinking, especially in living cells. This chapter will describe select examples reported over the last decade of compact photoprobes that have been used to answer an array of diverse biological questions, with an emphasis on the aspects of photoprobe design and workflow implementation.

3.2 Aryl Azide-Based Photoaffinity Probes

3.2.1 Photocrosslinking Properties of Aryl Azides

This section focuses on the development of photoprobes that incorporate an aryl azide photoreactive group. Typically, photoactivation of aryl azides results in the formation of a highly reactive nitrene intermediate that can undergo insertion into carbon–hydrogen (C–H) bonds. Advantages to utilizing the aryl azide include its moderate size and ease of synthetic installation, which is most fortuitous when the ligand scaffold contains an existing phenyl group that can tolerate an azide modification. In terms of disadvantages, the activation wavelength of the aryl azide is <300 nm, which has the potential to be absorbed by protein targets with damaging effects. Furthermore, conversion of the nitrene to alternate species can result in such problems as a decrease in photolabeling yields and nonspecific labeling. To circumvent this nonproductive decay of the nitrene, a common strategy relies on flanking the azide group with fluorine substituents to create a tetrafluorophenylazide group that results in a longer-lived singlet nitrene. Both the simple aryl azide and the more synthetically demanding tetrafluorophenylazide have been used in a variety of biological applications, some of which are described below.

3.2.2 *Design and Applications of Aryl Azide Photoprobes*

3.2.2.1 Ligand-Operated Molecular Chaperone (S1R)

Over the past decade, the laboratory of Arnold Ruoho developed aryl azide-based photoprobes to gain structural information on the ligand binding site of the sigma-1 receptor (S1R). S1R is a small (28 kDa) transmembrane protein whose molecular chaperone function can be regulated by an array of both endogenous and synthetic ligands. Because S1R promiscuously recognizes pharmacophores containing a phenylalkylamine, many compounds have been identified that can function as either S1R agonists or antagonists, including such examples as cocaine (psychostimulant), haloperidol (antipsychotic), fluvoxamine (antidepressant), fenpropimorph (antifungal), and the endogenous *N′,N′*-dimethyltryptamine (hallucinogen). The phenyl group moiety is thus amenable to conversion into an aryl azide photoreactive group, as well as modification with a radiolabeled iodine atom for detection of covalent small molecule–protein complexes. Such photoprobes were invaluable tools for defining ligand binding sites on S1R in advance of the recent determination of its three-dimensional structure (Schmidt et al. 2016).

For example, a photoprobe based on the cocaine scaffold (**1**) (Fig. 3.1a) in which the phenyl ring is substituted at the 3-position with an iodine atom and at the 4-position with an azido group ([^{125}I]-IACoc, **2**) (Fig. 3.1a) binds to S1R with an affinity 2–3 orders of magnitude higher than that of the parent compound, indicating that these additions to the molecule are well tolerated and do not disrupt binding (Kahoun and Ruoho 1992). In fact, later experiments demonstrated that the addition of an electron-withdrawing group at the *para* position of the phenyl ring is responsible for the increased affinity, most likely by stabilizing pi/pi interactions with aromatic residues in the binding site (Chen et al. 2007). Upon using [^{125}I]-IACoc to photolabel recombinant guinea pig S1R transiently expressed in a COS-7 cell proteome (i.e., cell lysate), subsequent cleavage and radiosequencing mapped the primary site of crosslinking to aspartate 188 within the steroid binding domain-like II (SBDLII) region located near the C-terminus; the specificity of photocrosslinking was demonstrated by competitive inhibition with preincubation with 1 μM of the high-affinity S1R ligand haloperidol (Chen et al. 2007). With the aim of probing the interaction region of the nitrogen atom in the tropane ring, photoprobe [^{125}I]-N-IACoc (**3**) (Fig. 3.1a) was synthesized and found to produce haloperidol-protectable labeling of both the N-terminus containing the transmembrane segment (TM) and the steroid binding domain-like I (SBDLI) region located within what was proposed to be a second transmembrane segment (TM2) (Fontanilla et al. 2008).

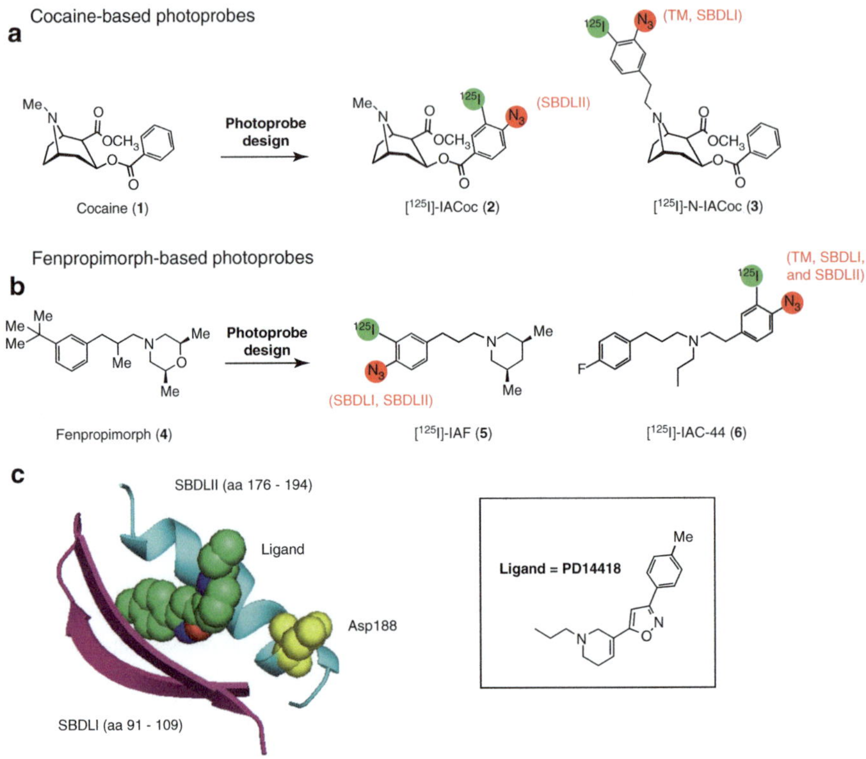

Fig. 3.1 Aryl azide-based sigma-1 receptor (S1R) photoprobes. (**a**) Cocaine-based photoprobes. (**b**) Fenpropimorph-based photoprobes. (**c**) S1R model from photolabeling experiments

An analogous series of experiments were performed with a photoprobe based on the fenpropimorph scaffold (**4**), also containing a 4-azido-3-iodo-phenyl group ([^{125}I]-IAF, **5**) (Fig. 3.1b) (Pal et al. 2007). Upon using [^{125}I]-IAF to photolabel S1R from rat liver and guinea pig liver membranes, specific photoinsertion was observed in both the SBDLI and SBDLII regions. The additional photoinsertion site observed with the fenpropimorph photoprobe, as compared to the cocaine photoprobe, was attributed to the innate flexibility of the fenpropimorph scaffold relative to the rigid tropane ring of the cocaine scaffold. Furthermore, by moving the aryl azide group to the nitrogen atom to produce photoprobe [^{125}I]-IAC44 (**6**) (Fig. 3.1b), labeling of the N-terminus, SBDLI, and SBDLII was detected (Fontanilla et al. 2008). Thus, by varying the position of the photoreactive group, the authors were able obtain multiple pieces of information about the ligand binding site, and thereby developed a model in which the juxtaposition of the SBDLI and SBDLII regions form a site for ligand recognition. The recent determination of the three-dimensional structure of S1R validated the role of amino acids in the SBDLI and SBDLII regions in ligand binding (Schmidt et al. 2016), even though the overall topology of the protein was different than in the proposed model (Fig. 3.1c). Thus, use of photoprobes provided critical and accurate insight into ligand binding by an important therapeutic target in the absence of structural information.

3.2.2.2 Monoamine Transporters

Inhibitors for the monoamine transporters for dopamine (DAT), norepinephrine (NET), and serotonin (SERT) block reuptake of the respective neurotransmitter into presynaptic neurons and are implemented to treat a variety of psychiatric disorders. Similar to the situation with S1R, solving the three-dimensional structures of these transporters has been challenging, leading researchers to rely on a combination of computational, chemical, and biological methods to decipher the functionally relevant binding modes for transport inhibitors. To this end, an array of reuptake inhibitors modified with an aryl azide crosslinking moiety has been developed to help gather information on the positioning of various ligands within the binding site(s).

Serotonin Transporter (SERT)

(*S*)-citalopram (**7**) (Fig. 3.2a) is an antidepressant of the selective serotonin reuptake inhibitor (SSRI) class that binds to the human serotonin transporter (hSERT). However, prior to the recent determination of the hSERT structure (Coleman et al. 2016), the molecular basis of the interaction between (*S*)-citalopram and hSERT was unknown. Elucidation of these molecular details is critical to understanding the high degree of potency and selectivity (*S*)-citalopram displays toward hSERT as compared to the dopamine transporter (DAT) and norepinephrine transporter (NET). To address this problem, three photoaffinity ligands based on the (*S*)-citalopram scaffold were synthesized (Kumar et al. 2015). Previous SAR analysis on citalopram suggested that the C-1 and C-5 positions could tolerate additional steric bulk without significant loss in hSERT binding affinity, and thus the aforementioned photoactivatable 4-azido-3-iodo-phenyl group was appended via variable linkers to generate ligands **8**–**10** (Fig. 3.2a). In order to test the pharmacological properties of these photoprobes, their ability to inhibit [^{3}H]5-HT uptake by the hSERT was determined. Compared to the parent compound, activity of the C-5 analogs were reduced by 11- and 17-fold for compounds **8** and **9**, respectively, while the C-1 analog (**10**) showed a greater reduction of 100 fold. Regardless, binding of all four compounds was negatively impacted by a Ser438Thr mutation commonly found to contribute to a significant loss in potency for many hSERT inhibitors. Successful photolabeling of HA-tagged hSERT in $LLCPK_1$ cells with compounds **8**–**10** was confirmed by immunoprecipitating with anti-HA or anti-hSERT antibodies and detection of a radiolabeled band at the expected molecular weight of ~100 kDa. Interestingly, while photolabeling with the C-5 analogs (**8** and **9**) was competed >99% with preincubation with 10 μM of the parent compound, labeling with the C-1 analog (**10**) was only reduced up to ~30%. These data suggested that the C-1 analog utilizes a different binding mode than the C-5 analogs, consistent with earlier studies suggesting the presence of a high-affinity primary binding site (S1) and a lower-affinity allosteric site (S2) (Banala et al. 2013). Thus, the photolabeling results suggested that compound **10** may bind through both the S1 and S2 sites. Remarkably, the recent crystal structure of hSERT demonstrated the presence of an allosteric site

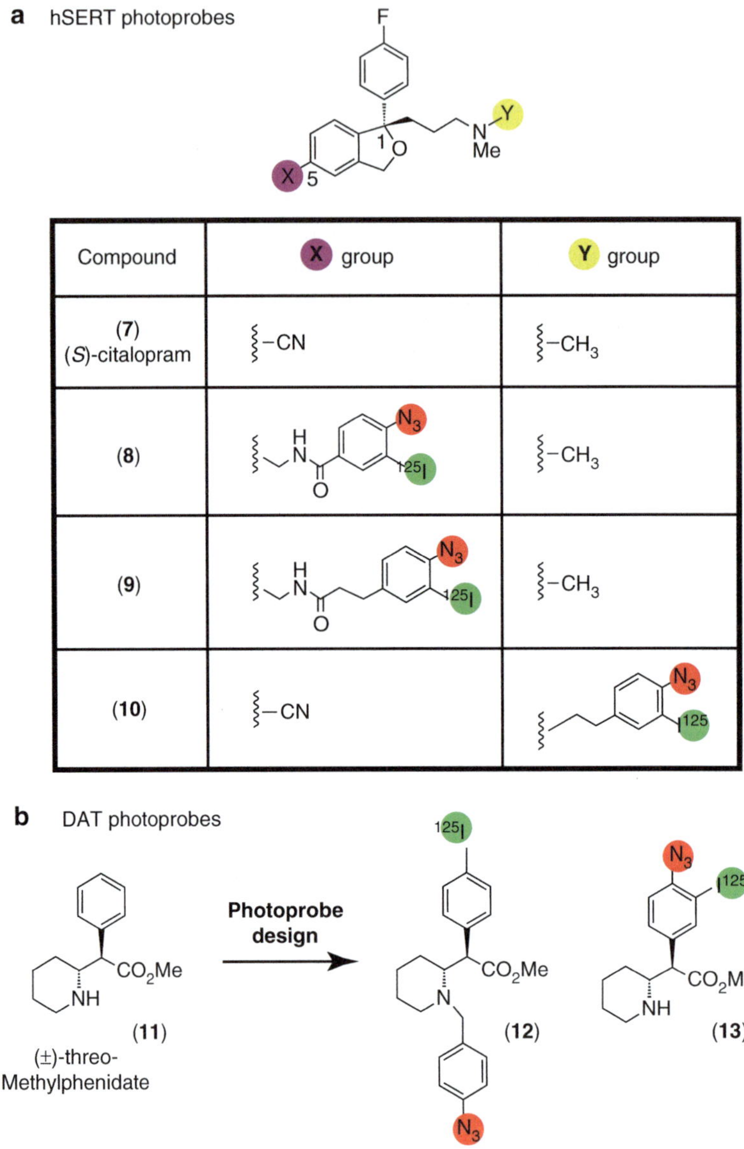

Fig. 3.2 Aryl azide-based monoamine transporter photoprobes. (**a**) Serotonin transporter (hSERT) photoprobes. (**b**) Dopamine transporter (DAT) photoprobes

adjacent to the primary (*S*)-citalopram binding site (Coleman et al. 2016), but whether compound **10** makes use of this allosteric site remains to be defined. Once again, placement of the photoreactive group at different positions on the drug scaffold provided differential information about the ligand binding pocket(s) and can be used as a strategy to characterize specific versus nonspecific drugs capable of binding to a target.

Dopamine Transporter (DAT)

The molecular basis by which the dopamine transporter (DAT) discriminates abused versus therapeutic compounds is also of interest. For example, structure–activity relationship (SAR) and site-directed mutagenesis (SDM) data indicates that the tropane and nontropane inhibitor classes may bind different domains or conformations of the transporter. While tropane-based photoprobes have been established, the photocrosslinking results suggested that the photoreactive azide group must be placed some distance away from the tropane pharmacophore by a variable linker in order to covalently modify the DAT protein at a residue near, but not directly within, the ligand binding domain. Therefore, recent efforts have focused on developing nontropane-based photocrosslinking probes. For example, a photoprobe based on the (±)-*threo*-methylphenidate scaffold (**11**) (Fig. 3.2b) was designed in which the photoreactive azide group and the radioactive iodine tag are on different parts of the molecule, based on the concern that the steric bulk of the iodine might detrimentally affect binding and, consequently, crosslinking efficiency (Lapinsky et al. 2011). To achieve this, the authors first used rational design from SAR data to create hybrid molecules of iodo-phenyl-substituted and *N*-benzyl-substituted methylphenidates, in which the azide group is systematically placed at different positions on the aromatic ring of the *N*-benzyl group. The authors were surprised to observe a significant decrease in DAT binding affinity with all of the hybrid molecules tested relative to (±)-*threo*-methylphenidate (~15-fold to 110-fold). While the exact position of the iodine atom in these hybrid molecules provided no advantage in terms of DAT binding affinity, positioning the azide group at either the *para* or *ortho* positions on the *N*-benzyl group seemed to be more advantageous as compared to the *meta* position. The photoprobe with the best affinity (**12**) (Fig. 3.2b) was carried forward in a photocrosslinking experiment against rat striatal membranes in the absence or presence of the competitor molecules β-CFT (a phenyltropane cocaine analog) or D-(+)-methylphenidate. DATs from the crosslinked samples were immunoprecipitated and separated by SDS-PAGE, upon which radiographic analysis revealed an ~80 kDa band that was reduced 75–90% with inclusion of 10–100 μM competitor. More recently, a more compact photoprobe based on (±)-*threo*-methylphenidate has been reported (Lapinsky et al. 2012). The design was based on a methylphenidate analog with hydrophobic chlorine atoms at positions 3 and 4 of the aromatic ring, which were replaced with hydrophobic iodine and azide groups in the photoprobe. In this case, the resulting probe (**13**) (Fig. 3.2b) displayed a fourfold higher hDAT affinity over (±)-*threo*-methylphenidate. A photocrosslinking experiment with compound **13** was carried out against $LLCPK_1$ cells expressing hexahistidine-tagged DAT in the absence or presence of 10 μM of the competitor molecules (−)-cocaine or D-(+)-methylphenidate. DATs from the crosslinked samples were immunoprecipitated and separated by SDS-PAGE, upon which radiographic analysis revealed an ~80 kDa band that was reduced >90% with inclusion of either of the competitors. The authors thus demonstrated that the nontropane class of inhibitors, unlike the tropane class, can effectively crosslink in the context of a compact photoprobe design. Notably, the recent determination of the three-dimensional structure of the

Drosophila melanogaster DAT is consistent with the photocrosslinking results, revealing that tropanes occupy a more tightly packed binding pocket than nontropanes, offering an explanation for the difficulty in functionalizing the tropane inhibitors (Wang et al. 2015).

3.2.2.3 Protease (γ-Secretase)

γ-Secretase (GS) is a membrane-bound enzyme complex, which catalyzes peptide bond hydrolysis of more than 20 reported substrates within the hydrophobic environment of the membrane. Presenilin (PS) is believed to contain the catalytic subunit, while nicastrin (NCT), presenilin enhancer 2 (Pen-2), and anterior pharynx defective 1 (Aph-1) are cofactors required for reconstituted GS activity. Selective modulation of γ-secretase activity has emerged as a potential treatment strategy for Alzheimer disease (AD), due to the fact that small molecule γ-secretase modulators (GSMs) can reduce formation of the pathogenic Aβ42, while not broadly inhibiting the cleavage of other substrates, such as Notch, that are critical for normal cellular signaling. GSM-1 (**14**) (Fig. 3.3a) is a second-generation NSAID-derived carboxylic acid-containing GSM, thus possessing improvements in potency and brain penetration over that of its predecessors. However, it was unknown whether the direct binding target for GSM-1 was a γ-secretase subunit, the amyloid precursor protein (APP) substrate, or both. Therefore, photoprobe GSM-5 (**15**) (Fig. 3.3a) was developed based on the GSM-1 scaffold to identify direct binding partners (Crump et al. 2011). To this end, the authors tried to incorporate moderate changes into the GSM-1 scaffold, such as substituting an alkyne handle into the alkyl side chain and substituting the chlorophenyl group with a perfluorophenylazide. When tested as an inhibitor against HeLa membranes with recombinant APP and Notch1 as substrates, photoprobe GSM-5 exhibited similar levels of potency (only reduced two to threefold) and selectivity (i.e., no activity against Notch) to that of the parent compound GSM-1. When carried forward in a photocrosslinking experiment, GSM-5 was capable of labeling endogenous PS-1 N-terminal fragment (NTF) from HeLa

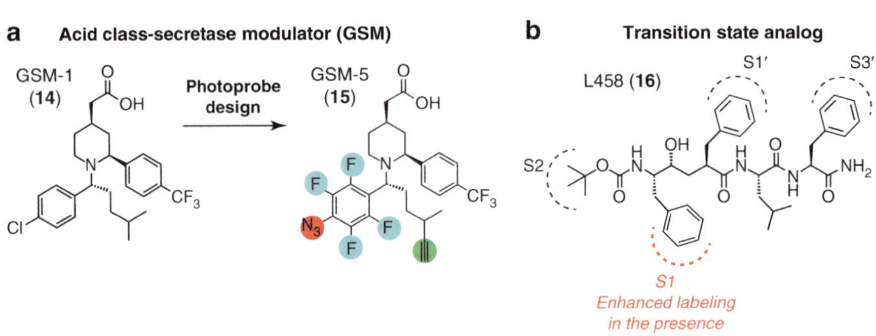

Fig. 3.3 Aryl azide-based γ-secretase modulator (GSM) protoprobe. (**a**) Acid class GSM protoprobe. (**b**) Transition state analog of γ-secretase

membranes (i.e., the active form of the catalytic subunit of γ-secretase), while no labeling occurred to other components such as nicastrin (NCT), Pen-2, and Aph-1. Therefore, GSM-1 binds directly to PS1 within the γ-secretase complex.

In addition to photoprobes based on γ-secretase modulators, PS1 photoprobes based on a peptide substrate analog have also been designed. To elucidate what specific effect the GSM-1 binding interaction was having on the PS1 active site, a panel of photoprobes (structures not shown) based on the transition-state analog L458 (**16**) (Fig. 3.3b) was developed. These photoprobes can photolabel different subpockets within PS1 (i.e., S2, S1, S1′, and S3′) via substitution of a benzophenone group at these corresponding positions (Li et al. 2000; Chun et al. 2004; Yang et al. 2009). GSM-1 binding to PS1 had a minimal effect on labeling of the L458-based photoprobes to the S2, S1′ and S3′ subsites. Rather, photocrosslinking to the S1 site was actually enhanced by inclusion of GSM-1 (Crump et al. 2011). This result suggests that GSM-1 functions through an allosteric binding mechanism that affects the S1 subsite, and which in turn leads to a reduction in the amount of Aβ42 produced.

3.2.2.4 Protein Toxin (Anthrax Lethal Toxin)

N[1]-(4-bromobenzylidene)-*N*[4]-(2,6-dimethylphenyl)semicarbazone, referred to as EGA (**17**) (Fig. 3.4), was identified in a high-throughput screen as a selective inhibitor of the pH-dependent endosomal trafficking pathway exploited by the anthrax lethal toxin (LT), with an IC_{50} value of 1.4 μM (Gillespie et al. 2013). However, to date the protein target(s) of this compound remain unknown, and thus the advent of a photoprobe would aid in this area. To achieve this, first an SAR study was conducted on the parent compound EGA to generate ~30 analogs (Jung et al. 2014). This study revealed that both the semicarbazone core and the 2,6-dimethylphenyl moiety are required for bioactivity, while the *N*[1]-phenyl ring could tolerate certain modifications. In particular, by adding a fluoro group to the

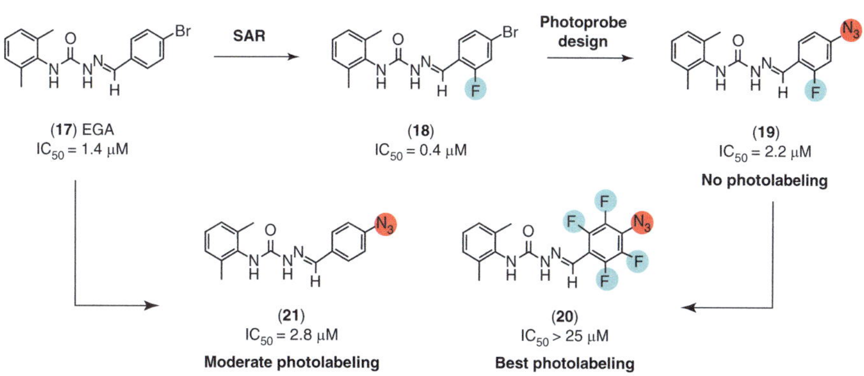

Fig. 3.4 Aryl azide-based EGA photoprobes

2-position of the N^1-phenyl ring, compound **18** (Fig. 3.4) was generated and found to be more potent than the parent compound, with an IC_{50} value of 0.4 μM. A photoaffinity probe was designed and synthesized based on compound **18**, in which the bromo group on the 4-position of the N^1-phenyl ring was replaced with an azide group (**19**) (Fig. 3.4) (Jung et al. 2014). Two comparison probes (**20** and **21**) (Fig. 3.4) were also synthesized. In probe **20**, the remaining hydrogen atoms on the N^1-phenyl ring of compound **19** were replaced with fluorine atoms to afford a tetrafluorophenylazide photoreactive group, while probe **21** was based on the original parent compound EGA that possesses no fluorine atoms on the N^1-phenyl ring. In the end, compounds **19** and **21** functioned as reversible inhibitors of LT intoxication (IC_{50} values of 2.2 and 2.8 μM, respectively), while compound **20** containing the tetrafluorophenylazide motif was an order of magnitude less protective (IC_{50} >25 μM). However, upon photoactivation of the aryl azide photocrosslinker prior to an exchange into fresh media (i.e., a washout step), only compounds **20** and **21** had an irreversible protective effect against LT, indicating that these photoaffinity probes were capable of covalently modifying their functionally relevant target proteins within the cell. Thus, activity and labeling efficiency are not directly correlated. Although the authors were unable to append an alkyne affinity handle to probe **21** and retain activity, they propose using a radiolabeled version of probe **20** as a means of identifying the bound proteins.

3.3 Benzophenone-Based Photoaffinity Probes

3.3.1 Photocrosslinking Properties of Benzophenones

This section will focus on the development of photoprobes that incorporate a benzophenone photoreactive group. Photoactivation of benzophenones results in the formation of a triplet diradical on the carbonyl that can undergo insertion into nearby X–H bonds (with a preference for C–H bonds over O–H bonds). An important advantage to incorporating a benzophenone into a photoprobe is that its activation wavelength is in the range of 350–360 nm, which lowers the risk of damaging biomolecules. However, in terms of disadvantages, the bulkiness of the benzophenone group has the potential to sterically hinder the interaction between the pharmacophore and its ligand binding site, which may lead to nonspecific labeling within the context of a native proteome. Furthermore, if the diradical species is unable to react with a nearby target, it can relax back to the ground state to regenerate the benzophenone. This can be viewed as an advantage—more opportunities for productive crosslinking—or a disadvantage, increased chance of nonspecific crosslinking. Regardless, the synthetic ease at which the benzophenone moiety can be installed has made it accessible to many research labs for photoprobe development.

3.3.2 Design and Applications of Benzophenone Photoprobes

3.3.2.1 Protease (γ-Secretase)

Another class of second-generation γ-secretase modulators (GSMs) is the non-NSAID imidazole-containing class. As with the acid class of GSMs described previously, the development of photoprobes has been an effective tool in elucidating the direct binding partners of these small molecules. To this end, the E2012-BPyne photoprobe (**23**) (Fig. 3.5) based on the E2012 scaffold (**22**) was reported in 2013. In the design of this probe, the *para*-fluorophenyl group of E2012 was substituted with a benzophenone photoreactive group and the methoxyphenyl moiety of E2012 was substituted with a propargyl group to allow for click chemistry (Pozdnyakov et al. 2013). E2012-BPyne could be used to specifically enrich PS1-NTF from HeLa cell membranes (known to possess high γ-secretase activity). Enrichment was dependent on E2012-BPyne concentration and could be competed with excess E2012. Labeling of other γ-secretase components such as PS1-CTF, Aph-1, Pen-2, or nicastrin couldn't be detected under these conditions. These results are analogous as to what was observed with the aryl azide probe for the acid class GSM-1. With these tools in hand, the authors could proceed to elucidate the overlap in binding sites between the different classes of GSMs (Pozdnyakov et al. 2013). For instance, excess GSM-1 (**14**) (Fig. 3.3a) had no effect on PS1-NTF labeling by E2012-BPyne, while analogously excess E2012 had no effect on PS1-NTF labeling by GSM-5 (**15**) (Fig. 3.3a). This indicates that these different classes of GSMs bind to distinct sites on PS1-NTF. Furthermore, the authors were able to show through photolabeling experiments that E2012 preferentially interacts with and labels only active γ-secretase, whereas GSM-1 appears to interact with both active and inactive γ-secretase complexes. Finally, *in situ* labeling reactions were performed with E2012-BPyne in both intact HeLa cells and primary cortical neurons. The membranes were precipitated, the crosslinked bands were modified by click chemistry with a biotin-azide, and the complexes enriched for immunoblot analysis. Prominent labeling of PS1-NTF was observed, but only in the presence of L458 (**16**) (Fig. 3.3b), which has been proven to increase binding affinity and/or labeling efficiency of E2012-BPyne in vitro. Taken together, photocrosslinking analysis with multiple photoprobes provides evidence for multiple binding sites within the active γ-secretase complex that confer the modulatory effects of different classes of compounds.

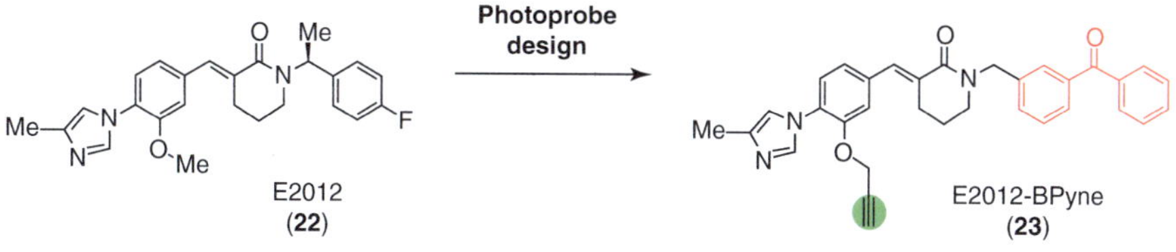

Fig. 3.5 Benzophenone-based γ-secretase modulator (GSM) protoprobe

3.3.2.2 Pyrophosphatase (DcpS)

DcpS is a scavenger mRNA-decapping enzyme, which hydrolyzes 7-methylguanylate (m^7G) from short 5′-capped oligonucleotides (m^7GpppN) produced by mRNA decay in the 3′ to 5′ direction. DcpS was identified on a protein microarray (containing >5000 proteins) to bind to C-5 substituted diaminoquinazolines (DAQs). In fact, in the "closed" active site conformation of DcpS, the 2,4-diaminquinazoline pharmacophore has been found to occupy the same binding pocket as the m^7G purine moiety of the m^7GpppG substrate (Singh et al. 2008). To this end, DcpS has emerged as a novel therapeutic target for small molecule modulation of gene expression, in particular for the treatment of spinal muscular atrophy (SMA) by upregulation of *SMN2*. With the C-5 position available to accommodate a bulky substituent, recent efforts have focused on developing covalent probes that can validate DcpS as a target in a more complex biological system. First, a *para*-sulfonyl fluoride group on a benzyl substituent at the C-5 position (SF-p1, **24**) (Fig. 3.6a) was able to covalently react with a nearby Tyr143 residue in the active site on recombinant DcpS, as determined by LC-MS peptide-mapping experiments (Hett et al. 2015). An alkyne-tagged version of this probe (SF-p1-yne) (structure not shown) was further able to purify DcpS from human primary cells after modification with an azido-biotin handle. Further, covalent reaction with SF-p1-yne could be competitively inhibited by preincubation of cells with the known inhibitor D156844 (**25**) (Fig. 3.6b) in a dose-dependent manner, thus validating DcpS as a cellular target. Next, in order to offset the amino acid bias associated with the sulfonyl fluoride electrophilic labeling approach, the authors also synthesized a library of 30 compounds each containing a photoreactive group (22 benzophenones, 6 aryl azides, and 2 diazirines) attached at the C-5 position by various linkers through either an amide or ether linkage (Xu et al. 2015). The authors were unable to detect covalent adducts with recombinant DcpS protein by intact mass spectrometric analysis, likely due a low yield or poor ionization. However, six photoaffinity probes containing both a benzophenone photoreactive group and a clickable handle could be tested for their ability to enable DcpS enrichment. Interestingly, compound **26** (Fig. 3.6b) was one of the most potent inhibitors in the library ($IC_{50} = 0.29$ nM), but a poor photoaffinity probe. In contrast, compound **27** (named DcpS-BPyne) was capable of enriching DcpS in a manner that could be effectively competed with SF-p1. But shortening the linker by just one methylene group resulted in a probe (**28**) whose crosslinking actually increased in the presence of SF-p1, indicating that labeling might be occurring through a nonspecific interaction, which correlates well with its appreciable loss in inhibitory activity ($IC_{50} = 5560$ nM). While no data was reported on the application of these photoprobes in a more complex proteome, this work gives a good comparison of the utility of different covalent labeling techniques and highlights the need to explore the optimal linker length that balances activity with specificity of labeling.

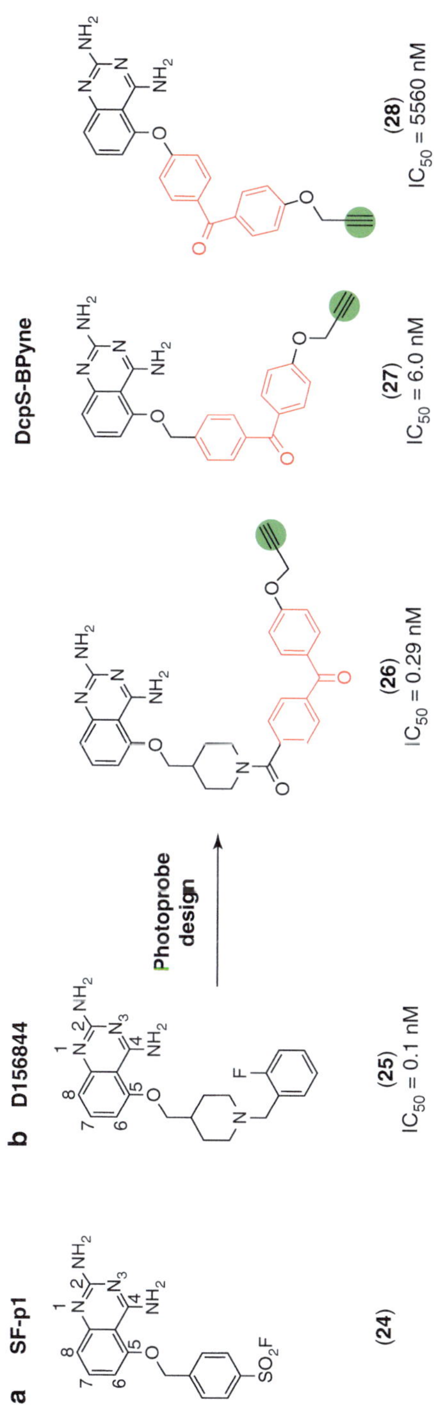

Fig. 3.6 Benzophenone-based DcpS photoprobes. (**a**) Electrophilic labeling. (**b**) Photoreactive labeling

3.3.2.3 Adenylation Enzymes

The biosynthesis of siderophores called mycobactins (i.e., iron chelating agents) is critical for the growth and virulence of *Mycobacterium tuberculosis* (*Mtb*). Therefore, small molecule inhibitors of the mycobactin biosynthetic pathway are currently being developed as novel antituberculosis (TB) agents. The adenylating enzyme MbtA catalyzes the initiating step in mycobactin production, in which it activates a salicylic acid to an acyladenylate intermediate (Sal-AMP) that remains non-covalently bound to the active site. MbtA then catalyzes the transfer of the salicyl moiety onto the thiolation domain of MbtB, whereby this tethered structure is further elaborated into the mycobactin core by a mixed nonribosomal peptide synthetase (NRPS) and polyketide synthetase (PKS) assembly line (MbtB-MbtF). An inhibitor of MbtA has been developed that mimics the acyladenylate, in which the labile acylphosphate linkage has been replaced with a non-hydrolyzable acylsulfamate (Sal-AMS, **29**) (Fig. 3.7a) (Ferreras et al. 2005; Somu et al. 2006). Interestingly, while Sal-AMS demonstrates potent antitubercular activity under iron-deficient conditions (MIC = 0.39 μM), it is also active under iron-rich conditions (MIC = 1.56 μM), suggesting that it may possess a secondary mechanism of action due to off-target binding events. In fact, the Mtb genome encodes over 60 putative adenylating enzymes responsible for catalyzing a multitude of essential biochemical processes. Therefore, in 2012 Aldrich and co-workers developed a photoprobe based on the Sal-AMS scaffold to try to identify its array of binding partners in *Mtb* (Duckworth et al. 2012). Previous structure–activity relationship (SAR) studies indicated that MbtA tolerates modification of Sal-AMS at the C-2 position of the adenosine ring (Neres et al. 2008). Based on these data, a benzophenone group tethered to a terminal alkyne was appended to this position in order to afford Sal-AMS ABP (**30**) (Fig. 3.7a) (Duckworth et al. 2012). Sal-AMS ABP retained an activity profile similar to the parent compound Sal-AMS, in which it was more active under iron-deficient conditions (MIC = 3–6 μM) versus iron-rich conditions (MIC = 50 μM) and was thus carried forward for photolabeling experiments. Sal-AMS ABP was able to label pure recombinant MbtA in vitro as visualized by in-gel fluorescence, as well as MbtA overexpressed in *E. coli* lysates, both of which could be competed with the inclusion of excess Sal-AMS. Furthermore, the Sal-AMS ABP was even effective at labeling overexpressed EntE, which is the adenylating enzyme involved in the biosynthesis of siderophores produced from *E. coli* called enterobactins; these data further expand the utility of this reagent to the study of other siderophore-producing pathogens which possess an aryl acid adenylating enzyme (AAAE). Finally, even endogenous levels of MbtA from mycobacterial cell lysate could be enriched by labeling with the Sal-AMS ABP and ultimately detected by LC-MS/MS. Thus, future work will entail the application of this photoprobe to the elucidation of additional binding partners that may account for its off-target effects, thereby aiding in the design of more selective inhibitors of MbtA.

Additional work has been focused on developing an array of "clickable" photoprobes that can directly label adenylation domains based on their cognate amino acid substrate specificities. This technology would complement genetic approaches

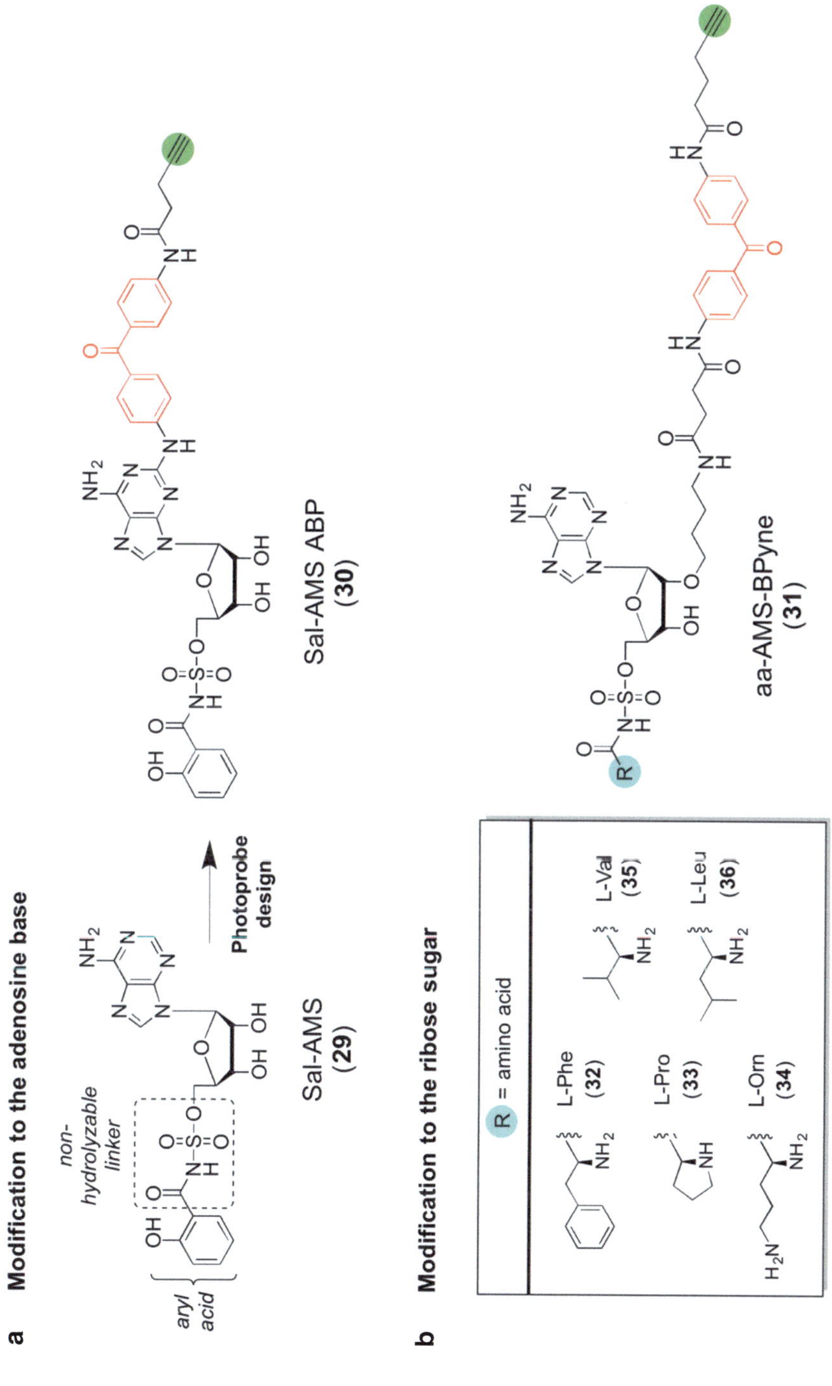

Fig. 3.7 Benzophenone-based NRPS photoprobes. **(a)** Modification to the adenosine base. **(b)** Modification to the ribose sugar

for characterizing NRPS modules in their native bacterial proteomes; for example, in sequenced producers the probes can be used to monitor expression dynamics and optimize bacterial culture conditions, while in unsequenced organisms the probes could facilitate the discovery of gene clusters. To this end, in 2003 the lab of Marahiel and co-workers developed an aminoacyladenylate inhibitor with the aforementioned acylsulfamate substitution (aa-AMS) targeted toward the gramicidin S biosynthetic enzyme, GrsA, which recognizes phenylalanine and AMP as substrates (Finking et al. 2003). Guided by the available crystal structure of GrsA with its substrates bound (Conti et al. 1997; Stachelhaus et al. 1999), the authors were able to incorporate a biotin functionality into the inhibitor via a long linker attached to the 2′-oxygen atom with no loss in inhibitory activity against the recombinantly expressed A domain fragment (Finking et al. 2003). From this work, the lab of Kakeya and co-workers were able to design an array of photoprobes collectively named aa-AMS-BPynes (**31**) (Fig. 3.7b), which incorporate a bifunctional probe onto the 2′-oxygen consisting of a benzophenone photoreactive group and terminal alkyne for click chemistry, as well as a variable amino acid component for targeting to different NRPS modules from bacterial cell lysates. As a model system, the authors similarly focused on characterizing the A-domains within the GrsA and GrsB modules, which together function to form the gramicidin S cyclic decapeptide antibiotic with the primary structure cyclo(-D-Phe_1-L-Pro_2-Val_3-L-Orn_4-L-Leu_5-$)_2$. To this end, the tridomain target protein GrsA (A ($_L$-Phe)-CP-E) could be labeled from crude lysate with L-Phe-AMS-BPyne (**32**) as determined by in-gel fluorescence, in which a ~120 kDa band was observed (and confirmed by LC-MS/MS to contain GrsA) that could be completely inhibited by pretreatment with L-Phe-AMS (Konno et al. 2015; Ishikawa et al. 2015a). Labeling specificity could also be observed by employing four inhibitors designed to target the A-domains of the GrsB module (C-A_2 ($_L$-Pro)-T-C-A_3 ($_L$-Val)-T-C-A_4 ($_L$-Orn)-T-C-A_5 ($_L$-Leu)-T-TE). For instance, GrsA labeling by L-Phe-AMS-BPyne (**32**) was inhibited by L-Leu-AMS (due to its known role as a miscognate substrate with lower catalytic efficiency), but not by L-Pro-AMS, L-Orn-AMS, and L-Val-AMS (Konno et al. 2015). Furthermore, endogenous GrsB could be labeled from crude lysate with L-Pro-AMS-BPyne (**33**), L-Orn-AMS-BPyne (**34**), L-Val-AMS-BPyne (**35**), or L-Leu-AMS-BPyne (**36**) as determined by in-gel fluorescence observation of a ~500 kDa fluorescent band (and confirmed by LC-MS/MS to contain GrsB), which could be inhibited by pretreatment with the aa-AMS analogs of the corresponding cognate substrate (Konno et al. 2015; Ishikawa et al. 2015a). As observed before, L-Leu-AMS was the most promiscuous inhibitor tested, also being able to prevent GrsB labeling with L-Val-AMS-BPyne (**35**), most likely due to their similar hydrophobic properties (Ishikawa et al. 2015a). Finally, when these photoprobes are used in combination in a cocktail format (i.e., allowing for up to quadruple labeling of GrsB), the detection and isolation of low-abundance multidomain NRPS enzymes from native proteome environments is improved, and further revealed the presence of an unknown inactive form of GrsB in a nonproducing strain (Ishikawa et al. 2015b).

3.3.2.4 Antibiotics (Vancomycin)

Vancomycin (**37**) (Fig. 3.8) is a glycosylated heptapeptide antibiotic that binds to structural motifs on peptidoglycans to perturb biosynthesis of the bacterial cell wall. In 2010, Gerard Wright and co-workers reported using rational design to develop a vancomycin photoaffinity probe (**38**) (Fig. 3.8) in order to determine the direct binding partners involved in the transcriptional activation of the *vanH*, *vanA*, and *vanX* resistance cassette (which results in the synthesis of peptidoglycans terminating in D-Ala-D-lactate instead of the usual D-Ala-D-Ala) (Koteva et al. 2010). Probe design was based on previous SAR data available on second-generation vancomycin derivatives currently in clinical trials. For instance, the authors predicted that the benzophenone group would be tolerated at the primary amino group on the vancosamine sugar moiety due to precedent of alkylation at this site and that direct tethering of a biotin handle at the C-terminus would not impact activity. In fact, probe **38** maintained comparable antibiotic activity to that of the parent compound when tested against various strains of Gram-positive bacteria (either inherently sensitive or resistant to the drug), and bound with a comparable affinity in vitro to the known target di-*N*-Ac-Lys-D-Ala-D-Ala (K_d of 17.1 ± 2.6 μM for probe **38** versus 10.5 ± 1.9 μM for vancomycin). Using photoaffinity probe **38**, the authors were able to biotin-label the 43 kDa transmembrane receptor histidine kinase from *Streptomyces coelicolor* (VanSsc) recombinantly expressed in *E. coli* and associated with purified membranes; further, labeling could be specifically blocked by free vancomycin (IC_{50} of 124 ± 12 μM). Similar results were obtained using isolates of membranes from *S. coelicolor*, in which the VanS photocrosslinked band was not detected in a strain deleted for the gene encoding VanS. The authors concluded that photoprobe **38** is a bioactive ligand for VanS (in fact binding directly to its N-terminal region), thus triggering its autophosphorylation and inducing glycopeptide antibiotic resistance. Overall, this work demonstrates that photoprobes can serve as important tools to discover ways in which bacteria can become resistant to antibiotics, information that is critical to the design of new analogs.

In 2011, Stephan Sieber and co-workers developed an additional series of benzophenone-based probes of vancomycin in order to determine the direct binding partners that may function as "noncanonical target sites" in clinically relevant bacterial strains (Eirich et al. 2011). In terms of probe design, the same sites were utilized as the lab of Wright and co-workers, but optimal placement of either the benzophenone or alkyne handle at these sites was explored, as well as attaching both functionalities on the glycan (**39–41**) (Fig. 3.8). In the end, all three probes maintained activities comparable to that of the unmodified vancomycin against the strains tested. In addition, all three probes were capable of labeling a single ~65 kDa membrane protein band within staphylococcal proteomes (e.g., *S. aureus* (SA) and methicillin-resistant *S. aureus* (MRSA)) and a single ~40 kDa membrane protein band within *enterococcal* proteomes (e.g., vancomycin-susceptible *E. faecalis* (VSE) and vancomycin-resistant *E. faecalis* (VRE)). Because there was no observed difference in labeling between SA and MRSA, or VSE and VRE, the labeled targets

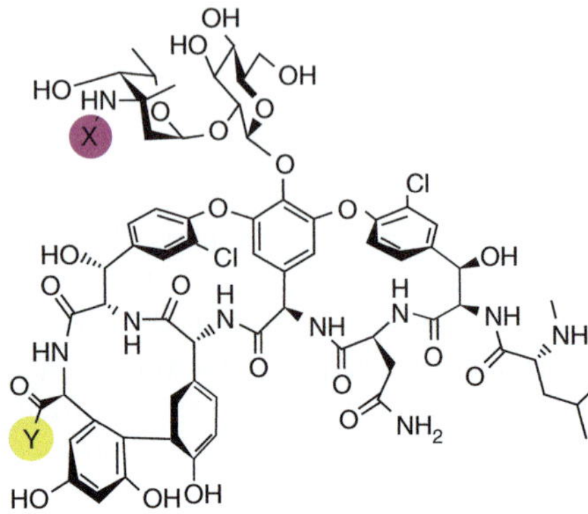

Compound	X group	Y group
(**37**) Vancomycin	–H	–OH
(**38**)	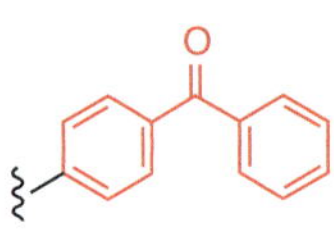	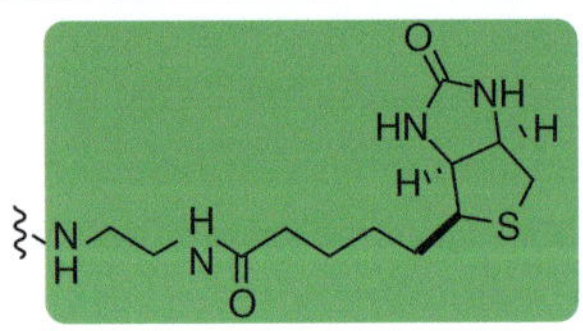
(**39**)	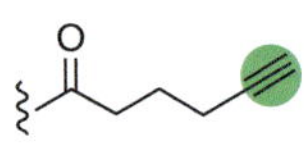	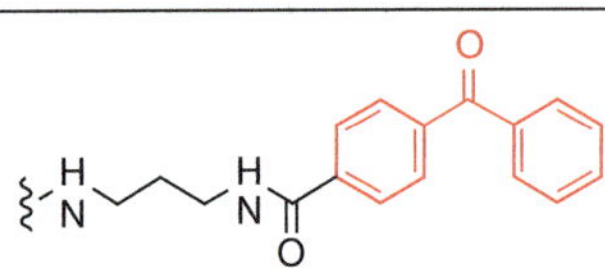
(**40**)	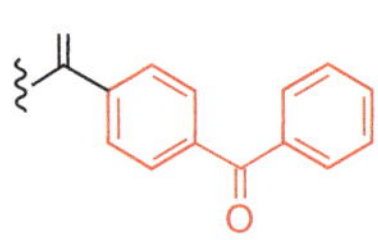	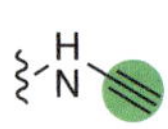
(**41**)	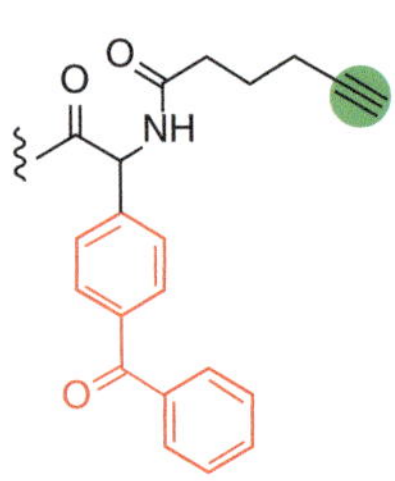	–OH

Fig. 3.8 Benzophenone-based vancomycin photoprobes

are most likely not directly responsible for vancomycin resistance. Reacting a trifunctional rhodamine-biotin-azide tag with the alkyne handle on probe **39** using click chemistry enabled both affinity purification of the crosslinked proteins and their visualization in a gel for subsequent band excision, tryptic digestion, and analysis by mass spectrometry. From these data, a strong candidate for the ~65 kDa band labeled in MRSA was revealed to be autolysin (ATL), an enzyme that functions to break down the peptidoglycan matrix to allow cell division to occur. Furthermore, the ~40 kDa band in VSE was tentatively identified as a peptide ABC transporter (pABC) that is important for the import of nutrients. Recombinant forms of the full-length ATL, the amidase domain of ATL, and pABC were expressed in *E. coli* and confirmed to photocrosslink to probe **39** in a manner that can be competed with excess unmodified vancomycin. From this work, the authors have determined two previously unknown targets that may likely contribute to the antibacterial activity of vancomycin. Thus, these results could facilitate the design of vancomycin variants with improved activity.

3.3.2.5 HDAC Inhibitors (SAHA)

Aberrant activity of histone deacetylases (HDACs) in cancer has inspired development of general inhibitors of class I/II HDACs. Chemical tools that can profile the HDAC activities in different human disease models would aid in the successful applications of these inhibitors. To this end, Benjamin Cravatt and co-workers have designed a chemical probe based on the scaffold of suberoylanilide hydroxamic acid (SAHA, **42**) (Fig. 3.9) that can be used to evaluate HDAC activity and to identify SAHA targets in both native complex proteomes (cell lysates) and in situ (live cells) (Salisbury and Cravatt 2007). In terms of probe design, the crystal structure of SAHA bound to an HDAC homologue from the thermophilic bacteria *Aquifex aeolicus* demonstrates that the phenyl ring resides at the "lip" of the substrate pocket, thus being exposed to solvent and providing room for modification. In order to create the photoprobe SAHA-BPyne (**43**) (Fig. 3.9), the authors therefore appended to this accessible position a bifunctional group consisting of both a benzophenone photoreactive group for HDAC protein capture and an azide to enable "click chemistry" modification with an affinity handle for enrichment of crosslinked complexes. Soluble proteomes prepared from human aggressive (MUM2B) and nonaggressive (MUM2C) melanoma cell lines were photocrosslinked with SAHA-BPyne in the absence or presence of excess SAHA competitor ligand. Three distinct

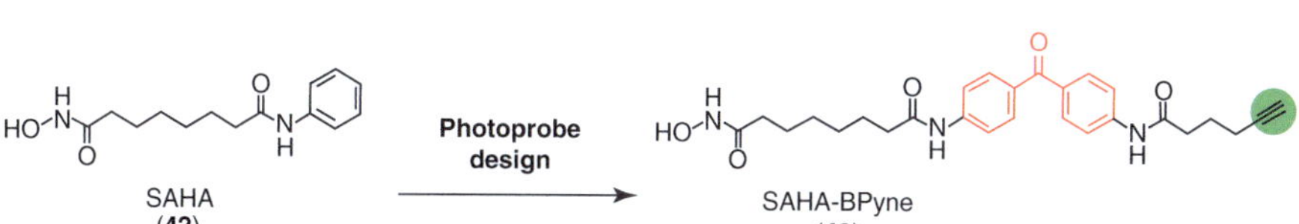

Fig. 3.9 Benzophenone-based SAHA photoprobe

SAHA-competed targets were identified, including HDACs 1 and 2 (~60 kDa) and the HDAC-associated protein MBD3 (methyl CpG binding protein 3) (~38 kDa), as determined by in-gel trypsin digestion and subsequent LC-MS/MS analysis. Lower abundance, but still specific, targets not clearly visible by a band on a gel could also be purified and identified through a process that required harsh protein-denaturing conditions, on-bead trypsin digestion, and a comparison to multiple control samples to be successful. HDAC6 was identified in this approach, as well as several non-HDAC proteins, including CoREST, p66β, and the metastasis-associated proteins MTA1 and MTA2, all of which have been reported to be components of functional HDAC complexes. Immunoblot analysis confirmed that the SAHA-BPyne probe was able to enrich for HDAC2, CoREST, and MTA2. Due to the stringent conditions used for purification of the crosslinked complexes, the authors hypothesize that the SAHA-BPyne probe is capable of directly crosslinking to proteins within the complex that are positioned close to the active site of the histone deacetylase. In addition, the authors demonstrate that cell type-dependent differences observed in spectral counts in the mass spectrometry analysis may reflect differences in the amount of assembled HDAC complexes rather than alterations in total protein abundance. Finally, the authors compared photolabeling in vitro (cell lysates) and in situ (live cells) in the human breast cancer cell line MDA-MB-231. While in-gel fluorescence scanning revealed SAHA-sensitive signals corresponding to HDAC1, HDAC2, and MBD3 under both conditions, multiple targets showed a stronger signal from the in situ crosslinking experiment. The authors speculate that this may be because the integrity of the functional states can be compromised upon cell homogenization, emphasizing the utility of probes that can enter and function in live cells.

3.4 Diazirine-Based Photoaffinity Probes

3.4.1 Photocrosslinking Properties of Diazirines

This section focuses on the development of photoprobes that incorporate a diazirine photoreactive group, consisting of either the aromatic or the aliphatic classes. Photoactivation of diazirines results in the formation of a highly reactive carbene that rapidly inserts into nearby X–H bonds (X = C, N, S, and O). Like the benzophenone, the activation wavelength for the diazirine (350–380 nm) is not strongly damaging to proteins. In particular, an important advantage to utilizing the alkyl diazirine is its compact size, being not much larger than a methyl group and allowing placement at the interface of a ligand-protein interaction. Furthermore, improved methods have been developed for synthesizing the alkyl diazirine from the ketone precursor. While factors such as rapid quenching by water or intramolecular rearrangements may decrease the yield of protein labeling, the labeling specificity remains high due to proximity limitations between the short-lived carbene and its bound protein targets. Aryl diazirines suffer less from intramolecular rearrangements than alkyl diazirines do, but also do not have the advantage of compact size.

3.4.2 *Design and Applications of Aryl Diazirine Photoprobes*

3.4.2.1 Natural Products

Cyclodepsipeptides

The natural product Hun-7293 (**44**) and related cyclodepsipeptides (**45** and **46**) (Fig. 3.10a) are capable of perturbing interactions between nascent ribosome-associated vascular cell adhesion molecule (VCAM) chains and the translocation channel to the endoplasmic reticulum (ER), resulting in reduced cell surface VCAM expression. The core of this protein-conducting channel is formed by a heterotrimeric membrane protein complex termed Sec61. The lab of Taunton and co-workers therefore designed a photoaffinity probe based on this class of inhibitors to decipher if a protein within the Sec61 complex is a direct binding target (MacKinnon et al. 2007). For the design of probe **47** (Fig. 3.10b), a conserved leucine at position 4 was replaced with a photocrosslinking analog, photo-Leu, to introduce a photoreactive diazirine, and a propargyl substituent was installed at the more variable position 1 to enable "click chemistry." Both photoprobe **47** and its competitor ligand (**48**) (Fig. 3.10b) were as potent as the natural product HUN-7293 at inhibiting VCAM expression (EC_{50} = ~25 nM). Photoprobe **47** was crosslinked to crude ER microsome fractions, and a major crosslinked band with an apparent molecular weight of

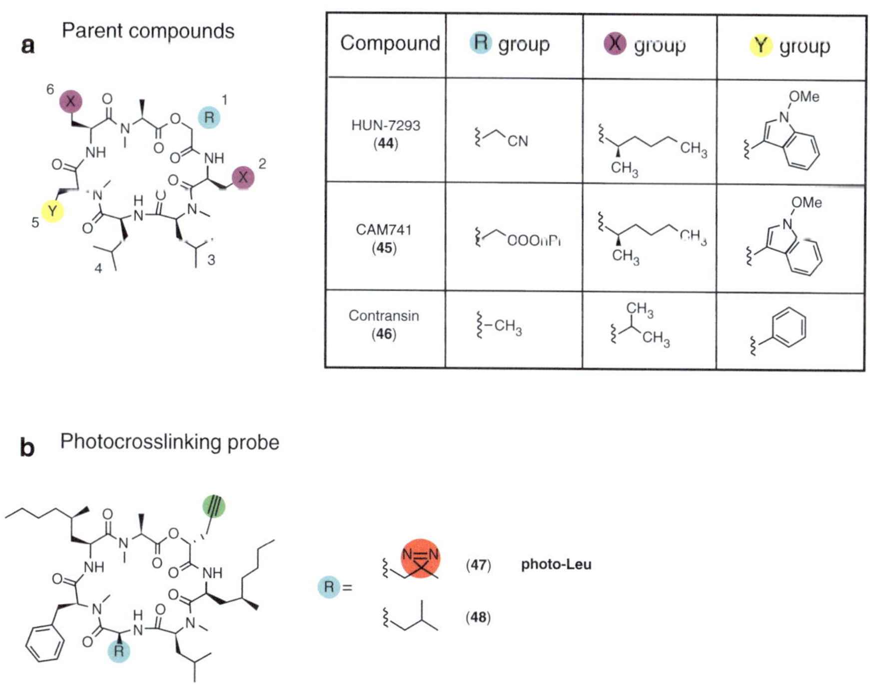

Fig. 3.10 Diazirine-based cyclodepsipeptide photoprobes. (**a**) Parent compounds. (**b**) Photo-crosslinking probes

~50 kDa was observed, which could be competed with excess ligand **48**. Furthermore, this fluorescently labeled band could be immunoprecipitated by an antibody against Sec61α. When the authors then crosslinked photoprobe **47** to proteoliposomes that had been reconstituted from a detergent extract of the ER microsomes, they observed that extracts that had been depleted of the intact Sec61 complex showed a decrease in intensity of the ~50 kDa band, which could be restored upon replenishment with purified Sec61 complex. These data led the authors to conclude that Sec61α is the primary target of this class of inhibitors. The authors also point out that the intimate placement of the diazirine group within the leucine side chain and its short lifetime (nanoseconds) may have aided in the identification of such a minor constituent of the proteome.

Polymyxins

Polymyxin antibiotics (namely, B and E) have reemerged in clinical settings as a last-resort treatment against multidrug-resistant Gram-negative bacteria. Polymyxins are polycationic decapeptides comprised of a heptapeptide ring and an N-terminal fatty acid tail, whose selectivity arises from their affinity for the negatively charged lipid A portion of lipopolysaccharide (LPS) within the outer leaflet of the asymmetric outer membrane (OM). In 2015, John Robinson and co-workers hypothesized that polymyxins might also interact with OM proteins in a stereospecific manner as part of their "self-promoted uptake" pathway. To address this hypothesis, the authors designed a photocrosslinking probe based on the polymyxin B3 scaffold (**49**) (Fig. 3.11) (van der Meijden and Robinson 2015). Based on previous structure–activity relationship (SAR) studies, it was known that

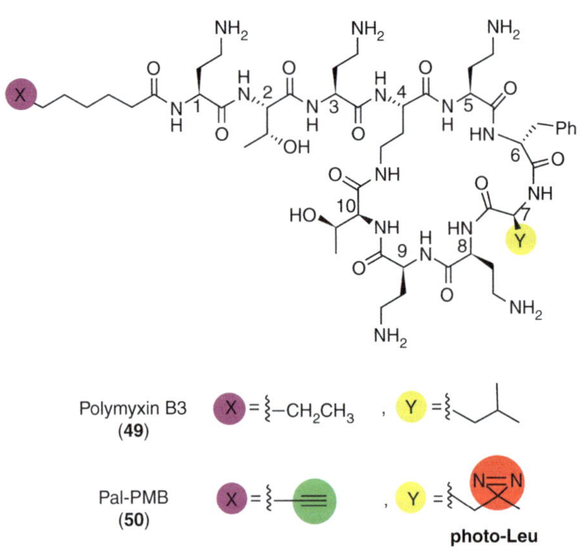

Fig. 3.11 Diazirine-based polymyxin photoprobes

conservative amino acid substitutions could be made within the D-Phe^{6}-L-Leu^{7} β-turn segment, such as an L-photo-Leu^{7} comprised of a diazirine photoreactive group. In addition, an alkyne for "click chemistry" was introduced into the fatty acyl chain at the N-terminus. The photoprobe retained potent antimicrobial activity against Gram-negative bacteria, with minimal inhibitory concentrations (MICs) for Pal-PMB (**50**) only twofold higher than those of the parental polymyxin B3 against both *E. coli* and *P. aeruginosa*. The test strain *E. coli* ATCC25922 was then photolabeled with Pal-PMB, and the OM proteins isolated by a modified carbonate extraction method for modification with a biotin tag and analysis by SDS-PAGE. The authors observed several distinct crosslinked bands in the size range 15–40 kDa, as well as a ~50 kDa and ~100 kDa band. Future work will entail identification of these crosslinked proteins by a proteomics approach and validation of their functional significance.

3.4.2.2 Histone Modifications

In 2010, Li and Kapoor set out to develop a chemical approach to "trap" proteins that recognize histone posttranslational modifications (PTMs) (Li and Kapoor 2010). This strategy relies on the use of histone peptide-based photoaffinity probes from the N-terminal "tail" region (i.e., amino acids 1–15). The authors focused on the trimethylation modification of lysine-4 of histone 3 (H3K4me3), which is typically associated with actively transcribed gene promoters. Available structures of protein "reader" domains in complex with H3K4me3 peptides provided a basis for probe design. For example, the plant homeodomain (PHD) finger ING2 interacts mainly through residues between P-3 and P+2 relative to lysine-4 (i.e., ARTK(me3)QT). Therefore, the authors synthesized a 15-amino acid photoprobe, in which a photoreactive benzoyl phenylalanine group (Bpa) replaced an Ala side chain at P+3 (so as to remain proximal to, but not sterically occlude, the binding site) and an alkyne was positioned at the C-terminus of the peptide (**51**) (Fig. 3.12a). Probe **51** was capable of labeling multiple recombinant proteins with folds that bind H3K4me3 in vitro, such as the PHD domain of ING2, the PHD domain of BPTF, and the JMJD2A double tudor domain. Furthermore, this labeling could be competed specifically by the native H3K4me3 peptide, but not with either the unmodified H3 peptide (H3K4me0) or the H3 peptide trimethylated at lysine-9 (H3K9me3). Specificity of crosslinking was further demonstrated through the observation that probe **51** was also able to affinity enrich endogenous ING2 from HeLa cells lysates, but not heterochromatin protein-1 (HP-1) known to preferentially interact with H3K9me3. In later work, the authors used probe **51** in combination with stable isotope labeling amino acid in cell culture (SILAC) to identify endogenous proteins from cell lysates that recognize the H3K4me3 modification (Li et al. 2012). While many of the identified hits were known previously to be H3K4me3 binders, the authors discovered a new H3K4me3 "reader" protein called MORC3, albeit one that possesses modest discrimination between the different methylation states of H3 peptides.

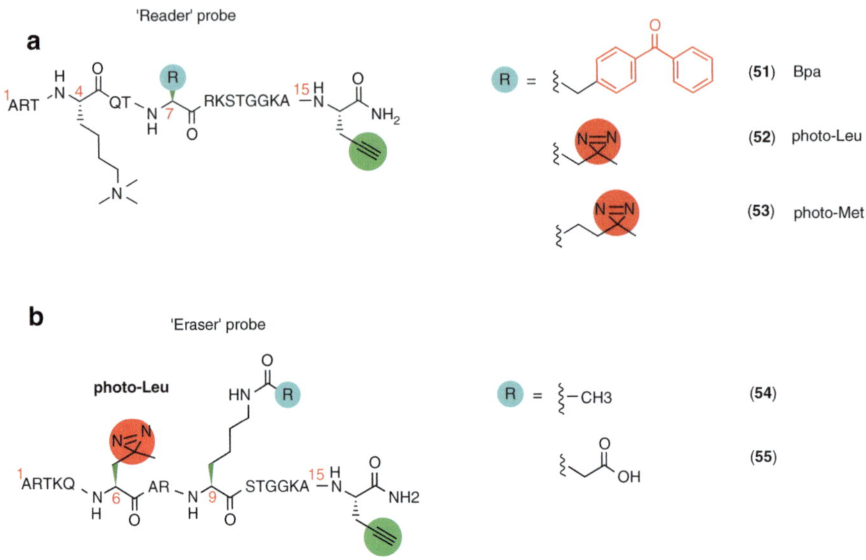

Fig. 3.12 Diazirine-based histone modification photoprobes. (**a**) Capturing histone 'readers' (**b**) Capturing histone 'erasers'

However, not all known "readers" of the H3K4me3 modification were enriched with probe **51**, possibly due to steric interference of the bulky benzophenone moiety. To address this issue, later work by Li and co-workers focused on the development of diazirine-based photoprobes (Yang et al. 2015). In the new probes **52** and **53** (Fig. 3.12a), the Ala side chain at P+3 was replaced by a diazirine-containing Leu or Met analog (photo-Leu or photo-Met), respectively, instead of Bpa found in the original probe **51**. As a comparison, photolabeling of recombinant SPIN1, a known "reader" of the H3K4me3 modification, was found to occur at a higher efficiency in vitro using the diazirine-based probes (**52** and **53**) over that of the benzophenone-based probe (**51**); analogous results were also observed for the PHD domain of ING2. Furthermore, when SPIN1 was added to HeLa S3 cell lysates and subjected to photocrosslinking to probes **51** (Bpa) and **52** (photo-Leu), the diazirine-based probe afforded not only a more robust signal, but a much higher "signal-to-noise" ratio over that of the benzophenone-based probe, indicative of a higher degree of specificity that can be achieved with the newly designed probes. Based on these data, the authors were inspired to design probes that could target proteins that remove histone modifications ("erasers"), a more technically challenging task due to the more transient nature of these interactions. To this end, the authors designed probes for the identification of "erasers" of histone lysine acetylation (e.g., Sirt3) and malonylation (e.g., Sirt5). In these constructs, the 15-amino acid H3 peptide is substituted with a photo-Leu at the position P+3 relative to the modified lysine-9, which is acetylated or malonylated in probes **54** and **55**, respectively (Fig. 3.12b). Probes **54** and **55** successfully labeled recombinant human Sirt3 and Sirt5, respectively, as well as when these proteins were added to HeLa S3 cell lysates.

Furthermore, probes **54** and **55** successfully enriched endogenous Sirt3 and Sirt5, respectively, from HeLa S3 cell lysates. Future work will entail SILAC proteomic analysis to profile "erasers" of newly identified histone PTMs.

3.4.2.3 Kinase Inhibitors

Minimalist Linker Design

Protein kinase (PK) inhibitors have merged as important drug candidates, but often suffer from unwanted side effects due to nonselective inhibition of multiple kinases. This arises from the fact that most kinase inhibitors target the highly conserved ATP-binding site of the enzyme. To identify these hazards early in drug development, there has been a growing emphasis in the field to conduct proteome-wide profiling of the potential targets of the compounds under native cellular settings. Photoprobes can aid in this endeavor. For example, dasatinib (**56**) (Fig. 3.13a) is a dual Src/Abl inhibitor that has been FDA approved for the treatment of imatinib-resistant chronic myelogenous leukemia (CML). Yao and co-workers have been successful in developing a photoprobe based on the dasatinib scaffold to effectively target interacting partners both in vitro (cellular lysates) and in situ (live cells) (Shi et al. 2012). To achieve this, the pyrimidine core was modified at the hydroethylpiperazinyl group, which functioned just for solubility purposes and doesn't make contacts with the kinase targets. A bifunctional linker was appended at this position, consisting of a both a photoreactive group (a benzophenone for DA-1 or an alkyl diazirine for DA-2) and an alkyne handle, which by modeling studies is expected to project into the solvent-exposed surface of the binding pocket. Both DA-1 (structure not shown) and DA-2 (**57**) (Fig. 3.13a) retain activity within an in vitro kinase

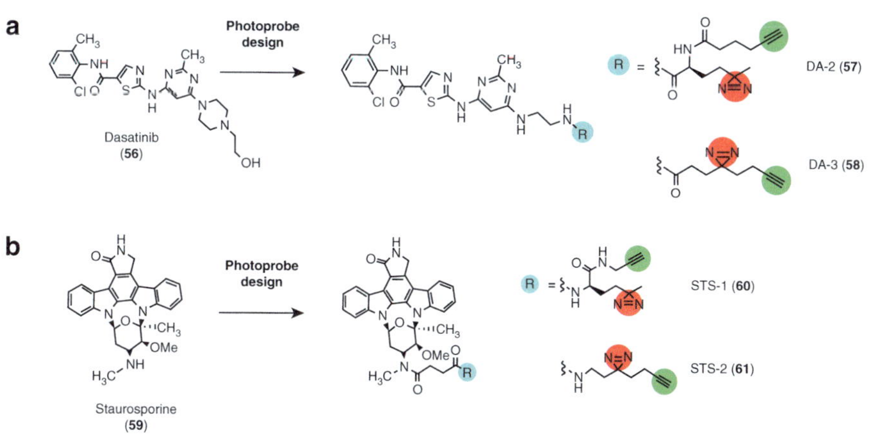

Fig. 3.13 Diazirine-based kinase inhibitor photoprobes. (**a**) Dasatinib photoprobes. (**b**) Staurosporine photoprobes

inhibition assay with recombinant purified c-Src and c-Abl kinase domains. However, only DA-2 retains the cell permeability of the parent compound, thereby retaining activity against full-length kinases transiently expressed in cell culture as determined through autophosphorylation and antiproliferative assays, and was therefore carried forward for additional photolabeling experiments.

DA-2 was able to photolabel recombinant purified c-Src and c-Abl kinase domains, but not a T338 M mutant of c-Src with a "deformed" ATP pocket. Analogous results were also observed against full-length c-Src constructs transiently expressed in cell culture, in a manner that can be blocked with dasatinib pretreatment. Based on these results, the authors optimized the DA-2 labeling conditions for comparison of endogenous protein targets present in either cellular lysates (in vitro) or live cells (in situ), striking a balance between productive crosslinking and minimal probe-induced cell death (<10%). Mass spectrometry analysis of purified crosslinked targets confirmed previous reports that a greater number of proteins can be identified that may be related to kinase activities/interactions by using photoprobes over conventional means of purification using resin-immobilized dasatinib (Fischer et al. 2011). Of the 84 potential targets identified by DA-2 photocrosslinking, a total of 50 kinases were identified (46 and 22 from in vitro and in situ settings, respectively, with 18 overlapping targets). The 13 serine/threonine kinases identified were the focus of further validation, only seven of which were previously shown to be targets of dasatinib. To this end, all six of the newly identified dasatinib targets (PCTK3, STK25, eIF-2A, PIM-3, PKA C-α, and PKN2) were successfully labeled by DA-2 in either cellular lysates or live cells, as determined by immunoanalysis of the purified crosslinked material. These results indicate that in addition to the Abl and Src family tyrosine kinases, serine/threonine kinases may be important direct binding partners of dasatinib and that the weak binding properties of these targets (as determined by kinase inhibition assays) have caused them to evade identification.

Yao and co-workers have also conducted analogous studies on the scaffold of the natural product staurosporine (**59**) (Fig. 3.13b), which is a potent general PK inhibitor with dozens of reported kinase targets. Previous work had indicated that the secondary amine position on staurosporine was highly tolerant to chemical modifications and thus could be appended with a bifunctional linker comprised of an alkyl diazirine and an alkyne handle to afford photoprobe STS-1 (**60**) (Fig. 3.13b) (Shi et al. 2011). Mass spectrometry analysis of endogenously expressed proteins that crosslinked to STS-1 in either cellular lysates (in vitro) or live cells (in situ) revealed a total of 43 kinases (35 and 21 from in vitro and in situ settings, respectively, with 13 overlapping targets). Of the eight kinases identified exclusively in situ, five were confirmed by immunoanalysis to be successfully labeled by STS-1 only in live cells and not lysates, suggesting the possibility that cell lysis may disrupt some complexes targeted by STS-1. More recently, these authors reported new reagents to streamline photoprobe production. They report first-generation "minimalist" linkers that incorporate a reactive functional group ($-CO_2H$, $-NH_2$, and –I) that can be conjugated to drug molecules through robust acylation and alkylation reactions (Li et al. 2013). In particular, these methods have been applied to create photoprobes

DA-3 (**58**) (Fig. 3.13a) and STS-2 (**61**) (Fig. 3.13b), which perform better than their predecessors, most likely due to further minimization of interference with the desired binding interactions. Second-generation "minimalist" linkers containing a novel cyclopropene handle and a diazirine have now been developed and can be used for both imaging and protein labeling (Li et al. 2014).

Type II Binding Mode

Protein kinases can interconvert between catalytically active and inactive states, but it remains challenging to know which states are represented inside of a living cell. The Maly group has designed photoprobes with the goal of globally profiling kinase functional states during specific signaling events. Utilizing two general scaffolds (Fig. 3.14a, b) based on type II inhibitors (**62** and **64**), two photocrosslinking probes were designed to capture the kinases that adopt the DFG-out inactive conformation (Ranjitkar et al. 2012). In order to poise a photoreactive group within the ATP-binding site without disrupting key interactions with the pharmacophore, the authors chose to substitute the trifluoromethyl groups of the inhibitors with trifluoromethyl diazirines (**63** and **65**). An alkyne tag was placed at a solvent-exposed region to allow access for derivatization. While the photoprobes do retain the inhibitory specificities of their parent compounds against a small panel of kinases known to be either sensitive (ABL, CSK, EPHA3, HCK, LCK, p38α, and SRC) or insensitive (IRAK4, MAPK3K5, PAK4, PAK5, PKA, and p38δ) to type II inhibitors, in general the IC_{50} values of the photoprobes were >5-fold higher than those of the parent molecules. Regardless, probes **63** and **65** were successful in photolabeling hexahistidine-tagged versions of kinases SRC, p38α, and EPHA3 (known to adopt the DFG-out conformation) when individually added to mammalian cell lysate, but

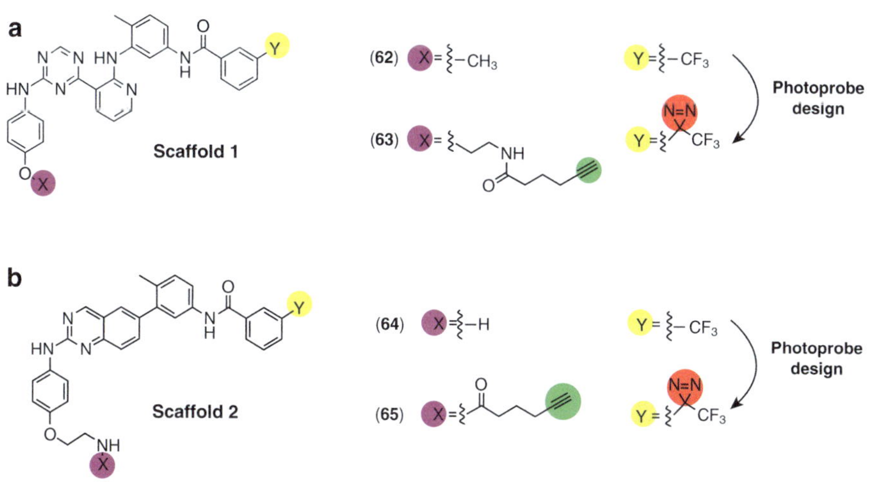

Fig. 3.14 Diazirine-based type II kinase inhibitor photoprobes. (**a**) Scaffold 1. (**b**) Scaffold 2

not in the presence of excess competitors **62** and **64**. Additionally, kinases not known to adopt the DFG-out conformation, MAP3K5 and PAK5, were not labeled by the photoprobes, suggesting that labeling occurred in a conformation-specific manner.

To identify endogenous (and possible low abundance) kinases capable of adopting the DFG-out conformation in mammalian cell lysate, SILAC methodology was employed for quantitative proteomic analysis. In this experiment, the photocrosslinking with probes **63** or **65** was performed with A431 lysates in the absence (heavy state) and presence (light state) of competitor ligand. The lysates were combined, conjugated to biotin-azide, and enriched with streptavidin-conjugated beads for on-bead trypsin digest. Proteins identified by mass spectrometric analysis with a heavy-to-light ratio (H/L) greater than twofold were classified as specific targets of the photoprobes. Of the specifically labeled proteins identified, 21 protein kinases were labeled by probe **63**, and 28 protein kinases were labeled by probe **65**, with an overlap of 15 protein kinases being labeled by both probes. From these data, a number of known as well as novel protein kinase targets were identified, which will require future characterization as to their ability to adopt the DFG-out conformation and thus susceptibility to type II inhibitors. In particular, the authors obtained a crystal structure of the catalytic domain of the STE kinase LOK bound to inhibitor **62** and confirmed that it adopts the DFG-out conformation in this context. Furthermore, an in situ (crosslinking performed in cells) quantitative proteomic analysis was carried out for probe **65**, in which 24 proteins were specifically enriched, with eight of these being protein kinases (all of which have been previously characterized to adopt the DFG-out conformation). Ongoing work will entail the in situ profiling of the active site conformations of these hits during ligand-stimulated signaling events.

3.4.2.4 O-linked β-*N*-acetylglucosamine (O-GlcNAc)

O-linked β-*N*-acetylglucosamine (O-GlcNAc) is a single sugar modification found on serine and threonine residues ubiquitously among proteins in higher eukaryotes. The O-GlcNAc modification is dynamic, being added to proteins by O-GlcNAc transferase (OGT) and hydrolytically removed by O-GlcNAcase (OGA). In 2012, Kohler and co-workers developed a method in which human cell lines are genetically engineered to metabolize a diazirine-modified form of GlcNAc-1-phosphate, thereby producing a diazirine-modified analog of the nucleotide sugar donor UDP-GlcNAc (Yu et al. 2012). This method relies on synthesis of a cell-permeable diazirine-modified analog of GlcNAc-1-phosphate and on mutation of the UDP-GlcNAc pyrophosphorylase (UAP1) to enable it to accept a diazirine-modified substrate. Under these conditions, cells produce diazirine-modified UDP-GlcNAc, known as UDP-GlcNDAz, which is accepted as a substrate by OGT, enabling in-cell production of diazirine-modified O-GlcNAc (O-GlcNDAz) (Fig. 3.15a). The authors used this technology to identify interaction partners of a subset of nuclear pore proteins (NUP153, NUP214, and NUP358) known to be heavily modified with O-GlcNAc residues. Multiple nuclear transport factors were shown to crosslink to nuclear pore

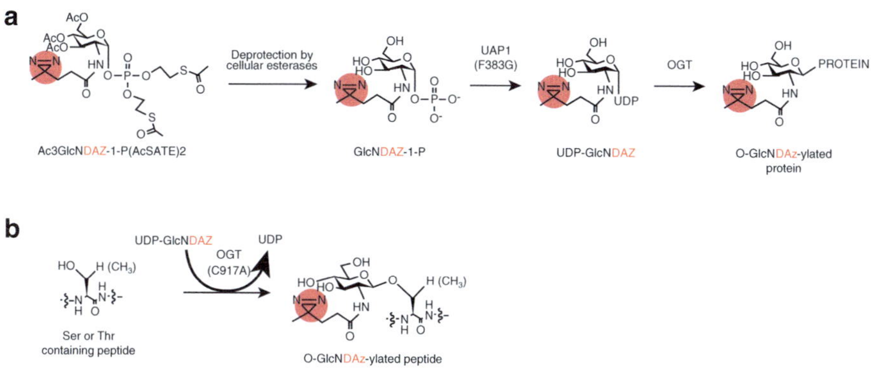

Fig. 3.15 Diazirine-based metabolically incorporated photocrosslinking sugars. (**a**) Cell-based GlcNDAz (**b**) *In-vitro* GlcNDAz

proteins, demonstrating that the photocrosslinking method was capable of capturing transient interactions that occur during nuclear transport. Following that success, the authors are working to expand this technology to enable mapping of glycan-dependent protein–protein interaction networks. Toward that goal, they have identified a mutant form of OGT that exhibits a substrate preference for UDP-GlcNDAz over UDP-GlcNAc and can be used for efficient chemoenzymatic synthesis of O-GlcNDAz-modified peptidic photoprobes (Rodriguez et al. 2015). In a proof-of-principle experiment, the authors were able to take a biotin-labeled 14-amino acid peptide sequence from the casein kinase II (CKII), modify it with O-GlcNDAz, and photocrosslink it to an O-GlcNAc-recognizing antibody (RL2) (Fig. 3.15b). Future work will entail constructing and crosslinking an array of peptidic O-GlcNDAz photoprobes to cell lysates to identify candidate binding partners, which can then be verified using the cell-based O-GlcNDAz photocrosslinking system.

3.5 Conclusions and Future Directions

This chapter describes select examples reported over the last decade illustrating how to convert a known ligand into a compact photoprobe. From these examples, it is apparent that researchers can utilize photoprobe technology for the following purposes: (1) identifying the direct binding partner of a ligand, (2) gaining structural information on the ligand binding site in the absence of a three-dimensional structure, (3) determining whether ligand binding is occurring through a primary or allosteric site, and (4) investigating the specificity determinants of ligand binding. The area in which the field of photocrosslinking has room to grow is the de novo identification of proteinaceous binding partners by mass spectrometry. This is typically performed on peptides released by trypsin digestion of affinity-enriched complexes, upon which stringent purification conditions should be implemented to truly take advantage of the covalent nature of the photoprobe approach. However, there is still difficulty in differentiating between specific versus nonspecific hits because of limitations on the abundance of biologically relevant crosslinked complexes within the sample, likely due to a combination of their low expression levels within a proteome and their substoichiometric labeling by the photoreactive group. Choosing the proper comparative sample(s) can be a key parameter to success in this area. For instance, inclusion of a competitor ligand in the photocrosslinking reaction can be a powerful tool for excluding binding of the photoprobe to the biologically active target, rather than to a non-relevant high abundant or "sticky" protein. Alternatively, Park and co-workers have recently utilized two-color fluorescence and two-dimensional gel electrophoresis (2DGE) to obtain a higher resolution picture of the bands that have been labeled with photoprobes that either do or do not contain the biologically relevant pharmacophore (Park et al. 2012). Continued efforts in this area will be needed to progress this technology to where it will be a common tool available in the arsenal of biochemists.

Looking toward the future, we anticipate that compact, "clickable" photoprobes will become essential tools in the process of identifying the protein targets of the many novel bioactive molecules that are now emerging from phenotypic screening approaches. Indeed, Cravatt and co-workers have pioneered an approach in which phenotypic screening is performed with molecules that already incorporate functionalities required for photoaffinity labeling and affinity purification. In 2012, this group reported a ~30-member photoprobe library constructed around two benzophenone-based scaffolds: the 5-benzoyl indole (BzIndole, **66**) (Fig. 3.16a)

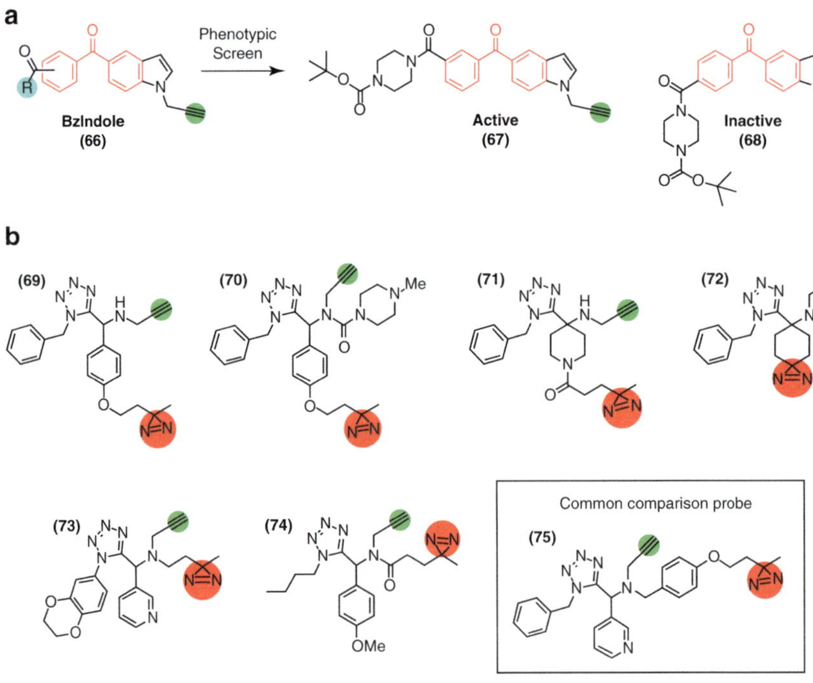

Fig. 3.16 Integration of phenotypic screening and target deconvolution. (**a**) Benzophenone-based library. (**b**) Diazirine-based library

and the 7-benzoyl-benzo-1,4-diazepin-2,5-dione (BzBD) (Cisar and Cravatt 2012). Interestingly, a compound with a *t*-butoxycarbmolpiperazine carbonyl group on the *meta* position of the BzIndole scaffold (**67**) demonstrated antiproliferative activity against a human breast cancer cell line MDA-MB-231 under low-glucose conditions that better mimic the nutrient-deprived state found in solid tumors, while placement of this substituent at the *para* position (**68**) abolished activity (Fig. 3.16a). Through comparative proteomics, the authors found that the differential target was MT-ND1, an integral membrane subunit of the ubiquinone oxidoreductase (complex 1) that is known to impair cancer cell viability under low-glucose conditions. Thus, photoprobes can be applied in a straightforward manner to find druggable targets for a given phenotypic effect, even if the photoprobe molecules are not themselves lead drug candidates. Indeed this strategy may be generalizable, as other examples are emerging where compact photoprobes are useful tools for identifying the targets of molecules identified in HTS campaigns (Theodoropoulos 2016). In 2014 Cravatt and co-workers further expanded on their goals of constructing a more structurally diverse photoprobe library that can be used to target multiple cellular pathways for phenotypic screening assays. To this end, they constructed a ~60-member library containing a diazirine photoreactive group by utilizing an Ugi-azide tri-component condensation reaction (Kambe et al. 2014). With this method, the authors were able to assemble 1,5-disubstituted tetrazoles from an array of isocyanide, aldehyde/ketone, and amine building blocks. Most probes showed distinct labeling patterns in situ within a human PC-3 prostate cancer cell line as determined by in-gel fluorescence, whereby six photoprobes were chosen for an in-depth MS analysis (**69**–**74**) (Fig. 3.16b). Because probe **75** was observed to possess a broad interaction profile, it was therefore chosen as a means for comparison. In the end, 24 preferred targets were identified that encompassed not only proteins generally considered to be druggable (i.e., enzymes and transmembrane channels) but also proteins believed to be less pharmacologically accessible such as those that function to mediate protein–protein interactions within signal transduction pathways (i.e., adaptor and scaffolding proteins). Thus, photoprobes have the potential to be important tools in future drug discovery efforts.

References

Banala AK, Zhang P, Plenge P, Cyriac G, Kopajtic T, Katz JL, Loland CJ, Newman AH (2013) Design and synthesis of 1-(3-(dimethylamino)propyl)-1-(4-fluorophenyl)-1,3-dihydroisobenzofuran-5-carbonitrile (citalopram) analogues as novel probes for the serotonin transporter S1 and S2 binding sites. J Med Chem 56(23):9709–9724. doi:10.1021/jm4014136

Chen Y, Hajipour AR, Sievert MK, Arbabian M, Ruoho AE (2007) Characterization of the cocaine binding site on the sigma-1 receptor. Biochemistry 46(11):3532–3542. doi:10.1021/bi061727o

Chun J, Yin YI, Yang G, Tarassishin L, Li YM (2004) Stereoselective synthesis of photoreactive peptidomimetic gamma-secretase inhibitors. J Org Chem 69(21):7344–7347. doi:10.1021/jo0486948

Cisar JS, Cravatt BF (2012) Fully functionalized small-molecule probes for integrated phenotypic screening and target identification. J Am Chem Soc 134(25):10385–10388. doi:10.1021/ja304213w

Coleman JA, Green EM, Gouaux E (2016) X-ray structures and mechanism of the human serotonin transporter. Nature 532(7599):334–339. doi:10.1038/nature17629

Conti E, Stachelhaus T, Marahiel MA, Brick P (1997) Structural basis for the activation of phenylalanine in the non-ribosomal biosynthesis of gramicidin S. EMBO J 16(14):4174–4183

Crump CJ, Fish BA, Castro SV, Chau DM, Gertsik N, Ahn K, Stiff C, Pozdnyakov N, Bales KR, Johnson DS, Li YM (2011) Piperidine acetic acid based gamma-secretase modulators directly bind to Presenilin-1. ACS Chem Nerosci 2(12):705–710. doi:10.1021/cn200098p

Duckworth BP, Wilson DJ, Nelson KM, Boshoff HI, Barry CE 3rd, Aldrich CC (2012) Development of a selective activity-based probe for adenylating enzymes: profiling MbtA involved in siderophore biosynthesis from Mycobacterium tuberculosis. ACS Chem Biol 7(10):1653–1658. doi:10.1021/cb300112x

Eirich J, Orth R, Sieber SA (2011) Unraveling the protein targets of vancomycin in living S. aureus and E. faecalis cells. J Am Chem Soc 133(31):12144–12153. doi:10.1021/ja2039979

Ferreras JA, Ryu JS, Di Lello F, Tan DS, Quadri LEN (2005) Small-molecule inhibition of siderophore biosynthesis in Mycobacterium tuberculosis and Yersinia pestis. Nat Chem Biol 1(1):29–32. doi:10.1038/nchembio706

Finking R, Neumuller A, Solsbacher J, Konz D, Kretzschmar G, Schweitzer M, Krumm T, Marahiel MA (2003) Aminoacyl adenylate substrate analogues for the inhibition of adenylation domains of nonribosomal peptide synthetases. Chembiochem 4(9):903–906. doi:10.1002/cbic.200300666

Fischer JJ, Dalhoff C, Schrey AK, Graebner OY, Michaelis S, Andrich K, Glinski M, Kroll F, Sefkow M, Dreger M, Koester H (2011) Dasatinib, imatinib and staurosporine capture compounds—complementary tools for the profiling of kinases by Capture Compound Mass Spectrometry (CCMS). J Proteomics 75(1):160–168. doi:10.1016/j.jprot.2011.05.035

Fontanilla D, Hajipour AR, Pal A, Chu UB, Arbabian M, Ruoho AE (2008) Probing the steroid binding domain-like I (SBDLI) of the sigma-1 receptor binding site using N-substituted photoaffinity labels. Biochemistry 47(27):7205–7217. doi:10.1021/bi800564j

Gillespie EJ, Ho CL, Balaji K, Clemens DL, Deng G, Wang YE, Elsaesser HJ, Tamilselvam B, Gargi A, Dixon SD, France B, Chamberlain BT, Blanke SR, Cheng G, de la Torre JC, Brooks DG, Jung ME, Colicelli J, Damoiseaux R, Bradley KA (2013) Selective inhibitor of endosomal trafficking pathways exploited by multiple toxins and viruses. Proc Natl Acad Sci U S A 110(50):E4904–E4912. doi:10.1073/pnas.1302334110

Hett EC, Xu H, Geoghegan KF, Gopalsamy A, Kyne RE Jr, Menard CA, Narayanan A, Parikh MD, Liu S, Roberts L, Robinson RP, Tones MA, Jones LH (2015) Rational targeting of active-site tyrosine residues using sulfonyl fluoride probes. ACS Chem Biol 10(4):1094–1098. doi:10.1021/cb5009475

Ishikawa F, Konno S, Suzuki T, Dohmae N, Kakeya H (2015a) Profiling nonribosomal peptide synthetase activities using chemical proteomic probes for adenylation domains. ACS Chem Biol 10(9):1989–1997. doi:10.1021/acschembio.5b00097

Ishikawa F, Suzuki T, Dohmae N, Kakeya H (2015b) A multiple-labeling strategy for nonribosomal peptide synthetases using active-site-directed proteomic probes for adenylation domains. Chembiochem 16(18):2590–2594. doi:10.1002/cbic.201500481

Jung ME, Chamberlain BT, Ho CL, Gillespie EJ, Bradley KA (2014) Structure-activity relationship of semicarbazone EGA furnishes photoaffinity inhibitors of anthrax toxin cellular entry. ACS Med Chem Lett 5(4):363–367. doi:10.1021/ml400486k

Kahoun JR, Ruoho AE (1992) (125I)iodoazidococaine, a photoaffinity label for the haloperidol-sensitive sigma receptor. Proc Natl Acad Sci U S A 89(4):1393–1397

Kambe T, Correia BE, Niphakis MJ, Cravatt BF (2014) Mapping the protein interaction landscape for fully functionalized small-molecule probes in human cells. J Am Chem Soc 136(30):10777–10782. doi:10.1021/ja505517t

Konno S, Ishikawa F, Suzuki T, Dohmae N, Burkart MD, Kakeya H (2015) Active site-directed proteomic probes for adenylation domains in nonribosomal peptide synthetases. Chem Commun 51(12):2262–2265. doi:10.1039/c4cc09412c

Koteva K, Hong HJ, Wang XD, Nazi I, Hughes D, Naldrett MJ, Buttner MJ, Wright GD (2010) A vancomycin photoprobe identifies the histidine kinase VanSsc as a vancomycin receptor. Nat Chem Biol 6(5):327–329. doi:10.1038/nchembio.350

Kumar V, Yarravarapu N, Lapinsky DJ, Perley D, Felts B, Tomlinson MJ, Vaughan RA, Henry LK, Lever JR, Newman AH (2015) Novel Azido-Iodo Photoaffinity Ligands for the Human Serotonin Transporter Based on the Selective Serotonin Reuptake Inhibitor (S)-Citalopram. J Med Chem 58(14):5609–5619. doi:10.1021/acs.jmedchem.5b00682

Lapinsky DJ, Velagaleti R, Yarravarapu N, Liu Y, Huang Y, Surratt CK, Lever JR, Foster JD, Acharya R, Vaughan RA, Deutsch HM (2011) Azido-iodo-N-benzyl derivatives of threo-methylphenidate (Ritalin, Concerta): Rational design, synthesis, pharmacological evaluation, and dopamine transporter photoaffinity labeling. Bioorg Med Chem 19(1):504–512. doi:10.1016/j.bmc.2010.11.002

Lapinsky DJ, Yarravarapu N, Nolan TL, Surratt CK, Lever JR, Tomlinson M, Vaughan RA, Deutsch HM (2012) Evolution of a compact photoprobe for the dopamine transporter based on (+/−)-threo-methylphenidate. ACS Med Chem Lett 3(5):378–382. doi:10.1021/ml3000098

Li X, Kapoor TM (2010) Approach to profile proteins that recognize post-translationally modified histone "tails". J Am Chem Soc 132(8):2504–2505. doi:10.1021/ja909741q

Li YM, Xu M, Lai MT, Huang Q, Castro JL, DiMuzio-Mower J, Harrison T, Lellis C, Nadin A, Neduvelil JG, Register RB, Sardana MK, Shearman MS, Smith AL, Shi XP, Yin KC, Shafer JA, Gardell SJ (2000) Photoactivated gamma-secretase inhibitors directed to the active site covalently label presenilin 1. Nature 405(6787):689–694. doi:10.1038/35015085

Li X, Foley EA, Molloy KR, Li Y, Chait BT, Kapoor TM (2012) Quantitative chemical proteomics approach to identify post-translational modification-mediated protein-protein interactions. J Am Chem Soc 134(4):1982–1985. doi:10.1021/ja210528v

Li Z, Hao P, Li L, Tan CY, Cheng X, Chen GY, Sze SK, Shen HM, Yao SQ (2013) Design and synthesis of minimalist terminal alkyne-containing diazirine photo-crosslinkers and their incorporation into kinase inhibitors for cell- and tissue-based proteome profiling. Angewandte Chemie 52(33):8551–8556. doi:10.1002/anie.201300683

Li Z, Wang D, Li L, Pan S, Na Z, Tan CY, Yao SQ (2014) "Minimalist" cyclopropene-containing photo-cross-linkers suitable for live-cell imaging and affinity-based protein labeling. J Am Chem Soc 136(28):9990–9998. doi:10.1021/ja502780z

MacKinnon AL, Garrison JL, Hegde RS, Taunton J (2007) Photo-leucine incorporation reveals the target of a cyclodepsipeptide inhibitor of cotranslational translocation. J Am Chem Soc 129(47):14560–14561. doi:10.1021/ja076250y

van der Meijden B, Robinson JA (2015) Synthesis of a polymyxin derivative for photolabeling studies in the gram-negative bacterium Escherichia coli. J Pept Sci 21(3):231–235. doi:10.1002/psc.2736

Neres J, Labello NP, Somu RV, Boshoff HI, Wilson DJ, Vannada J, Chen L, Barry CE 3rd, Bennett EM, Aldrich CC (2008) Inhibition of siderophore biosynthesis in Mycobacterium tuberculosis with nucleoside bisubstrate analogues: structure-activity relationships of the nucleobase domain of 5′-O-[N-(salicyl)sulfamoyl]adenosine. J Med Chem 51(17):5349–5370. doi:10.1021/jm800567v

Pal A, Hajipour AR, Fontanilla D, Ramachandran S, Chu UB, Mavlyutov T, Ruoho AE (2007) Identification of regions of the sigma-1 receptor ligand binding site using a novel photoprobe. Mol Pharmacol 72(4):921–933. doi:10.1124/mol.107.038307

Park J, Oh S, Park SB (2012) Discovery and target identification of an antiproliferative agent in live cells using fluorescence difference in two-dimensional gel electrophoresis. Angew Chem 51(22):5447–5451. doi:10.1002/anie.201200609

Pozdnyakov N, Murrey HE, Crump CJ, Pettersson M, Ballard TE, Am Ende CW, Ahn K, Li YM, Bales KR, Johnson DS (2013) gamma-Secretase modulator (GSM) photoaffinity probes reveal distinct allosteric binding sites on presenilin. J Biol Chem 288(14):9710–9720. doi:10.1074/jbc.M112.398602

Ranjitkar P, Perera BG, Swaney DL, Hari SB, Larson ET, Krishnamurty R, Merritt EA, Villen J, Maly DJ (2012) Affinity-based probes based on type II kinase inhibitors. J Am Chem Soc 134(46):19017–19025. doi:10.1021/ja306035v

Rodriguez AC, Yu SH, Li B, Zegzouti H, Kohler JJ (2015) Enhanced transfer of a photocross-linking N-acetylglucosamine (GlcNAc) analog by an O-GlcNAc transferase mutant with converted substrate specificity. J Biol Chem 290(37):22638–22648. doi:10.1074/jbc.M115.667006

Salisbury CM, Cravatt BF (2007) Activity-based probes for proteomic profiling of histone deacetylase complexes. Proc Natl Acad Sci U S A 104(4):1171–1176. doi:10.1073/pnas.0608659104

Schmidt HR, Zheng S, Gurpinar E, Koehl A, Manglik A, Kruse AC (2016) Crystal structure of the human sigma1 receptor. Nature 532(7600):527–530. doi:10.1038/nature17391

Shi H, Cheng X, Sze SK, Yao SQ (2011) Proteome profiling reveals potential cellular targets of staurosporine using a clickable cell-permeable probe. Chem Commun 47(40):11306–11308. doi:10.1039/c1cc14824a

Shi H, Zhang CJ, Chen GY, Yao SQ (2012) Cell-based proteome profiling of potential dasatinib targets by use of affinity-based probes. J Am Chem Soc 134(6):3001–3014. doi:10.1021/ja208518u

Singh J, Salcius M, Liu SW, Staker BL, Mishra R, Thurmond J, Michaud G, Mattoon DR, Printen J, Christensen J, Bjornsson JM, Pollok BA, Kiledjian M, Stewart L, Jarecki J, Gurney ME (2008) DcpS as a therapeutic target for spinal muscular atrophy. ACS Chem Biol 3(11):711–722. doi:10.1021/cb800120t

Somu RV, Wilson DJ, Bennett EM, Boshoff HI, Celia L, Beck BJ, Barry CE, Aldrich CC (2006) Antitubercular nucleosides that inhibit siderophore biosynthesis: SAR of the glycosyl domain. J Med Chem 49(26):7623–7635. doi:10.1021/jm061068d

Stachelhaus T, Mootz HD, Marahiel MA (1999) The specificity-conferring code of adenylation domains in nonribosomal peptide synthetases. Chem Biol 6(8):493–505. doi:10.1016/S1074-5521(99)80082-9

Theodoropoulos PC, et al (2016) Discovery of tumor-specific irreversible inhibitors of stearoyl CoA desaturase. Nat Chem Biol 12:218–225

Wang KH, Penmatsa A, Gouaux E (2015) Neurotransmitter and psychostimulant recognition by the dopamine transporter. Nature 521(7552):322–327. doi:10.1038/nature14431

Xu H, Hett EC, Gopalsamy A, Parikh MD, Geoghegan KF, Kyne RE Jr, Menard CA, Narayanan A, Robinson RP, Johnson DS, Tones MA, Jones LH (2015) A library approach to rapidly discover photoaffinity probes of the mRNA decapping scavenger enzyme DcpS. Mol Biosyst 11(10):2709–2712. doi:10.1039/c5mb00288e

Yang G, Yin YI, Chun J, Shelton CC, Ouerfelli O, Li YM (2009) Stereo-controlled synthesis of novel photoreactive gamma-secretase inhibitors. Bioorg Med Chem Lett 19(3):922–925. doi:10.1016/j.bmcl.2008.11.117

Yang TP, Liu Z, Li XD (2015) Developing diazirine-based chemical probes to identify histone modification 'readers' and 'erasers'. Chem Sci 6(2):1011–1017. doi:10.1039/c4sc02328e

Yu SH, Boyce M, Wands AM, Bond MR, Bertozzi CR, Kohler JJ (2012) Metabolic labeling enables selective photocrosslinking of O-GlcNAc-modified proteins to their binding partners. Proc Natl Acad Sci U S A 109(13):4834–4839. doi:10.1073/pnas.1114356109

Chapter 4
Utilities of Anti-ligand Antibody and Mass Spectrum to Elucidate Photolabeled Sites

Kohichi Kawahara, Katsuyoshi Masuda, Masaji Ishiguro, and Hitoshi Nakayama

Abstract In studies of photoaffinity labeling, combinatorial use of mass spectrometry (MS) analysis has been increasing. Among the increasing numbers of such reports, however, major use of the method is for mapping labeled peptides, and application for identifying photolabeled peptides and their labeled amino acid residues is still limited. This is mainly caused that separation of the labeled peptide fragment(s) by HPLC, for example, is tedious and labour intensive. Therefore, in order to locate the ligand binding sites precisely even when photolabeling efficiency is low, a general and an efficient method to separate the photolabeled peptide(s) is required for identifying not only photolabeled peptides but also their labeled amino acid residues. As an example to fulfil this requirement, we here describe a method using simple purification step using anti-ligand antibody followed by MS analysis to characterize efficiently the binding sites for a calcium antagonist semotiadil in human serum albumin as the drug-binding protein.

Keywords Photoaffinity labeling • Anti-ligand antibody • Mass spectrometry • Drug-binding site • Human serum albumin • Semotiadil • FNAK

K. Kawahara (✉) • M. Ishiguro
Niigata University of Pharmacy and Applied Life Sciences, Niigata, Niigata 956-8603, Japan
e-mail: kkawa@nupals.ac.jp; ishiguro@nupals.ac.jp

K. Masuda
Suntory Institute for Bioorganic Research, Shimamoto-cho, Mishima-gun, Osaka 618-8503, Japan
e-mail: Katsuyoshi_Masuda@suntory.co.jp

H. Nakayama
Kumamoto University, Kumamoto, Kumamoto 862-0973, Japan
e-mail: jin@gpo.kumamoto-u.ac.jp

Y. Hatanaka, M. Hashimoto (eds.), *Photoaffinity Labeling for Structural Probing Within Protein*, DOI 10.1007/978-4-431-56569-7_4

4.1 Introduction

Photoaffinity labeling deserves a versatile method to identify chemically not only target molecules in biological preparations but also amino acid residues that confer the ligand binding site(s) within the target molecules. In addition, recent progress in mass spectrometry enables us to detect peptides and determine their sequences in sub-picomolar quantities (for reviews, see Chaurand et al. 1999; Jensen et al. 1999; Kuster and Mann 1998). Matrix-assisted laser desorption ionization-time-of-flight mass spectrometry (MALDI-TOF MS) provides accurate information about peptide size, even when the sample is a peptide mixture, and mass-fit analysis gives a possible amino acid sequence when the primary structure is known. To analyse the sequence of a peptide fragment precisely, several methods, including post-source decay (PSD) MALDI-TOF MS and tandem MS (MS/MS), have been used.

In studies of photoaffinity labeling, introduction of MS analysis has been increasing (for reviews, see Sinz 2014; Lee et al. 2014; Cheng et al. 2010; Robinette et al. 2006). Among the increasing numbers of such reports, however, major use of the method is for mapping labeled peptides (Dianoux et al. 2000; Pinke et al. 2009; Rowland et al. 2011; Pettelkau et al. 2012; Chen et al. 2012), and application for identifying photolabeled peptides and their labeled amino acid residues is still limited. This is mainly caused by separation of the labeled peptide fragment(s) from unlabeled peptide(s) and/or large numbers of other peptide fragments by HPLC, for example, which is tedious and time-consuming. Therefore, in order to locate the ligand binding sites precisely even if photolabeling efficiency may happen to be low, a general and an efficient method to separate the photolabeled peptide(s) is required for identifying not only photolabeled peptides but also their labeled amino acid residues.

As an example to fulfil this requirement, we here describe a method using simple purification step using anti-ligand antibody followed by MS analysis to characterize efficiently the binding sites for a calcium antagonist semotiadil (Fig. 4.1) in human serum albumin (HSA) as the drug-binding protein (Kawahara et al. 2002).

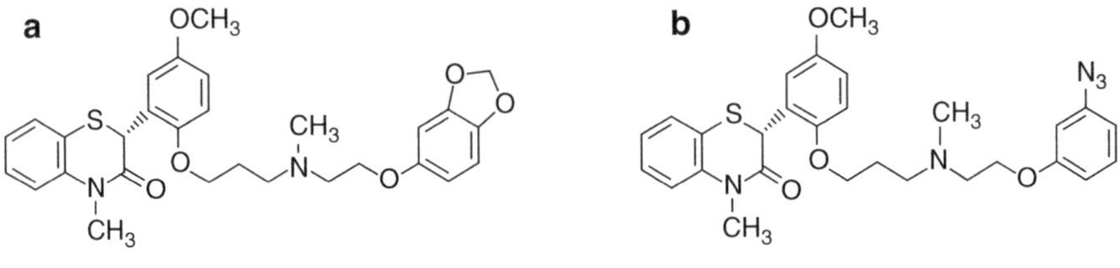

Fig. 4.1 Chemical structure of semotiadil derivatives. (**a**) Semotiadil. (**b**) FNAK

4.2 Conjugate Preparation and Generation of Anti-ligand Antibody

We prepared norsemotiadil–thyroglobulin conjugate as antigen and generated the anti-ligand antibody as follows. (±)-2-[5-Methoxy-2-[3-[2-[3,4-(methylenedioxy) phenoxyl]-ethylamino]propoxy]phenyl]-4-methyl-2*H*-1, 4-benzothiazin-3-(4*H*)-one (norsemotiadil) was reacted with equimolar N-succinimidyl (4-idoacetyl)-aminobenzoate (SIAB) and triethylamine in dimethylformamide (DMF) for 24 h at room temperature, and the residual mixture was purified on a silica gel column. The purified material in DMF was incubated with thyroglobulin for 1.5 h at room temperature. After dialysis against PBS, a norsemotiadil–thyroglobulin conjugate was obtained (Fig. 4.2). Spectral analysis of the conjugate indicated 112 moles of norsemotiadil/mole of thyroglobulin. Then, Japanese white rabbits were immunized with the conjugate (1.6 mg/rabbit) emulsified in Freund's complete adjuvant with a biweekly schedule of injection. After 4 weeks, the immunization was repeated an additional four times with the conjugate (0.8 mg/rabbit/immunization) in Freund's incomplete adjuvant. Titres were monitored by ELISA assay using norsemotiadil–HSA conjugate and horseradish peroxidase-labeled goat anti-rabbit IgG. The norsemotiadil–HSA conjugate (49 moles of norsemotiadil/mole of HSA) was prepared in a manner similar to the norsemotiadil–thyroglobulin.

To assess antibody specificity, a competitive ELISA assay was performed by addition of a preincubated mixture of antiserum and test drugs at various concentrations

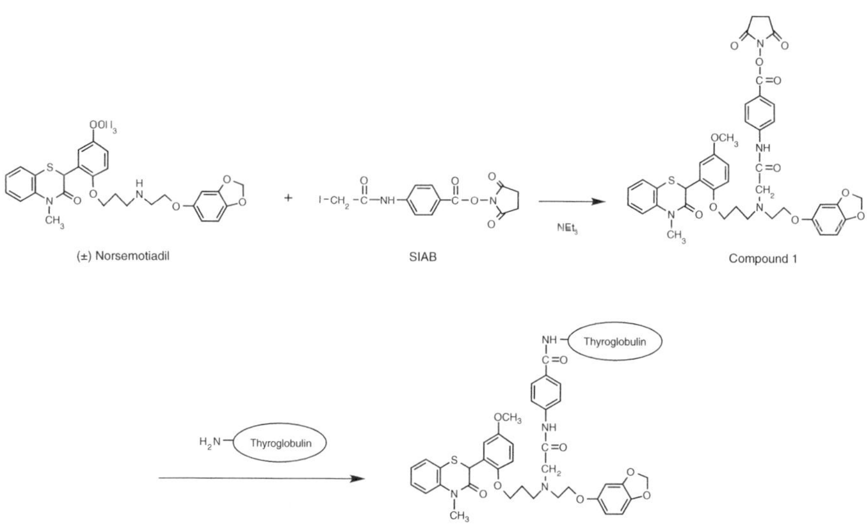

Fig. 4.2 Preparation of a norsemotiadil–thyroglobulin conjugates

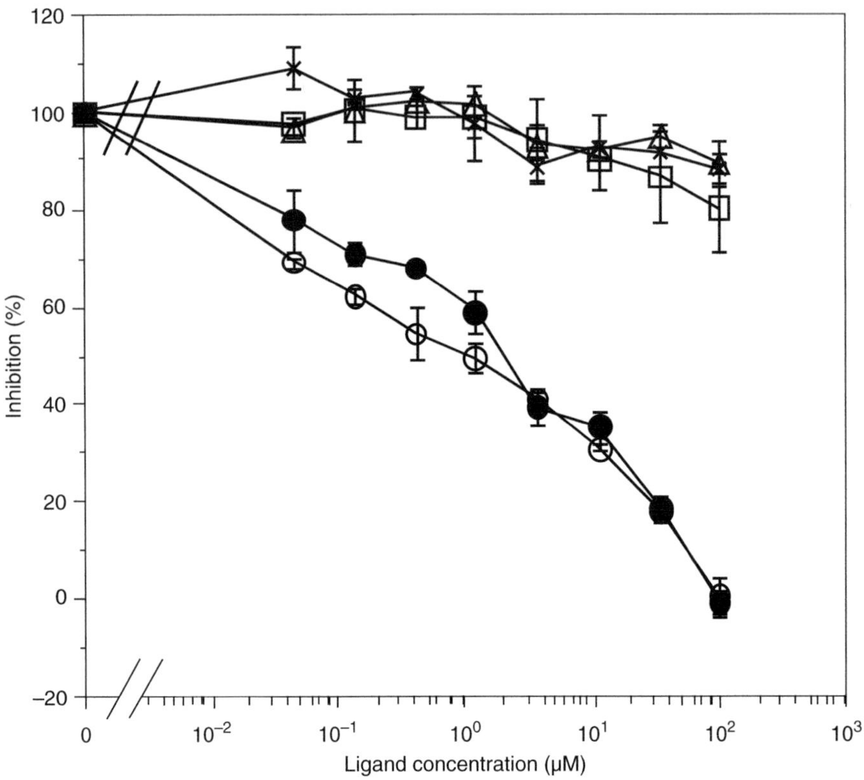

Fig. 4.3 Binding specificity of anti-semotiadil antibody for semotiadil derivatives. Cross-reactivity was determined by the competitive ELISA method. Before the antibody was added to the 96-well plates that were coated by norsemotiadil–HSA conjugates, it was preincubated with or without test compounds at various concentrations. Semotiadil (*open circle*), FNAK (*filled circle*), diltiazem (*triangle*), verapamil (*square*) and SIAB (×) were used as test drugs. Results are shown as means ± SD ($n = 4$) [Transferred from Kawahara et al. (2002)]

to 96-well plates coated with the norsemotiadil–HSA conjugate. The results showed that the antibody selectively recognized semotiadil and its phenylazido derivative FNAK (Fig. 4.3). SIAB, the crosslinker used to prepare the antigenic conjugate of norsemotiadil with thyroglobulin, was not recognized by the antibody. Semotiadil is a hybrid compound of diltiazem and verapamil, since not only the benzothiazine ring of semotiadil is homologous to the benzothiazepine ring of diltiazem, but also semotiadil has a structural component, Ar-C(R_1R_2)-$CH_2CH_2CH_2$-N(Me)-CH_2CH_2-Ar, similar to verapamil (Watanabe et al. 1996). However, the results indicated that the antibody showed no cross-reactivity with diltiazem and verapamil. In addition, the antibody recognized FNAK-labeled HSA in a Western blot, while unmodified HSA was not recognized, indicating that the antibody recognizes the semotiadil moiety.

4.3 Photolabeling of HSA

To determine photolabeling efficiency, labeling was performed at various ratios of the [^{3}H-FNAK]/[HSA], and the photoincorporated radioactivity into HSA was measured after SDS-PAGE (Kawahara et al. 2002). Typically, HSA (0.67 mg, 10 nmol) in 1 mL of 50 mM Tris/HCl (pH 7.4) was incubated with 10–100 μM [^{3}H]FNAK at 30 °C for 1 h in the presence and absence of 1 mM semotiadil. The incubation mixture was kept on ice and irradiated for 20 min with a 100 W black light/blue lamp (Ultra-Violet Products, Inc., San Gabriel, CA) at a distance of 10 cm. After photolysis, the sample was concentrated to 0.1 mL in a SpeedVac concentrator (Savant), and cold acetone (0.8 mL) was added. The sample was allowed to stand overnight at −30 °C and then centrifuged in a microfuge (18,000 *g*) at 4 °C for 10 min to obtain photolabeled HSA in the pellet. Photolabeled HSA was separated by SDS-PAGE (10% gels). Photolabeling efficiency was determined by calculation of the incorporated radioactivity (mol) compared with HSA (mol). Photolabeled samples for mass spectroscopic analysis were separately prepared using cold FNAK (20 μM) and 20 μM HSA.

The photolabeling efficiency was progressively enhanced by increasing the [^{3}H-FNAK]/[HSA] ratios and reached a plateau (20%) at a ratio of 3. In the presence of excess semotiadil, a parent compound of FNAK, labeling was effectively inhibited (Kawahara et al. 2002). The stoichiometry of photoincorporation was measured by nanoelectrospray ionization (nanoESI) MS using a Q-TOF instrument and MassLynx data acquisition. All HSA samples photolabeled with a 1:10 molar ratio of unlabeled FNAK/HSA showed FNAK-modified HSA bearing only 1 mol of FNAK (*m/z* 66945.0, 67062.0 and 67158.0) and unmodified HSA (*m/z* 66439.0, 66557.0 and 66653.0) (Masuda et al. 2005).

4.4 Immunoaffinity Purification of the Photolabeled Peptides

Photolabeled and acetone-precipitated HSA (1 mg/0.1 mL) was reduced with 25 mM dithiothreitol in 0.2 M Tris/HCl (pH 8.6) containing 10 M urea and 1 mM EDTA at 37 °C for 2 h and reacted with 0.1 M iodoacetamide for an additional hour. After gel filtration on a Sephadex G-50 column, labeled HSA was digested with endoprotease Lys-C ([substrate]/[enzyme] = 20) in 50 mM Tris/HCl (pH 9.0) at 37 °C for 12 h. The digest (≈30 pmol equivalent) was diluted with 50 mM Tris/HCl (pH 7.4) containing 0.5 M NaCl (0.1 mL) and incubated with the antibody-immobilized gel (30 μL) at 4 °C for 12 h. The gel was washed twice with 50 mM Tris/HCl (pH 7.4) containing 0.5 M NaCl, once with 1 mL of 50 mM Tris/HCl (pH 7.4) and twice with 1 mL of distilled water. The adsorbed materials were eluted three times with 0.1 mL of 50% acetonitrile/10 mM HCl, and the combined eluates were concentrated to 10 μL using a SpeedVac concentrator. Yields of the adsorbed and eluted products were determined using the Lys-C digest of [^{3}H]FNAK-labeled HSA. A control sample was prepared similarly using ethanolamine-treated Affi-Gel 10.

4.5 MALDI-TOF MS Analysis of the Photolabeled Peptides

The eluted and concentrated sample from the antibody-immobilized gel was mixed (1:1) with an α-cyano-4-hydroxycinnamic acid solution [40% acetonitrile/60% water containing 0.06% trifluoroacetic acid]. An aliquot (≈100 fmol) of the mixture was put on a probe tip, air-dried, and then MALDI-TOF MS measurements taken using a Bruker's Reflex MALDI-TOF mass spectrometer.

The MALDI-TOF MS spectrum showed three ion peaks at *m/z* 2557.54, 1322.44 and 1279.55 in the eluate from the antibody-immobilized column, while no apparent peaks were observed in the eluate from the control column (Fig. 4.4a). Among the three ion peaks described above, the *m/z* 1279.55 ion

a

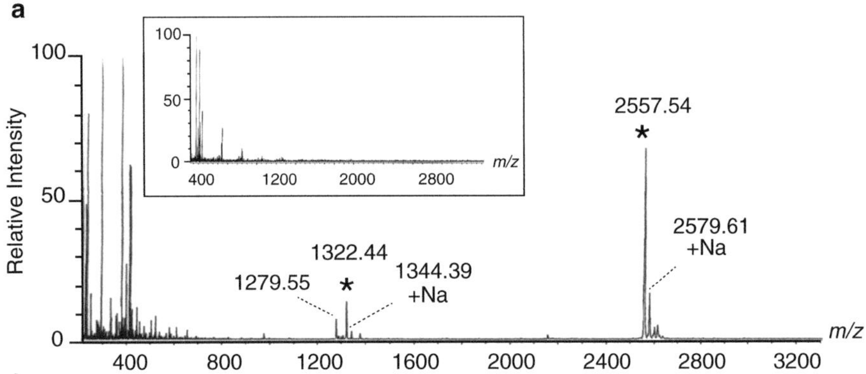

b

```
401  YKFQNALLVR  YTKKVPQVST  PTLVEVSRNL  GKVGSKCCKH  PEAKRMPCAE
451  DYLSVVLNQL  CVLHEKTPVS  DRVTKCCTES  LVNRRPCFSA  LEVDETYVPK
501  EFNAETFTFH  ADICTLSEKE  RQIKKQTALV  ELVKHKPKAT  KEQLKAVMDD
551  FAAFVEKCCK  ADDKETCFAE  EGKKLVAASQ  AALGL
```

c

Residues no.	Putative fragments with the 505-u label	Calculated mass	Observed mass
414-432	KVPQVSTPTLVEVSRNLGK	2557.38	2557.54
539-545	ATKEQLK	1322.68	1322.44

Fig. 4.4 MALDI-TOF MS analysis of the photolabeled peptide mixture. (**a**) After photolabeled HSA was digested with Lys-C ([substrate]/[enzyme] = 20) at 37 °C for 12 h, photolabeled peptide fragments were immunoprecipitated with Sepharose beads on which either anti-semotiadil antibody or control IgG (*inset*) was immobilized and analysed by MALDI-TOF MS. Ion peaks marked with an asterisk were candidates for FNAK-labeled fragments. Ion peaks at *m/z* 1344.39 and 2579.61 are sodium ion adducts to the corresponding molecular ions, and ion peak at *m/z* 1279.55 is assigned as doubly charged ion of the *m/z* 2557.54. (**b**) The partial amino acid sequence of HSA (residues 401–585) is shown, and putative photolabeled peptide sequences are doubly underlined. (**c**) Photolabeled peptides that were detected by MALDI-TOF MS are summarized [Transferred from Kawahara et al. (2002)]

peak was assigned as a doubly charged ion of the *m/z* 2557.54 peak. Therefore, the eluted fraction contained two photolabeled fragments at *m/z* 2557.54 (major) and 1322.44 (minor). As shown in Fig. 4.4b, c, mass-fit analysis of the two fragments in the HSA sequence revealed that *m/z* 2557.54 and 1322.44 were in complete agreement with the calculated mass of the peptides Lys-414 to Lys-432 and Ala-539 to Lys-545, respectively, when 505 was taken as the mass of the photoincorporated phenylnitrene generated from FNAK during photolysis.

4.6 NanoESI-MS/MS Analysis of the Photolabeled Amino Acid

To identify photolabeled amino acid residues in the two fragments, nanoESI-MS/MS analysis was performed after capillary RP-HPLC separation (Kawahara et al. 2002). Figure 4.5 shows the MS/MS spectrum of the triply charged ion (*m/z* 853.46) for the labeled peptide of *m/z* 2557.54. Two significant peaks were observed at *m/z* 399.22 and 1080.64. The *m/z* 399.22 ion peak was determined to be the fragment derived from cleavage at the C–O bond adjacent to the substituted aniline moiety of the labeled peptide (structure is shown in Fig. 4.5a), and the *m/z* 1080.64 was the doubly charged ion for the resultant labeled peptide fragment. In the expanded spectra 400–1075 (Fig. 4.5b) and 1100–2100 (Fig. 4.5c), we assigned the sequence of the 2nd through 15th amino acids as VPQVSTPTLVEVSR, based on the observation of contiguous y ion series. In addition, b2 and b4 ions were also detected at *m/z* 733.46 and 958.57, respectively. Taken together with the observed y ion series, we conclude that N-terminus Lys was the photolabeled amino acid residue in the peptide. This observation is consistent with the fact that the Lys residue remained at the N-terminus of the labeled peptide after Lys-C digestion.

In the second labeled peptide of *m/z* 1322.44, the MS/MS spectrum of doubly charged ion (*m/z* 661.85) showed two strong peaks of *m/z* 399.20 and 924.50 (Fig. 4.6). Although the C-terminal short sequence QLK was assigned, based on the part of a contiguous b and b* (=b-399) ion series, the entire sequence and the labeled amino acid residue were not determined. This may be due to the sensitivity of the labeled peptide to ionization at or near the photolabeled residue. Nevertheless, the Lys residue that lies in the third position of the peptide is a strong candidate for the labeled site, since it was resistant to Lys-C digestion. The photolabeled amino acid was not identified by ESI-MS/MS but could be determined by amino acid sequence analysis, in which six amino acid residues of the hepta-peptide were determined except for the third Lys residue.

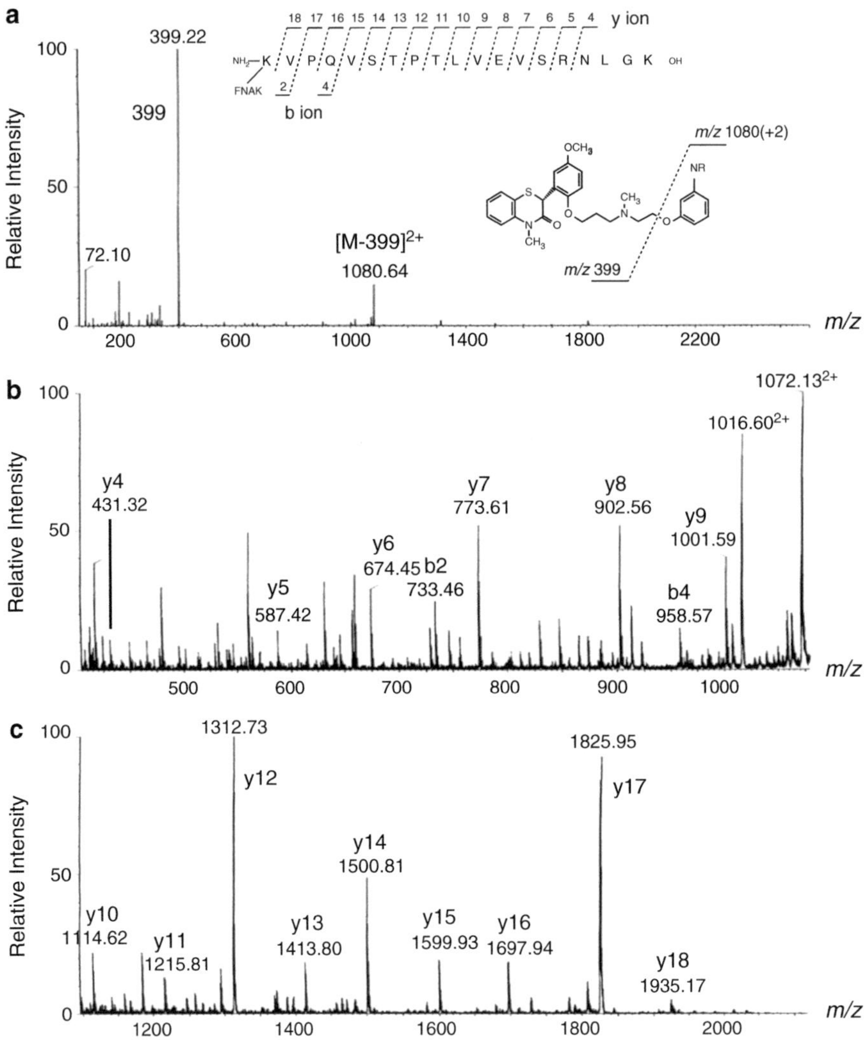

Fig. 4.5 NanoESI-MS/MS spectra of the photolabeled peptide Lys-414–Lys-432. The peptide was obtained by immunoprecipitation followed by fractionation by capillary RP-HPLC. (**a**) NanoESI-MS/MS of the triply charged ion (*m/z* 853.46) using a collision energy of 45 eV and argon as collision gas. y and b ion series of the peptide fragments corresponding to the internal cleavages in Lys-414–Lys-432 are indicated. Fragment ion series are defined using standard nomenclature (Biemann 1990). (**b**) and (**c**) show expanded spectra of (**a**) [Transferred from Kawahara et al. (2002)]

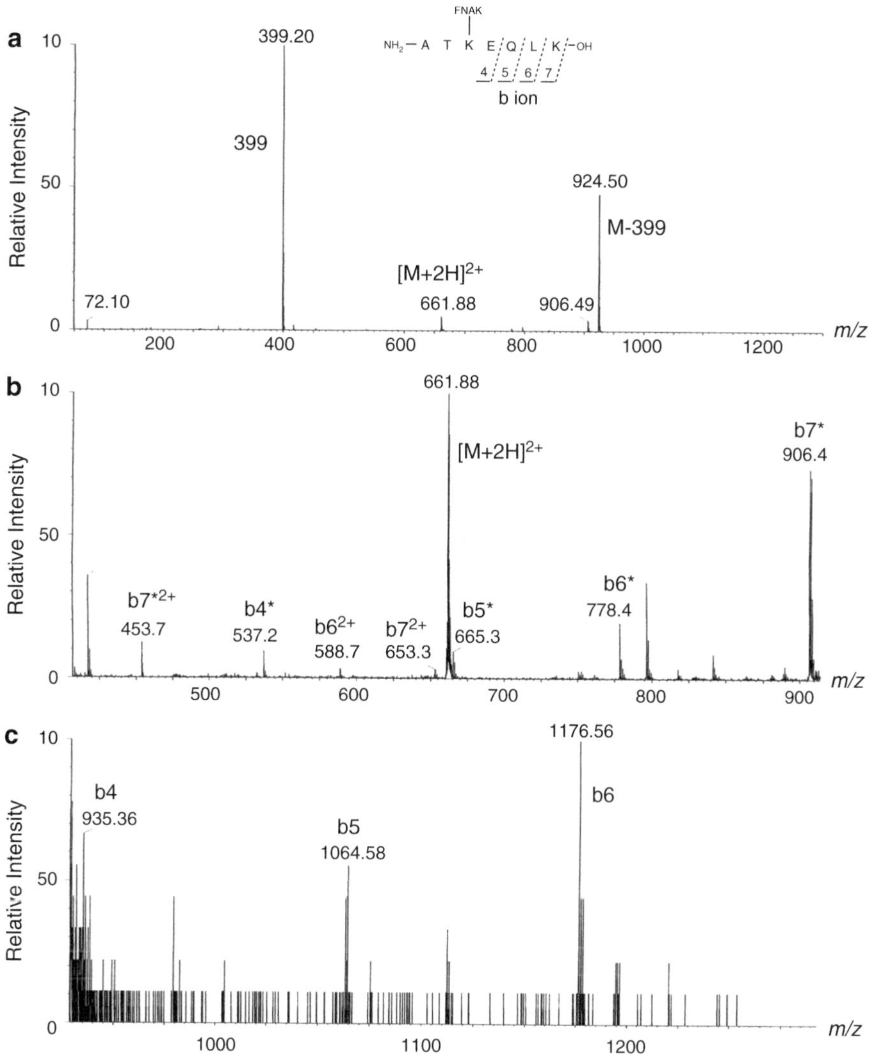

Fig. 4.6 NanoESI-MS/MS spectra of the photolabeled peptide Ala-539–Lys-545. (**a**) NanoESI-MS/MS of doubly charged ion (*m/z* 661.88) using a collision energy of 25 eV and argon as collision gas. (**b**, **c**) show expanded spectra of (**a**). b* ion indicates b-399 [Transferred from Kawahara et al. (2002)]

4.7 Different Regulation of Photolabeling on Lys-414 and Lys-541 by Addition of Myristate

HSA is a major binding protein for many drugs and fatty acids, and binding sites for several drugs have been defined (Sudlow et al. 1976 and reviewed in Kragh-Hansen 1990). It has also been reported that semotiadil strongly binds to HSA in serum

(Rosas et al. 1997). However, no data was available on the semotiadil binding site in HSA and its relation to fatty acid binding.

The effect of myristate (Myr) on FNAK photolabeling of HSA was investigated by isolating and quantifying two photolabeled fragments by HPLC. Photoincorporation into Lys-414 increased in the condition of 1 < [Myr]/[HSA] but decreased when the ratio was greater than 1 (Kawahara et al. 2002). In the case of photoincorporation into the Lys-541 containing peptide, however, a mild decrease was observed up to a ratio of 1, followed by a steep decrease with further addition of Myr. The results indicate that Myr binding affected FNAK photolabeling in two phases: Myr addition up to [Myr]/[HSA] = 1 increased labeling in Lys-414, while it decreased that in Lys-541, but further addition of Myr uniformly decreased labeling in both lysine residues. Therefore, FNAK labeling is closely associated with Myr binding.

4.8 FNAK Binding Site in HSA and Its Relevance to Those for Myr

We identified two photolabeled sites by FNAK on Lys-414 and Lys-541 of HSA. Under photolabeling conditions, HSA binds one mole of FNAK. A binding stoichiometry of 1:1 was also reported for semotiadil binding to HSA (Skammelsrud et al. 1999). Therefore, the FNAK binding site is located near Lys-414 and Lys-541. When we constructed the complex model of semotiadil with HSA, there were two plausible pockets for semotiadil binding. At a fairly large pocket that accommodates two myristates (Myr-3 and Myr-4) in subdomain IIIA, the 2-methoxyphenyl-1,4-benzothiazine-3-one moiety of semotiadil replaced the two myristates. The other binding pocket, however, allowed semotiadil to bind only the 1,4-benzothiazine-3-one moiety at the hydrophobic hole in subdomain IIIB, while the remaining moiety was exposed at the cleft formed by subdomains IIIA and IIIB. A definite conclusion was not obtained which model is more favoured one. However, based on the photolabeled results as described above, we propose a complex model of FNAK with HSA, as shown in Fig. 4.7.

The binding site is located primarily in subdomain IIIA, which is composed of helices 1, 2, 3 and 6, and helix 9 is also in close apposition. In the model, the aromatic ring of the azidophenyl group interacts with the aromatic ring of Tyr-411 almost in parallel arrangement, and the hydroxyl group of Tyr-411 forms a hydrogen bond interaction with the carbonyl group of the 1,4-benzothiazine-3-one moiety. The methylamino group of FNAK is bonded by a salt bridge to the side chain of Glu-492, which is held in place by a second hydrogen bond interaction with the side chain of Lys-414. The azidophenyl group of FNAK can take two distinct conformations, with respect to the orientation of the m-substituted azide moiety. In one conformation, the azide moiety is located 3 Å from ε-amino group of Lys-414, as shown in Fig. 4.7. In the other conformation, the azide moiety orientates Lys-541 to reside

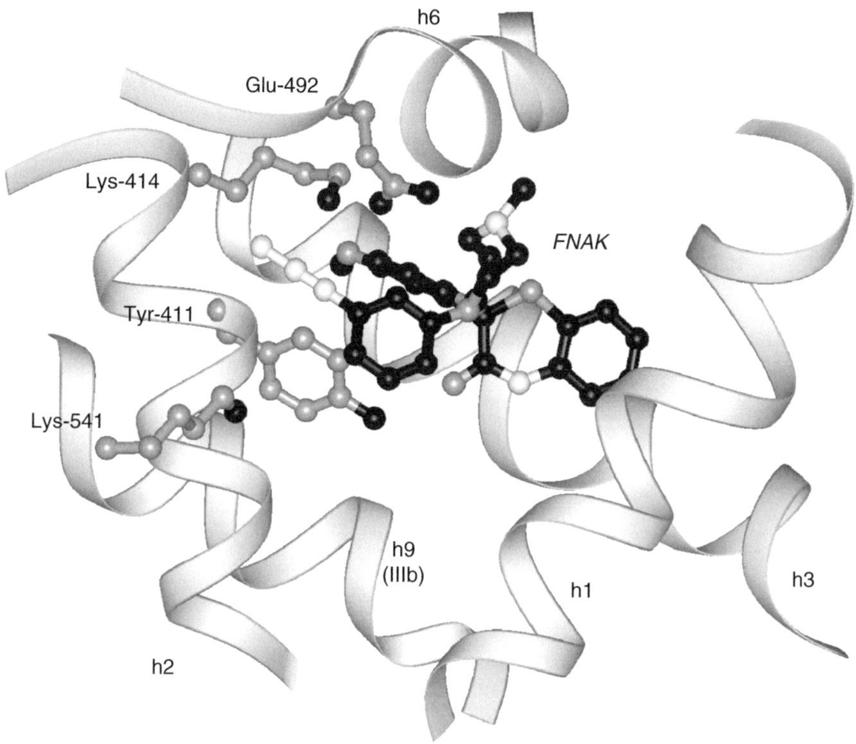

Fig. 4.7 Model of FNAK-HSA complex in the absence of myristate. FNAK bound to subdomain IIIA of HSA. A local main-chain structure is shown by a ribbon. Particular residues (side chains only) and helices (h) that noted in the text are shown for clarity. For amino acid residues of HSA, carbon atoms are coloured *grey*, and heteroatoms are *black*. For FNAK, carbon atoms are coloured *black*, nitrogen atoms are *white* and oxygen and sulphur atoms are *grey* [Transferred from Kawahara et al. (2002)]

within 3 Å from its ε-amino group when the side chain of Lys-541 forms an extended conformation. Alternatively, a more flexible conformation of the Lys-541 side chain is consistent with the results that Lys-541 was photolabeled less efficiently than Lys-414.

4.9 Conclusion

The method utilizing anti-ligand antibody and modern mass spectrometry has proven efficient and sensitive to identify photolabeled amino acids. It is significant that both labeled amino acid residues were lysine. Although determination of the photoproduct structure remains to be solved, we proposed a mechanism for this reaction: ring expansion of the nitrene of FNAK to a dehydroazepine followed by nucleophilic addition of ε-amino group of Lys-414 or Lys-541, because *m/z* 399 was easily observed in ESI-MS/MS (Figs. 4.5 and 4.6) and MALDI-PSD analyses (unpublished observation). The detailed mechanism will be reported elsewhere in combination with the NMR analysis (K. Masuda, K. Kawahara and H. Nakayama, unpublished work).

4.10 Additional Concluding Remarks

As the first example for the efficient isolation of photolabeled peptides followed by identification of their peptide sequences and the labeled amino acid residues, we described above a combinatorial use of anti-ligand antibody and highly sensitive mass spectrometry method.

The work was designed and carried out as a model system for identifying the binding site for a calcium antagonist semotiadil on L-type Ca^{2+} channel. The model system has proved to work nicely as expected. As the second stage of utility for this novel method, we wait for coming results in the L-type Ca^{2+} channel system using membrane preparations that are described in Chap. 7. We mention that the subsequent example of revealing photolabeled site using anti-ligand antibody was reported by Shinohara et al. (2007), suggesting that the method presented here is also useful for the system between peptide ligand and receptor kinase.

If we remind of accumulated knowledge on the structural basis of antibody specificity starting in 1960s (a representative book by Kabat (1968)), anti-hapten antibodies with high affinity, such as anti-phenylarsenate antibody (Marrack and Smith 1932) and anti-dinitrophenyl antibody (Metzger and Singer 1963), have been well contributed in the chemical or biochemical understanding of the molecular recognition mechanism of antigen-antibody interactions. In addition, utilizing antibodies against antigens of low molecular weight, namely, anti-ligand antibodies, is a logical extension to isolate the photolabeled proteins or their peptide fragments, since we use similar ligand reagent, anyway, as a probe for photoaffinity labeling, in which a suitable photosensitive group is attached onto the same ligand moiety.

Another contrivance for efficient isolation of the photolabeled peptides is the application of avidin–biotin system, introducing biotin moiety before (Hatanaka et al. 1996) or after (Hashimoto and Hatanaka 2005) photolabeling in the photoactivatable ligand reagent (*see* also Chaps. 3 and 6). You are now able to select either the method of anti-ligand antibody or avidin–biotin system, as you like.

References

Biemann K (1990) Nomenclature for peptide fragment ions (positive ions). Methods Enzymol 193:886–888

Chaurand P, Leutzenkirchen F, Spengler P (1999) Peptide and protein identification by matrix-assisted laser desorption ionization (MALDI) and MALDI-post-source decay time-of-flight mass spectrometry. J Am Soc Mass Spectrom 10:91–103

Chen ZW, Manion B, Townsend RR, Reichert DE, Covey DF, Steinbach JH, Sieghart W, Fuchs K, Evers AS (2012) Neurosteroid analog photolabeling of a site in the third transmembrane domain of the β3 subunit of the GABA(A) receptor. Mol Pharmacol 82:408–419

Cheng KW, Wong CC, Wang M, He QY, Chen F (2010) Identification and characterization of molecular targets of natural products by mass spectrometry. Mass Spectrom Rev 29:126–155

Dianoux AC, Noel F, Fiore C, Trezeget V, Kiefer S, Jaquinod M, Lauquin GJM, Brandolin G (2000) Two distinct regions of the yeast mitochondrial ADP/ATP carrier are photolabeled by a new ADP analogue: 2-azido-3′-O-naphthoyl-[β-^{32}P]ADP. Identification of the binding segments by mass spectrometry. Biochemistry 39:11477–11487

Hashimoto M, Hatanaka Y (2005) Post-biotinylation of photocrosslinking by Staudinger-Bertozzi ligation of preinstalled alkylazide tag. Chem Pharm Bull 53:1510–1512

Hatanaka Y, Hashimoto M, Nishihara S, Narimatsu H, Kanaoka Y (1996) Synthesis and characterization of a carbene-generating biotinylated N-acetylglucosamine for photoaffinity labeling of β-(1→4)-galactosyltransferase. Carbohydr Res 294:95–108

Jensen ON, Wilm M, Shevshenko A, Mann M (1999) Peptide sequencing of 2-DE gel-isolated proteins by nanoelectrospray tandem mass spectrometry. Methods Mol Biol 112:571–588

Kabat EA (1968) Structural concepts in immunology and immunochemistry. Holt, Rinehart and Winston, New York

Kawahara K, Kuniyasu A, Masuda K, Ishiguro M, Nakayama H (2002) Efficient identification of photolabelled amino acid residues by combining immunoaffinity purification with MS: revealing the semotiadil-binding site and its relevance to binding sites for myristates in domain III of human serum albumin. Biochem J 363:223–232

Kragh-Hansen U (1990) Structure and ligand binding properties of human serum albumin. Dan Med Bull 37:57–84

Kuster B, Mann M (1998) Chemical protein synthesis. Curr Opin Biotechnol 9:412–426

Lee S, Braun CR, Bird GH, Walensky LD (2014) Photoreactive stapled peptides to identify and characterize BCL-2 family interaction sites by mass spectrometry. Methods Enzymol 544:25–48

Marrack J, Smith FC (1932) Quantitative aspects of immunity reactions: the combination of antibodies with simple haptenes. Br J Exp Pathol 13:394–402

Masuda K, Kawahara K, Kuniyasu A, Ishiguro M, Nakayama H (2005) Stoichiometry and binding site analysis of ligand-human serum albumin complex by photoaffinity labeling and mass spectrometry. J Mass Spectrom Soc Jpn 53:18–24

Metzger H, Singer SJ (1963) Binding capacity of reductively fragmented antibodies to the 2,4-dinitrophenyl group. Science 142:674–675

Pettelkau J, Schröder T, Ihling CH, Olausson BE, Kölbel K, Lange C, Sinz A (2012) Structural Insights into retinal guanylylcyclase-GCAP-2 interaction determined by cross-linking and mass spectrometry. Biochemistry 51:4932–4949

Pinke MW, Rijkers DT, Dostman WR, Heck AJ (2009) Mode of action of cGMP-dependent protein kinase-specific inhibitors probed by photoaffinity cross-linking mass spectrometry. J Biol Chem 284:16354–16368

Robinette D, Neamati N, Tomer KB, Borchers CH (2006) Photoaffinity labeling combined with mass spectrometric approaches as a tool for structural proteomics. Expt Rev Proteomics 3:399–408

Rosas MER, Shibukawa K, Ueda K, Nakagawa T (1997) Enantioselective protein binding of semotiadil and levosemotiadil determined by high-performance frontal analysis. J Pharm Biomed Anal 15:1595–1601

Rowland MM, Bostic HE, Gong D, Speers AE, Lucas N, Cho W, Cravatt BF, Best MD (2011) Phosphatidylinositol 3,4,5-trisphosphate activity probes for the labeling and proteomic characterization of protein binding partners. Biochemistry 50:11143–11161

Shinohara H, Ogawa M, Sakagami Y, Matsubayashi Y (2007) Identification of ligand binding site of phytosulfokine receptor by on-column photoaffinity labeling. J Biol Chem 282:124–131

Sinz A (2014) The advancement of chemical cross-linking and mass spectrometry for structural proteomics: from single proteins to protein interaction networks. Expert Rev Proteomics 11:733–743

Skammelsrud N, Martin ER, Murphy P, Khuu C, Frengen E, Kolstø AB (1999) The gene for human transcription factor TCF11 is located telomeric to D17S1827, BTR and HP1Hsbeta on chromosome 17q22. Anal Sci 15:217–222

Sudlow G, Birkett DJ, Wade DN (1976) Further characterization of specific drug binding sites on human serum albumin. Mol Pharmacol 12:1052–1061

Watanabe Y, Osanai K, Nishi T, Miyakawa N, Shii D, Honda T, Shibano T (1996) Synthesis of azido derivatives of semotiadil, a novel 1,4-benzothiazine calcium antagonist, for photoaffinity probes of calcium channels. Bioorg Med Chem Lett 6:1923–1926

Chapter 5
Combination of Photoaffinity Label and Site-Directed Antibody for Target Proteins

Akihiko Kuniyasu and Hitoshi Nakayama

Abstract Photoaffinity labeling is a useful method for studying the molecular interaction of ligands with their target proteins, including membrane receptor proteins, by locating or identifying the binding sites of the ligands in the receptor molecules. The ligand binding site is ideally determined by the direct identification of photolabeled amino acid residues via the complete isolation of photolabeled peptide fragments using high-performance liquid chromatography (HPLC) followed by microsequencing. However, the low efficiency of photolabeling often hampers this aim.

The photochemical identification of ligand binding sites through the use of a site-directed antibody provides and simplifies the determination of the photolabeled site in its receptor without requiring the complete purification of labeled peptides. This method is based on the immunological detection of the photolabeled peptides using prepared anti-peptide antibodies that can specifically recognize proteolytic digests of the target receptor protein if the primary sequence of the protein is known.

Here, we describe the principle and protocol of this method as the convenient identification of the ligand binding site(s) in transmembrane receptor proteins. We focus on successful applications of the method, which have identified the ligand and/or toxin binding sites on ionic channels and drug transporter proteins

Keywords Photoaffinity labeling • Site-directed antibody • Immunoprecipitation • Antibody mapping • Ligand binding site • Membrane receptor protein

---

A. Kuniyasu (✉)
Sojo University, Kumamoto, Kumamoto 860-0082, Japan
e-mail: kuniyasu@ph.sojo-u.ac.jp

H. Nakayama
Kumamoto University, Kumamoto, Kumamoto 862-0973, Japan
e-mail: jin@gpo.kumamoto-u.ac.jp

Y. Hatanaka, M. Hashimoto (eds.), *Photoaffinity Labeling for Structural Probing Within Protein*, DOI 10.1007/978-4-431-56569-7_5

Abbreviations

BTZs	Benzothiazepines
DHPs	1,4-Dihydropyridines
LqTX	Toxin V from *Leiurus quinquestriatus*
PAAs	Phenylalkylamines
SDS-PAGE	Sodium dodecyl sulfate-polyacrylamide gel electrophoresis
TM	Transmembrane domain
TTX	Tetrodotoxin

5.1 Introduction

Photoaffinity labeling is a useful technique for studying the interaction of ligands with their target proteins, such as receptor proteins and enzymes (Hatanaka and Sadakane 2002). The identification of ligand binding sites provides information about the structure and conformational changes of allosteric proteins.

In general, a primary target for many drugs is a membrane protein such as G protein-coupled receptors, ion channels, and membrane transporters, which all have multiple membrane-spanning domains. For example, the α1 subunit of the calcium channel acts as a receptor for the antihypertensive drug 1,4-dihydropyridines (DHPs), which consists of more than 1800 amino acid residues (molecular weight >180 kDa). In photoaffinity labeling, ligand binding is identified by both the complete purification of the labeled peptides following proteolytic or chemical digestion and the determination of the amino acid sequence by Edman degradation or mass spectrometry (MS) analysis (Hatanaka and Sadakane 2002; Robinette et al. 2006). However, in most cases of multiple transmembrane proteins, it is not easy to analyze the labeled peptide fragments by using these proteomic methods. First, the membrane proteins with high molecular weights (>150 kDa) are commonly less abundant in cells and tissues. Second, many drugs of small molecular weight bind to a pocket within a transmembrane domain. As a consequence, the low abundance and hydrophobicity of the labeled peptide would be difficult to purify. Third, the low efficiency of photoaffinity labeling (0.1–5%) sometimes makes it challenging to detect photolabeled peptide fragments by automated MS analysis, especially when the structure of the photoproduct is unknown. The conditions for analyzing small quantities of photolabeled peptides of hydrophobic character through the use of MS alone have not been well developed.

The immunochemical detection of photolabeled peptides by anti-peptide antibodies, which specifically recognize each of the proteolytic digests derived from the target receptor proteins, allows the rapid identification of the photolabeled domains without the complete purification of the labeled peptides.

This method was first employed for the domain mapping of α-scorpion toxin in voltage-dependent sodium channels (Tejedor and Catterall 1988). It was subsequently applied to the identification of covalently labeled receptor sites for calcium antagonists (Nakayama et al. 1991; Striessnig et al. 1990, 1991; Watanabe et al. 1993; Kraus et al. 1996; Kuniyasu et al. 1998; Ii et al. 1998) on L-type calcium channels and neurotoxin binding sites (Tejedor and Catterall 1988, 1990; Nakayama et al. 1992; Trainer et al. 1991, 1996) on sodium channels. The subsequent analysis by site-directed mutagenesis identified individual amino acid residues that are required for binding of dihydropyridines, benzothiazepines (Peterson et al. 1997; Motoike et al. 1999), and phenylalkylamines (Hockerman et al. 1997) to L-type calcium channels. Indeed, such site-directed mutagenesis has proved that the combinatorial use of photoaffinity labeling and antibody-based peptide mapping strategies provides a satisfying method to locate these receptor sites, although site-directed mutagenesis provided a finer resolution, as expected.

This chapter will focus on the method and protocols using anti-peptide antibodies to identify the photolabeled site(s) of drug binding on L-type calcium channels and other examples of ligand binding sites in membrane channels and receptors.

5.2 Principle and Experimental Strategy

5.2.1 Outline

The typical steps involved in photoaffinity labeling experiments combined with antibody mapping to identify the ligand binding sites are shown in Scheme 5.1. There are three steps: (1) the photolabeled target protein is purified, in the course of which radioactivity is incorporated; (2) the labeled protein is digested with highly specific proteases and/or chemicals; and (3) the photolabeled peptide fragment(s) are analyzed by antibody mapping using the site-directed antibodies.

First, a receptor preparation such as a membrane fraction and/or the solubilized and partially purified receptor is incubated with the radiolabeled photoaffinity ligand and irradiated by UV light to form a covalent link between the ligand reagent and the receptor. Removed from the free ligand reagent and contaminated with

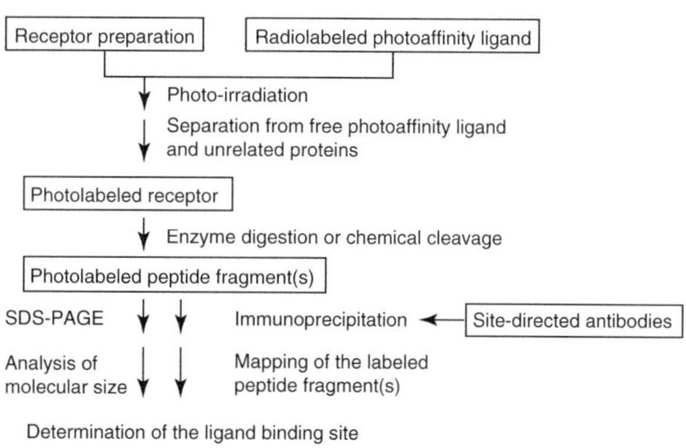

Scheme 5.1 Identification of the ligand binding site on the target receptor

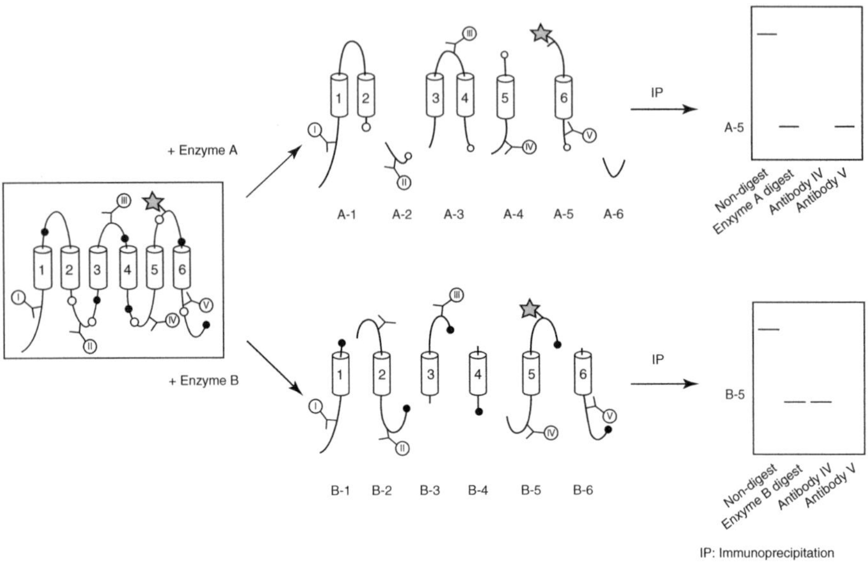

Fig. 5.1 Analysis of the photolabeled fragment(s)

unrelated proteins, the photolabeled receptor proteins are then purified by appropriate procedures, such as affinity chromatography and high-performance liquid chromatography. Alternatively, photolabeled protein with radioactivity can be extracted from the gel after SDS-PAGE for purification. Next, isolated receptor proteins are fragmented with proteases or chemicals, and the digests are separated by SDS-PAGE or tricine SDS-PAGE and visualized by radioactivity detection methods such as autoradiography, fluorography, or radioluminography. Counting radioactivity of the sliced gels is another simple method that is used to detect the molecular size of the fragments.

Figures 5.1 and 5.2 show outlines of a photolabeled site that was determined using a site-directed antibody. For this identification, the sample digested by enzymes and/or cleaved by chemicals is subject to immunoprecipitation of the labeled fragments by the site-directed antibodies that recognize the particular sequence of the receptor (Fig. 5.1). The immunoprecipitates by the specific antibody are then separated by conventional SDS-PAGE or tricine SDS-PAGE systems, and the radioactivity is determined to estimate the molecular weight of the labeled peptide fragments. The labeled peptide fragments are mapped to the primary structure of the target receptor protein. As shown in Fig. 5.2, the photolabeled site(s) can be deduced by comparing the immunoprecipitated peptide fragment sizes with those predicted from the cleavable site. A combination of multiple digestion samples produced by different proteases (e.g., enzymes A and B in Fig. 5.2) allows one to reach a fragment of the smallest size containing the photolabeled amino acid residues. If the precipitated fragment size is larger than predicted, the contribution of glycosylation should be considered in the case of glycosylated receptor proteins.

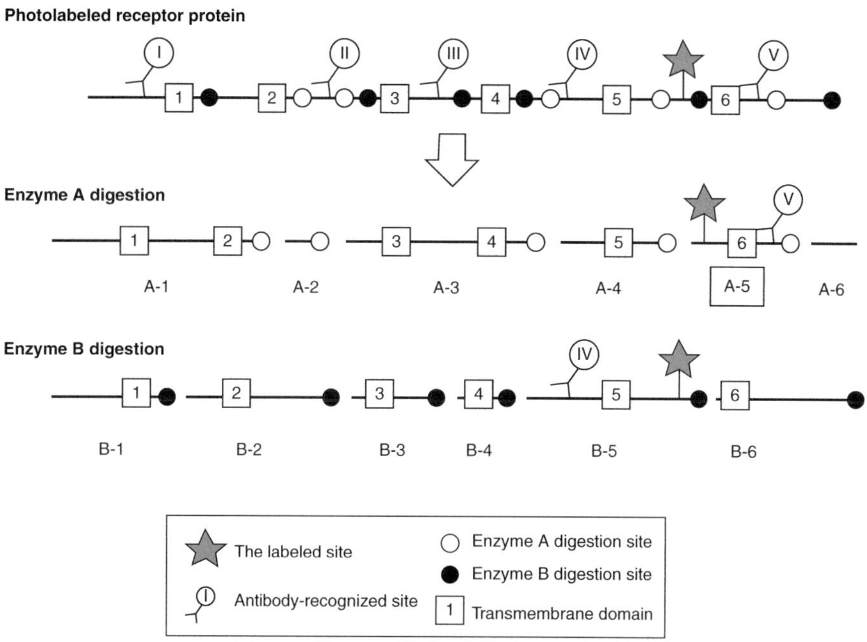

Fig. 5.2 Antibody mapping of the photolabeled fragment(s)

5.2.2 Photoaffinity Ligand

The employment of photoaffinity ligands is required to introduce not only a photoactivating group but also radioisotopes into the structure. The frequently used photoactivating groups are the arylazido group, aryldiazirine group, and benzophenones. Photoirradiation produces highly reactive intermediates, namely, nitrenes, carbenes, and triplet ketones, respectively, which form covalent bonds with receptor proteins (see Chap. 1, in more details). The photoaffinity ligand contains radioisotope such as iodine-125 (^{125}I) and tritium (^{3}H). The incorporation of the radioisotope into the photoaffinity ligand allows the easy detection of the labeled proteins/peptides with high sensitivity. SDS-PAGE analysis allows the size of labeled proteins and peptides to be measured.

5.2.3 Site-Specific Proteolysis

The site-specific cleavage of receptor proteins is an important step in this method to estimate the molecular weight of photolabeled peptide fragments. Site-specific proteases such as trypsin, endoproteinase Lys-C, and endoproteinase Glu-C (V8 protease) are frequently employed. When trypsin is utilized, the sequencing grade and TPCK-treated enzyme should be used. Chemical cleavage is also useful for the specific fragmentation of the photolabeled proteins. Cyanogen bromide is the most popular reagent that cleaves the bond to the carboxyl-terminal side of methionyl residues with high specificity and few side reactions.

5.2.4 Site-Directed Antibody

The site-directed antibody plays a critical role in this method. A series of polyclonal antibodies directed against synthetic peptides that correspond to defined segments of the target protein are prepared. In general, synthetic peptides (10–20 amino acids) with a selected sequence are used as haptens to immunize host animals. Antigen peptide sequences should be carefully selected. Notably, the hydrophobic transmembrane domain and the region containing the proteolytic cleavage sites of the target membrane receptor should be avoided due to low antigenicity and reduced immunoreactivity after proteolysis, respectively. The immunoreactivity of the obtained antibody (or antisera) is examined if the antibody can precipitate at least the denatured target protein.

5.3 Identification of the Photolabeled Sites

5.3.1 Calcium Antagonist Binding Sites on the L-Type Calcium Channel

L-type calcium channels with a total mass of ~450 kDa are composed of five distinct subunits (α_1, α_2, β, γ, and δ) and allow calcium flux into cells in response to membrane depolarization (Striessnig et al. 1998). The α_1 subunit forms the ion-conducting pore and comprises four (I–IV) characteristic homologous domains. Each domain contains six transmembrane α-helices (S1–S6) and an additional pore loop, which is located between the S5 and S6 segments.

Calcium antagonists (also called calcium channel blockers) are used as antihypertensive drugs, which decrease blood pressure in patients with hypertension. Calcium antagonists bind with high affinity to the α_1 subunit of L-type calcium channels and block the entry of extracellular Ca^{2+}. Three particular classes of calcium antagonists have been identified and include 1,4-dihydropyridines (DHPs: nitrendipine, isradipine, etc.), phenylalkylamines (PAAs: verapamil, etc.), and benzothiazepines (BTZs: diltiazem, etc.). These drugs bind to different sites on the α_1 subunit of calcium channels (Striessnig et al. 1998) and have been pharmacologically explained to show allosteric interactions with one another, as shown in Fig. 5.3. Combination analysis of photoaffinity labeling and antibody mapping revealed that all three drugs bind to different regions in more than one motif. Photoreactive derivatives of DHPs (diazipine, azidopine, and (+)PN200-110), PAAs (LU49,888), and BTZs (azidobutyryl clentiazem and benziazem) were photoincorporated into their

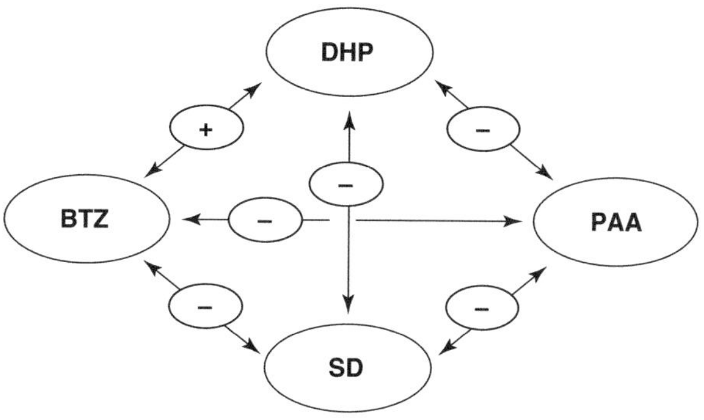

Fig. 5.3 Allosteric interactions of the different classes of calcium antagonists on the binding domain of α_1 subunits

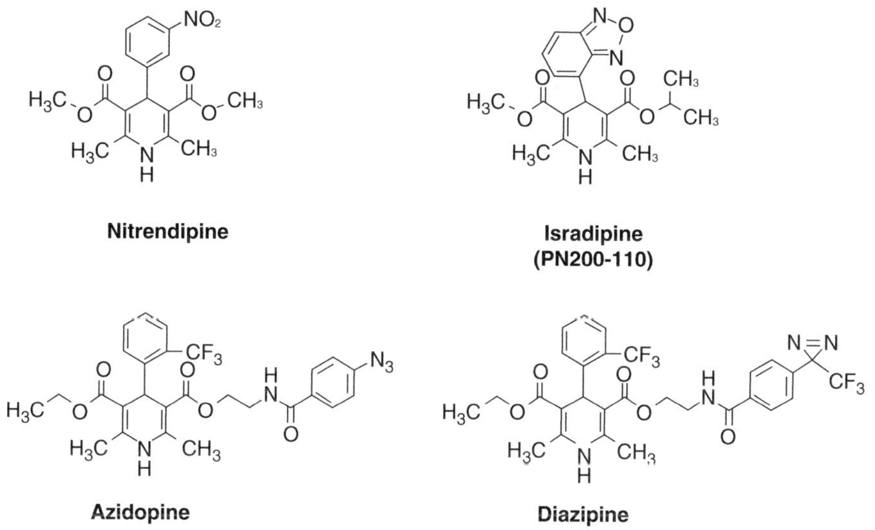

Fig. 5.4 Dihydropyridines and the photoreactive probes

respective binding domains in purified skeletal muscle L-type calcium channels, and the photolabeled sites were identified by immunoprecipitation with site-directed antibodies after proteolysis of the labeled α_1 subunit.

The identification of the DHP binding site was first achieved by photoaffinity labeling of the purified skeletal muscle calcium channels using diazipine (Fig. 5.4), a new DHP analog with a phenyldiazirine moiety as the photosensitive group (Nakayama et al. 1991). Diazipine has considerable advantages over azidopine as a photoaffinity probe for L-type calcium channels. The advantages include a higher yield of photoincorporation and the formation of more stable photoproduct(s), though its binding affinities are similar to those of azidopine (Taki et al. 1991a, b). The photoaffinity labeling results (Nakayama et al. 1991) show that two sites form the DHP binding site: the extracellular location of the connecting loop between transmembrane segments of S5 and S6 in repeat III and the extracellular end of the transmembrane segment S6 in repeat IV, as shown in Fig. 5.10. It was subsequently reported that two extracellular regions adjacent to the IIIS6 and IVS6 were also identified as the photolabeled sites with another DHP, isradipine (PN200-110) (Striessnig et al. 1991). Because isradipine, unlike diazipine or azidopine, has no spacer arm to connect the 1,4-dihydropyridine ring with the photosensitive group of phenyldiazirine or phenylazide at a 14-Å length in diazipine or azidopine, respectively, it can be inferred that isradipine labeled a core region of the binding site.

Regulla et al. reported that the peptide fragments photolabeled by azidopine were located in hydrophilic intracellular parts near the carboxyl- and amino-terminal regions (Regulla et al. 1991). They analyzed the reversed-phase HPLC-purified peptides using a microsequencer. However, if the DHP binding site is strongly hydrophobic, as expected, it seems unlikely that the component peptides can be efficiently recovered from a hydrophobic reversed-phase HPLC column. By contrast, the antibody-mapping method after protease digestion actually showed that it was able to recover the extracellular end of peptide fragments, even if they contained hydrophobic transmembrane segments.

A more detailed analysis was provided for the IIIS6 and IVS6 helices in which individual residues were systematically replaced by alanine (alanine-scanning mutagenesis). The mutation of five residues, four amino acids in IIIS6 and one in IVS6, decreased the apparent binding affinity of DHP by more than fivefold, demonstrating that the five residues are the most important contributors to drug binding in these segments (Peterson et al. 1997). Therefore, two extracellular regions located near the transmembrane contribute to the DHP binding site.

The PAA binding site in skeletal muscle calcium channels has been identified by ludopamil (LU49888), a photoreactive phenylazide derivative of PAA (Fig. 5.5a) (Striessnig et al. 1990). The peptide of 42 amino acid residues extending from 1349 to 1391, which includes the transmembrane segment IVS6 and several adjacent intracellular and extracellular amino acid residues, was identified as the photolabeled fragment. The binding site for quaternary PAAs consists of the intracellular end of IVS6.

Azidobutyryl clentiazem (Watanabe et al. 1993), a photoreactive diltiazem analog with higher affinity than azidobutyryl diltiazem (Naito et al. 1989), an earlier photoaffinity ligand, labels benzothiazepine binding sites on the α_1 subunit. The binding site is in domain IV but at a location different from DHPs and PAAs (Fig. 5.5b). Another potent diltiazem-related photoreactive probe, benziazem with a photoactivating benzophenone, allowed the specific photoincorporation into the BTZ-binding domain of α1 subunits with high labeling efficiency (<60%). The benziazem-labeled sites were determined to be the transmembrane segments S6 in the repeats III and IV together with extracellular and intracellular sequence stretches (Kraus et al. 1996).

We further identified the binding site of semotiadil (SD-3211), a new calcium antagonist with a unique 1,4-benzothiazine ring structure (Fig. 5.6) (Kuniyasu et al. 1998). Prior to the photoaffinity labeling experiment, we prepared a series of site-directed antibodies (Ab-1~Ab-7) to recognize the defined region of the skeletal muscle calcium channel primary structure (Fig. 5.7).

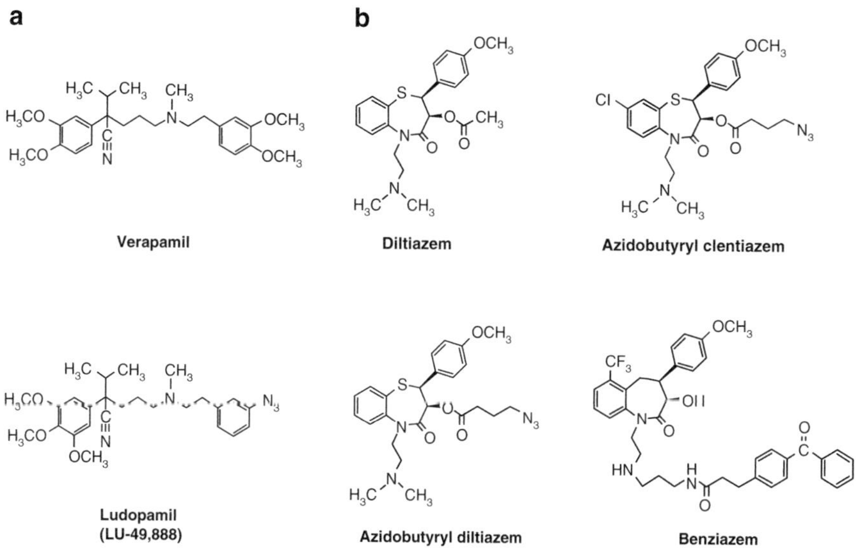

Fig. 5.5 (**a**) Phenylalkylamines, (**b**) Benzothiazepines, and their photoreactive probes

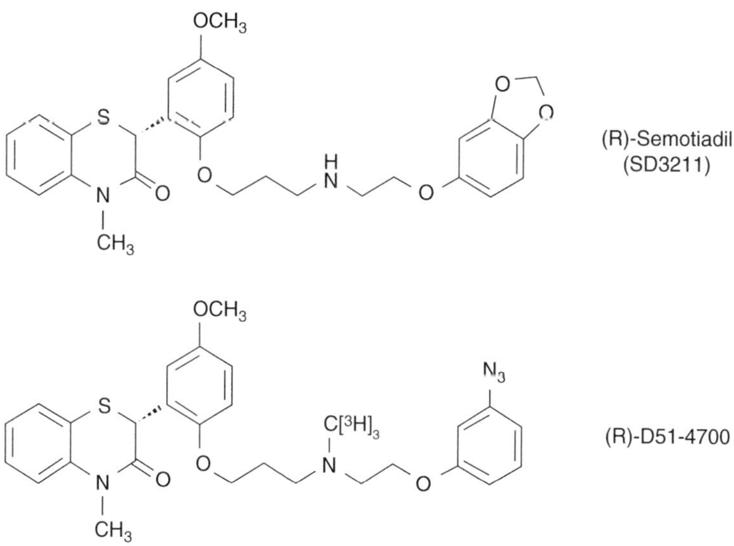

Fig. 5.6 A new calcium antagonist semotiadil and its photoreactive probe

Tritium-labeled D51-4700, a photoaffinity probe of semotiadil, selectively labeled the α_1 subunit of calcium channels in skeletal triad membranes (Fig. 5.8a). In the absence of unlabeled semotiadil, the photoaffinity probe labeled several polypeptides in addition to the α_1 subunit. However, the photoincorporation was observed only in the α1 subunit (170 kDa) after the photolabeled sample was purified by

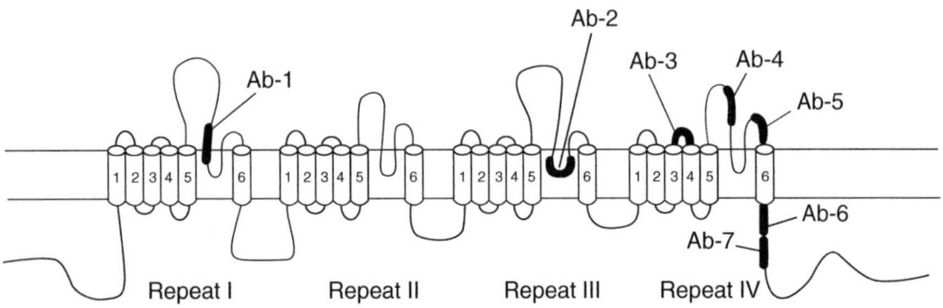

Fig. 5.7 Location of site-directed antibodies against the calcium channel α_1 subunit

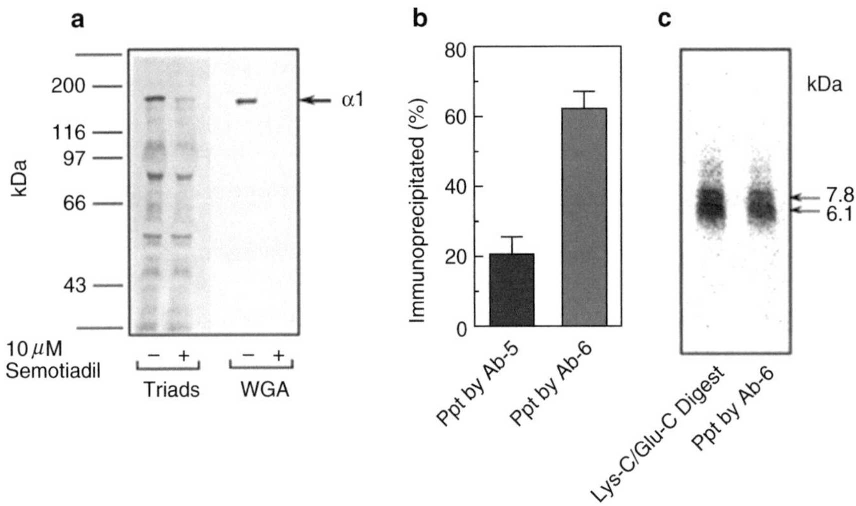

Fig. 5.8 Identification of the semotiadil binging sites (modified from Kuniyasu et al., J. Biol. Chem., 273, 4635–4641, 1997). (**a**) Autoradiography, (**b**) Precipitation rate, (**c**) Autoradiography

WGA column, and excess unlabeled semotiadil entirely blocked photoincorporation into the α1 subunit. The [^{3}H]D51-4700-labeled α1 was digested with the site-specific proteases (Lys-C, trypsin, and Glu-C) and cyanogen bromide. After the treatments, the sample was separated on a tricine-SDS-polyacrylamide gel (Schagger and von Jagow 1987) and visualized using radioluminography. Because the Lys-C fragment contains potentially cleavable sites by Glu-C, the labeled Lys-C fragment was subsequently digested with endoprotease Glu-C to restrict the photolabeled sites further. After the digestion, the D51-4700-labeled α1 subunits produced 7.8- and 6.1-kDa fragments (Fig. 5.8c, left lane). We examined whether a series of site-directed antibodies would recognize the photolabeled peptide fragments. Approximately 20 and 60% of the α_1-associated labeling were observed in the Lys-C/Glu-C fragments that were recognized by antibody Ab-5 (corresponding to amino acids 1338–1351) and antibody Ab-6 (corresponding to amino acids 1382–1400), respectively (Fig. 5.8b).

Further, tricine SDS-PAGE analysis revealed that the photolabeled peptides that were immunoprecipitated by the site-directed antibody Ab-6 are composed of a primary 6.1-kDa band and a small amount of a 7.8-kDa fragment (Fig. 5.8c, right lane). This observation indicates that the 6.1-kDa peptide was the smallest labeled fragment after the successive digestion with Lys-C and Glu-C. From searching the overlapped peptide sequences of the peptides produced by the Lys-C/Glu-C digestion and the CNBr cleavage, the smallest labeled fragments were deduced to be Tyr^{1350}-Met^{1366} and Leu^{1367}-Met^{1381}, respectively (Fig. 5.9). In the case of cardiac calcium channels, the most probable labeled site was the repeat IV containing peptide between Thr^{1475} and Met^{1507}, which is the corresponding peptide (Tyr^{1350}-Met^{1381}) (Ii et al. 1998) identified in the skeletal muscle counterpart. Semotiadil binding site is configured with the homologous segment containing transmembrane IVS6 and adjacent extracellular and/or intracellular stretches in both calcium channels. These data indicate that IVS6 is an important region for semotiadil binding (Fig. 5.9).

Collectively, Fig. 5.10 shows an overview illustration of the binding sites for the four typical calcium antagonists within the primary structure of skeletal muscle

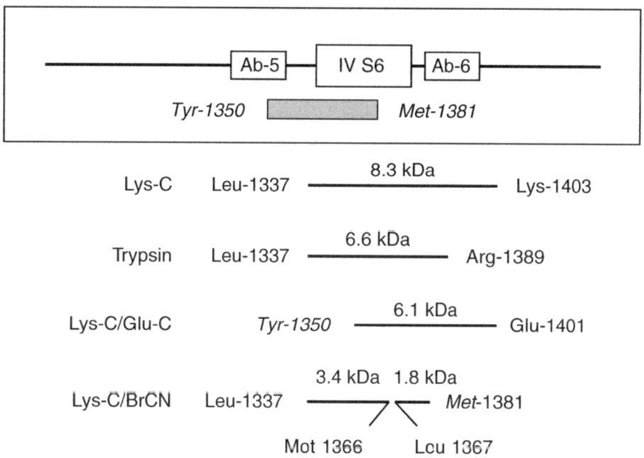

Fig. 5.9 Identification of the semotiadil binging sites (summary) (modified from Kuniyasu et al., J. Biol. Chem., 273, 4635–4641, 1997)

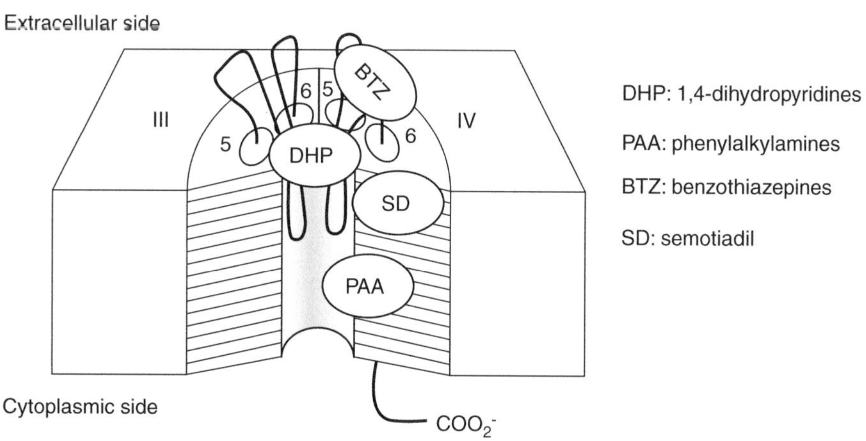

Fig. 5.10 Calcium antagonist binding sites on the α_1 subunit of L-type calcium channel

calcium channel. DHP binds the extracellular site composed of the IIIS5-IIIS6 loop and the extracellular mouth of IVS6, and BTZ binds another extracellular site, part of which consists of the IVS5-IVS6 loop. By contrast, PA binds the intracellular site, which consists of the intracellular mouth of IVS6, and the semotiadil site is located in the transmembrane IVS6 and adjacent extracellular and/or intracellular stretches.

Photoaffinity labeling results show that the pore-forming regions of L-type calcium channel α_1 subunits in repeats III and IV provide a "hot spot" for the binding and action of chemically unrelated calcium antagonists. These findings not only explain the pharmacological allosteric binding of the four calcium antagonists in more details based on the calcium channel structure but also shed light on the molecular characteristics of the extracellular and intracellular constituents of the channel pore, to which the calcium antagonists bind and block calcium ion permeation.

5.3.2 Neurotoxin Binding Sites on the Sodium Channel

Voltage-dependent sodium channels consist of α subunit, which is approximately 260 kDa, associated with auxiliary β subunits (Catterall et al. 2005, 2007). The α subunit is sufficient for the functional expression of the sodium channel, but the β subunits modify the voltage dependence and kinetics of channel gating. Six different neurotoxin receptor sites (site 1~site 6) have been identified on voltage-dependent sodium channels.

Tetrodotoxin (TTX) is a potent toxin, which specifically binds to the neurotoxin receptor site 1 within the channel pore region with high affinity. TTX blocks the influx of sodium ions across the membrane (Catterall et al. 2005, 2007). To identify the TTX binding site on the sodium channel (Fig. 5.11), a (diazirino) trifluoroethylbenzoyl derivative of 11-(aminoethylamino)methyl-TTX (TTXenDTB) was employed. It specifically labeled the 250-kDa sodium channel protein from the electroplax of *Electrophorus electricus* (Yoshida et al. 1990).

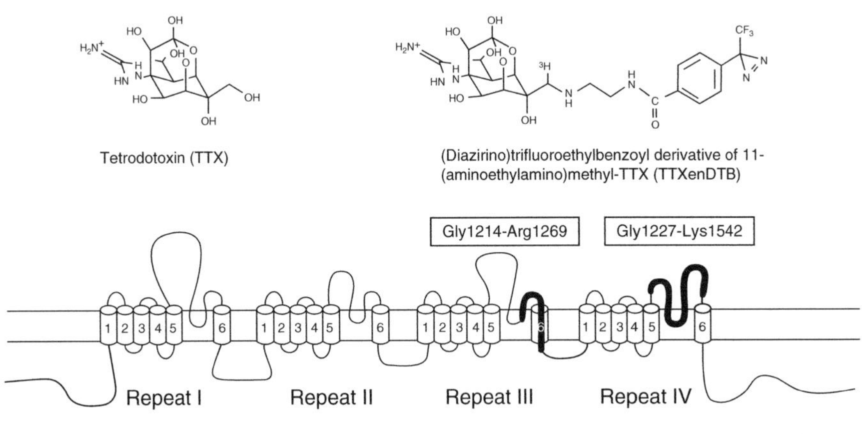

Fig. 5.11 TTX and its photoreactive probe

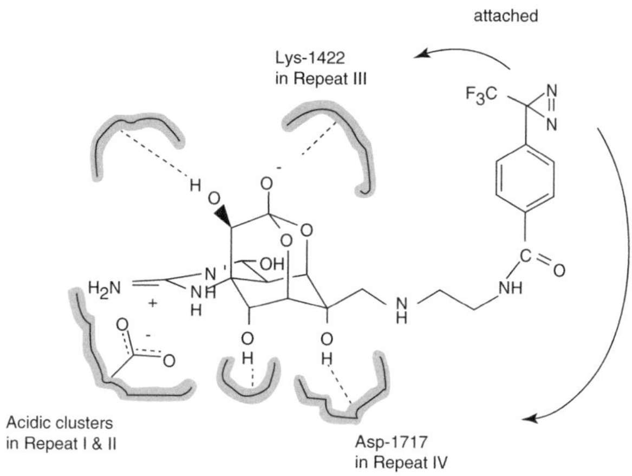

Fig. 5.12 A proposed TTX binding site model on the sodium channel

However, low photoincorporation efficiency (up to 2%) resulted in a failure to isolate the labeled peptide fragments and identify them by protein chemical methods. The introduction of antibody mapping allowed us to determine the photolabeled sites, which are located in the loop between segments S5 and S6 of repeat IV, besides the regions containing transmembrane segment S6 and adjacent extracellular and cytoplasmic sequences in repeat III (Nakayama et al. 1992). Site-directed mutagenesis experiments have determined the amino acid residues that are required for the high-affinity binding of tetrodotoxin and saxitoxin at neurotoxin receptor site 1 of the α subunit (Catterall et al. 2005). No photolabeled fragments were detected in the corresponding region of domain I. The results suggest that C-11 of tetrodotoxin, where the photoreactive moiety is attached, orients to the linker region between S5 and S6 in repeat III and IV (Fig. 5.12).

The binding site for batrachotoxin, an alkaloid toxin acting neurotoxin receptor site 2, is determined using a tritium-labeled batrachotoxinin-A-ortho-azidobenzoate ([^{3}H] BTX-OAB) to covalently label purified and reconstituted rat brain sodium channels. Antibody mapping indicates that the batrachotoxin binding site is located on the S6 transmembrane region of repeat I of the sodium channel α subunit (Trainer et al. 1996).

Brevetoxins are lipid-soluble polyether neurotoxins that are produced by the marine dinoflagellate *Ptychodiscus brevis* and bind to the neurotoxin receptor site 5 on voltage-dependent sodium channels. The brevetoxin binding site was identified using a photoaffinity ligand, *p*-azidobenzoyl tritium-labeled brevetoxin (*p*-AB [^{3}H] PbTx-3) (Trainer et al. 1991). Brevetoxin interacts with the transmembrane interface between repeats I and IV near the S5 segment of repeat IV and the S6 segment of repeat I. These results provide direct evidence for the participation of distant segments of the α subunit in the formation of the brevetoxin receptor site.

The identification of the interaction site for the α-scorpion toxin (toxin V) from *Leiurus quinquestriatus* (LqTx) was first used to successfully map sodium channel α subunits. A site of covalent attachment of ANB-LqTx has been identified in an extracellular loop of repeat I of the α subunit (Tejedor and Catterall 1988, 1990). In contrast to small organic ligands including brevetoxin and batrachotoxin, photolabeling with the larger polypeptide toxin can occur at sites distant from the core of the binding domain. The IS5–S6 linker is the toxin-labeled region, whereas the site mutagenesis residues essential for toxin sensitivity were located in the IVS3–IVS5 linker. In this case, photoaffinity labeling has not provided a direct hint to the regions containing the amino acids that are crucial for toxin sensitivity.

5.3.3 Substrate Binding Sites on the Multidrug Resistance Transporter

P-glycoprotein, the product of the human *MDR1* gene, is an energy-dependent transport protein, which has broad specificity for the structurally diverse hydrophobic antitumor drugs that do not have a common structure or intracellular target (Gottesman et al. 1996; Safa 2004). P-glycoprotein is a large 1276-amino-acid plasma membrane protein that consists of 12 transmembrane segments (TMs) and two nucleotide-binding domains (NBDs). It remains unclear how P-glycoprotein can recognize and transport many various structurally different substances, including therapeutic drugs. Combination analysis of the photoaffinity labeling and antibody mapping has been applied to the determination of drug binding sites. A photoaffinity ligand of the α_1-adrenergic receptor agonist [^{125}I]iodoarylazidoprazosin (IAAP) was initially used to identify the substrate binding sites (Greenberger et al. 1991). Two major labeled sites for the IAAP are located in each half of the P-glycoprotein (TM6 and TM12), and one minor labeled site is found within TM4 up to, but not including, TM6 (Greenberger 1993). These data indicate that IAAP interacts with an intracellular region just beyond TM6 and TM12 and in close proximity to the ATP binding site (Safa 2004). The genetic approach demonstrates that mutations in TM12 and the loop between TM11 and TM12 of *MDR1* can alter the specificity of P-glycoprotein for anticancer drugs (Zhang et al. 1995). In contrast, two photoreactive analogs of paclitaxel, containing a *p*-benzoyldihydrocinnamoyl substituent at either the C-3 or C-7 positions (Wu et al. 1998), are photoincorporated into a region of the transporter, which includes half of TM12 and terminates just after the Walker A motif in the second nucleotide-binding fold and a region that includes all of TM 7 and half of TM 8 plus the intervening extracellular loop, respectively. These data indicate that paclitaxel interacts with the carboxyl-terminal half of the mouse *mdr1b* P-glycoprotein. Subsequent site-directed mutagenesis studies have identified the amino acids that are critical for the binding of some of these agents to P-glycoprotein.

5.3.4 Ligand Binding Sites on Some Other Receptors

5.3.4.1 Calcium Release Channel/Ryanodine Receptor

The calcium release channels/ryanodine receptors (RyRs) play a role in the regulation of intracellular free calcium concentrations in muscles. Ryanodine, a neutral plant alkaloid, binds to the calcium release channel of RyRs and modulates their function.

Dantrolene used as a therapeutic drug to malignant hyperthermia also targets RyR. By using a photoreactive ryanodine derivative, [^{3}H]10-*O*-[3-(4-azidobenzamido)propionyl] ryanodine, the 76-kDa carboxyl-terminal fragment was identified as the high-affinity binding site for ryanodine (Witcher et al. 1994). In contrast, a photoreactive dantrolene derivative [^{3}H]azidodantrolene labeled the amino-terminal region corresponding to amino acids 590–509 in skeletal isoform RyR1 (Paul-Pletzer et al. 2001) and amino acids 601–620 in cardiac isoform RyR2 (Paul-Pletzer et al. 2005).

5.3.4.2 Arginine Vasopressin Receptor

Neurohypophysial hormones, such as arginine-vasopressin (AVP) 1, bind to the G protein-coupled receptor superfamily, V1a receptor, which contributes to vasoconstriction. The V1a receptor antagonist is used for the treatment of hyponatremia, particularly in congestive heart failure patients. By using a photoreactive linear peptide AVP1 antagonist, [^{125}I]3-azidophenylpropionyl-D-Tyr(Me)-Phe-Gln-Asn-Arg-Pro-Arg-Tyr-NH_2 (3-N_3-Phpa-LVA), the antagonist binding site has been identified as the amino acid residues Asn^{327}-Lys^{370}, including transmembrane segment VII (Phalipou et al. 1997). A subsequent photoaffinity labeling experiment indicated that the linear peptide antagonist could interact with a transmembrane binding site overlapping with the cyclic peptide AVP1.

5.3.4.3 Sodium Bile Salt Co-transporter

Ileal sodium bile salt co-transporter (IBAT), a 93-kDa integral membrane protein, plays a role in the sodium-dependent reabsorption of bile acids from the lumen of the small intestine and in cholesterol metabolism. To identify the bile acid binding site of IBAT, brush border membrane vesicles from rabbit ileum were labeled with a radioactive 7-diazirine bile acid analog, [^{3}H]7-azi-derivative of cholyltaurine (Kramer et al. 2001). The site-directed antibody generated against the carboxyl-terminus of IBAT recognized the chymotryptic peptide fragment of the carboxyl-terminal 67–56 amino acids downstream of Ser^{280} or possibly even downstream of Pro^{291}. Consistent with this finding, Pro^{290}-to-Ser, Thr^{262}-to-Met, or Leu^{243}-to-Pro mutations led to a complete loss of transporter function (Oelkers et al. 1997). These results suggest that the bile acid binding site is localized to the intracellular regions of the carboxyl-terminus of IBAT.

5.3.4.4 Dopamine Transporter

Dopamine transporter (DAT) is a 12-membrane-spanning protein, which takes up the neurotransmitter dopamine from the synaptic cleft into the cytosol. To identify the binding site of psychotropic drugs such as cocaine and amphetamine, rat DATs were labeled with photoaffinity ligands of two classes of uptake blockers: a GBR analog, [^{125}I]DEEP, and a cocaine analog, [^{125}I]RTI 82 (Vaughan 1995; Vaughan and Kuhar 1996). Antibody-mapping analysis of the labeled fragments revealed that

DEEP and RTI 82 were attached to the amino-terminal half of the DAT near the first two transmembrane segments (TM1 and TM2) and the carboxyl-terminal region between TM4 and TM12, respectively.

5.4 Conclusion

Over the past two decades, photoaffinity labeling combined with antibody mapping has been successfully applied to the identification of ligand binding site on the membrane receptor, especially multiple membrane-spanning proteins with high molecular weights. Advances employing MS techniques have contributed to the further utility of photoaffinity labeling (Robinette et al. 2006). By replacing microsequencing with Edman degradation, LC–MS and LC–MS/MS have been used to determine the labeled amino acid residues. However, these techniques have not been fully expanded to multiple transmembrane proteins such as ion channels and drug transporters. Therefore, photoaffinity labeling combined with site-directed antibodies is still useful for identifying drug binding sites in membrane protein receptors (the main target proteins for drugs). The structural information obtained by photoaffinity labeling will shed light on the potential mechanisms of drug action and thereby assist in the rational design of new therapeutic agents.

References

Catterall WA, Perez-Reyes E, Snutch TP, Striessnig J (2005) International Union of Pharmacology. XLVIII. Nomenclature and structure-function relationships of voltage-gated calcium channels. Pharmacol Rev 57(4):411–425

Catterall WA, Cestele S, Yarov-Yarovoy V, FH Y, Konoki K, Scheuer T (2007) Voltage-gated ion channels and gating modifier toxins. Toxicon 49(2):124–141

Gottesman MM, Pastan I, Ambudkar SV (1996) P-glycoprotein and multidrug resistance. Curr Opin Genet Dev 6(5):610–617

Greenberger LM (1993) Major photoaffinity drug labeling sites for iodoaryl azidoprazosin in P-glycoprotein are within, or immediately C-terminal to, transmembrane domains 6 and 12. J Biol Chem 268(15):11417–11425

Greenberger LM, Lisanti CJ, Silva JT, Horwitz SB (1991) Domain mapping of the photoaffinity drug-binding sites in P-glycoprotein encoded by mouse mdr1b. J Biol Chem 266(31):20744–20751

Hatanaka Y, Sadakane Y (2002) Photoaffinity labeling in drug discovery and developments: chemical gateway for entering proteomic frontier. Curr Top Med Chem 2(3):271–288

Hockerman GH, Peterson BZ, Johnson BD, Catterall WA (1997) Molecular determinants of drug binding and action on L-type calcium channels. Annu Rev Pharmacol Toxicol 37:361–396

Ii N, Kuniyasu A, Kawahara K, Shibano T, Schwartz A, Nakayama H (1998) Photochemical localization of the semotiadil binding region within the cardiac Ca^{2+} channel alpha1 subunit. Comparison with the skeletal muscle counterpart. FEBS Lett 441(1):83–87

Kramer W, Girbig F, Glombik H, Corsiero D, Stengelin S, Weyland C (2001) Identification of a ligand-binding site in the Na^+/bile acid cotransporting protein from rabbit ileum. J Biol Chem 276(38):36020–36027

Kraus R, Reichl B, Kimball SD, Grabner M, Murphy BJ, Catterall WA, Striessnig J (1996) Identification of benz(othi)azepine-binding regions within L-type calcium channel alpha1 subunits. J Biol Chem 271(33):20113–20118

Kuniyasu A, Itagaki K, Shibano T, Iino M, Kraft G, Schwartz A, Nakayama H (1998) Photochemical identification of transmembrane segment IVS6 as the binding region of semotiadil, a new modulator for the L-type voltage-dependent Ca^{2+} channel. J Biol Chem 273(8):4635–4641

Motoike HK, Bodi I, Nakayama H, Schwartz A, Varadi G (1999) A region in IVS5 of the human cardiac L-type calcium channel is required for the use-dependent block by phenylalkylamines and benzothiazepines. J Biol Chem 274(14):9409–9420

Naito K, McKenna E, Schwartz A, Vaghy PL (1989) Photoaffinity labeling of the purified skeletal muscle calcium antagonist receptor by a novel benzothiazepine, [^{3}H]azidobutyryl diltiazem. J Biol Chem 264(35):21211–21214

Nakayama H, Taki M, Striessnig J, Glossmann H, Catterall WA, Kanaoka Y (1991) Identification of 1,4-dihydropyridine binding regions within the alpha 1 subunit of skeletal muscle Ca^{2+} channels by photoaffinity labeling with diazipine. Proc Natl Acad Sci U S A 88(20):9203–9207

Nakayama H, Hatanaka Y, Yoshida E, Oka K, Takanohashi M, Amano Y, Kanaoka Y (1992) Photolabeled sites with a tetrodotoxin derivative in the domain III and IV of the electroplax sodium channel. Biochem Biophys Res Commun 184(2):900–907

Oelkers P, Kirby LC, Heubi JE, Dawson PA (1997) Primary bile acid malabsorption caused by mutations in the ileal sodium-dependent bile acid transporter gene (SLC10A2). J Clin Invest 99(8):1880–1887

Paul-Pletzer K, Palnitkar SS, Jimenez LS, Morimoto H, Parness J (2001) The skeletal muscle ryanodine receptor identified as a molecular target of [^{3}H]azidodantrolene by photoaffinity labeling. Biochemistry 40(2):531–542

Paul-Pletzer K, Yamamoto T, Ikemoto N, Jimenez LS, Morimoto H, Williams PG, Ma J, Parness J (2005) Probing a putative dantrolene-binding site on the cardiac ryanodine receptor. Biochem J 387(Pt 3):905–909

Peterson BZ, Johnson BD, Hockerman GH, Acheson M, Scheuer T, Catterall WA (1997) Analysis of the dihydropyridine receptor site of L-type calcium channels by alanine-scanning mutagenesis. J Biol Chem 272(30):18752–18758

Phalipou S, Cotte N, Carnazzi E, Seyer R, Mahe E, Jard S, Barberis C, Mouillac B (1997) Mapping peptide-binding domains of the human V1a vasopressin receptor with a photoactivatable linear peptide antagonist. J Biol Chem 272(42):26536–26544

Regulla S, Schneider T, Nastainczyk W, Meyer HE, Hofmann F (1991) Identification of the site of interaction of the dihydropyridine channel blockers nitrendipine and azidopine with the calcium-channel alpha 1 subunit. EMBO J 10(1):45–49

Robinette D, Neamati N, Tomer KB, Borchers CH (2006) Photoaffinity labeling combined with mass spectrometric approaches as a tool for structural proteomics. Expert Rev Proteomics 3(4):399–408

Safa AR (2004) Identification and characterization of the binding sites of P-glycoprotein for multidrug resistance-related drugs and modulators. Curr Med Chem Anticancer Agents 4(1):1–17

Schagger H, von Jagow G (1987) Tricine-sodium dodecyl sulfate-polyacrylamide gel electrophoresis for the separation of proteins in the range from 1 to 100 kDa. Anal Biochem 166(2):368–379

Striessnig J, Glossmann H, Catterall WA (1990) Identification of a phenylalkylamine binding region within the alpha 1 subunit of skeletal muscle Ca^{2+} channels. Proc Natl Acad Sci U S A 87(23):9108–9112

Striessnig J, Murphy BJ, Catterall WA (1991) Dihydropyridine receptor of L-type Ca^{2+} channels: identification of binding domains for [^{3}H](+)-PN200-110 and [^{3}H]azidopine within the alpha 1 subunit. Proc Natl Acad Sci U S A 88(23):10769–10773

Striessnig J, Grabner M, Mitterdorfer J, Hering S, Sinnegger MJ, Glossmann H (1998) Structural basis of drug binding to L Ca^{2+} channels. Trends Pharmacol Sci 19(3):108–115

Taki M, Kuniyasu A, Nakayama H, Kanaoka Y (1991a) Synthesis of diazipine and [^{3}H]diazipine: novel dihydropyridines as photoaffinity probes of calcium channels. Chem Pharm Bull(Tokyo) 39(7):1860–1862

Taki M, Nakayama H, Kanaoka Y (1991b) Diazipine, a novel photoaffinity probe for dihydropyridine receptors of calcium channels. FEBS Lett 283(2):259–262

Tejedor FJ, Catterall WA (1988) Site of covalent attachment of alpha-scorpion toxin derivatives in domain I of the sodium channel alpha subunit. Proc Natl Acad Sci U S A 85(22):8742–8746

Tejedor FJ, Catterall WA (1990) Photoaffinity labeling of the receptor site for alpha-scorpion toxins on purified and reconstituted sodium channels by a new toxin derivative. Cell Mol Neurobiol 10(2):257–265

Trainer VL, Thomsen WJ, Catterall WA, Baden DG (1991) Photoaffinity labeling of the brevetoxin receptor on sodium channels in rat brain synaptosomes. Mol Pharmacol 40(6):988–994

Trainer VL, Brown GB, Catterall WA (1996) Site of covalent labeling by a photoreactive batrachotoxin derivative near transmembrane segment IS6 of the sodium channel alpha subunit. J Biol Chem 271(19):11261–11267

Vaughan RA (1995) Photoaffinity-labeled ligand binding domains on dopamine transporters identified by peptide mapping. Mol Pharmacol 47(5):956–964

Vaughan RA, Kuhar MJ (1996) Dopamine transporter ligand binding domains. Structural and functional properties revealed by limited proteolysis. J Biol Chem 271(35):21672–21680

Watanabe T, Kalasz H, Yabana H, Kuniyasu A, Mershon J, Itagaki K, Vaghy PL, Naito K, Nakayama H, Schwartz A (1993) Azidobutyryl clentiazem, a new photoactivatable diltiazem analog, labels benzothiazepine binding sites in the alpha 1 subunit of the skeletal muscle calcium channel. FEBS Lett 334(3):261–264

Witcher DR, McPherson PS, Kahl SD, Lewis T, Bentley P, Mullinnix MJ, Windass JD, Campbell KP (1994) Photoaffinity labeling of the ryanodine receptor/Ca2+ release channel with an azido derivative of ryanodine. J Biol Chem 269(18):13076–13079

Wu Q, Bounaud PY, Kuduk SD, Yang CP, Ojima I, Horwitz SB, Orr GA (1998) Identification of the domains of photoincorporation of the 3′- and 7-benzophenone analogues of taxol in the carboxyl-terminal half of murine *mdr1b* P-glycoprotein. Biochemistry 37(32):11272–11279

Yoshida E, Nakayama H, Hatanaka Y, Kanaoka Y (1990) Photoaffinity labeling of the electroplax sodium channel with tetrodotoxin derivatives. II. Comparison of the photoreactivity of different photoactivable groups in the tetrodotoxin binding site. Chem Pharm Bull(Tokyo) 38(4):982–987

Zhang X, Collins KI, Greenberger LM (1995) Functional evidence that transmembrane 12 and the loop between transmembrane 11 and 12 form part of the drug-binding domain in P-glycoprotein encoded by MDR1. J Biol Chem 270(10):5441–5448

Chapter 6
Synthesis of Diazirinyl Photophore and Optically Pure Diazirinylphenylalanines for Photoaffinity Labeling

Yuta Murai, Lei Wang, and Makoto Hashimoto

Abstract As photophores in photoaffinity labeling, three major types, arylazide, benzophenone, and diazirine have been used and contributed to the understanding of molecular biology processes. In particular, diazrine is a significantly useful photophore because it has some advantages such as its relatively small size, stability, low rate of rearrangement, and generating a high reactive carbene with longer wavelength which reduces damage to the target molecules. Nevertheless, despite these many advantages in diazirine, there are reports in which arylazide and benzophenone are generally utilized for photoaffinity labeling compared with diazirine due to its synthetic difficulty onto ligands. To overcome this problem, recently many researchers have addressed to establish the efficient synthetic route of diazirinyl compounds. Particularly, α-aromatic amino acids possess not only biological activities themselves but also basic component of protein (bioactive peptide). In addition, a number of heterocyclic bioactive compounds have variety of attractive biological activities. Therefore, their diazirinyl compounds promise the molecular biology processes. In this chapter, firstly, it is demonstrated the simple and efficient preparation of diazirine on aromatics and aliphatic chains. Furthermore post-functional derivatization of them is also described to diversify bioactive compounds.

Keywords Photophore • Diazirine • α-Aromatic amino acid • Heterocyclic compound • Deuterium

---

Y. Murai (✉)
Faculty of Advanced Life Science, Frontier Research Center for Post-Genome Science and Technology, Hokkaido University, Kita 21, Nishi 11, Kita-ku, Sapporo 001-0021, Japan
e-mail: ymurai@sci.hokudai.ac.jp

L. Wang • M. Hashimoto
Division of Applied Science, Graduate School of Agriculture, Hokkaido University, Kita 9, Nishi 9, Kita-ku, Sapporo 060-8589, Japan

Y. Hatanaka, M. Hashimoto (eds.), *Photoaffinity Labeling for Structural Probing Within Protein*, DOI 10.1007/978-4-431-56569-7_6

6.1 Introduction

Photoaffinity labeling is a crucial chemical method to understand biological processes such as ligand–receptor, nucleic acid–protein, lipid–protein, and protein–protein interactions. So far, typically three photophores, arylazides, benzophenones, and diazirines, are widely used in this field. Arylazides are easily prepared, however the photoactivated wavelength of them is at 280 nm which often damages biomolecules. Besides generating active-species nitrenes, they sometimes rearrange to ketenimines as undesired by-products. Benzophenones require a long period irradiation to generate active-species biradical (triplet carbonyl states), which sometimes occur as a risk for nonspecific labeling. On the other hand, diazirines have great advantages such as thermodynamic stability, generating high reactive species carbenes within a short time at 360 nm wavelengths. Therefore, diazirines are most promising to investigate biological processes. However despite these utilities, the application of diazirines is less due to its synthetic difficulty onto bioactive compounds as compared with other photophores. To overcome this big problem, recently many researchers have addressed to improve the efficient synthesis of diazirinyl compounds. In this chapter, it is described about simple and efficient methods of a diazirine synthesis and its post-functional derivatives.

6.2 General Methods to Prepare Diazirine Photophore

6.2.1 3-Trifluoromethyl-3-Phenyl-Diazirines

Among all of the photoreactive groups used in photoaffinity labeling arylazides, benzophenones, and diazirines (Fleet et al. 1969; Galardy et al. 1974; Hashimoto and Hatanaka 2008; Dubinsky et al. 2013), 3-trifluoromethyl-3-phenyl-diazirine (TPD) (Brunner et al. 1980) has emerged as the most effective photoreactive group in photoaffinity labeling due to its relatively small size, thermal and chemical stability, low rate of rearrangement, and high reactivity of its intermediate. Furthermore, its activated wavelength is around 360 nm that are safe for cells and tissues (Smith and Knowles 1973; Hatanaka et al. 1989; Hiramatsu et al. 2007; Qiu et al. 2008; Das 2011).

To prepare TPD derivatives, the corresponding α,α,α-trifluoroacetophenones **1** are widely used as the starting material (Fig. 6.1). Under the oximation with hydroxylammonium chloride at 60–85 °C, the ketone can be converted to oxime derivatives. Next, alternative strategies are available: mesylation (Delfino et al. 1993) and tosylation (Song and Zhang 2009). Under the certain reagents (*p*-TsCl or MsCl), corresponding tosyl or mesyl oximes **2** are readily constructed. Due to the unstability and relatively high toxicity of MsCl, tosylation with *p*-TsCl is the widely used strategy for this purpose. Then, the tosyl or mesyl oximes are subjected to ammonia to afford diaziridines **3** which are the key intermediates to construct a diazirinyl ring. Finally, diazrines **4** are synthesized by oxidation of the diaziridines with various oxidants such as $(CH_3)_3COCl$ (Carrera et al. 2005), Ag_2O (Al-Omari et al. 2006), oxalyl chloride (Kumar and Young 2009), and I_2/NEt_3 (Burkard et al. 2010). Electrochemical oxidation of diaziridines in the presence of Pt/Ti anode was also reported (Vedenyapina et al. 2006). Obviously, the formation of diaziridines from tosyloxime derivatives is the crucial step to further construct TPD. It was well known that the tedious synthetic procedure of TPD derivatives was a major consideration for researchers. Furthermore, the inevitable isolation of diaziridines, preparation of fresh oxidant, and post-treatment after oxidation were time-consuming and yield-diminishing. To shorten the synthetic procedure of TPD as well as TPD based photoaffinity labeling probes, authors developed an effective one-pot strategy to afford TPD **4** from tosyloximes **2** (Fig. 6.2). Under liquid ammonia at 80 °C or with addition of $LiNH_2$ at room temperature in a sealed tube, authors could directly

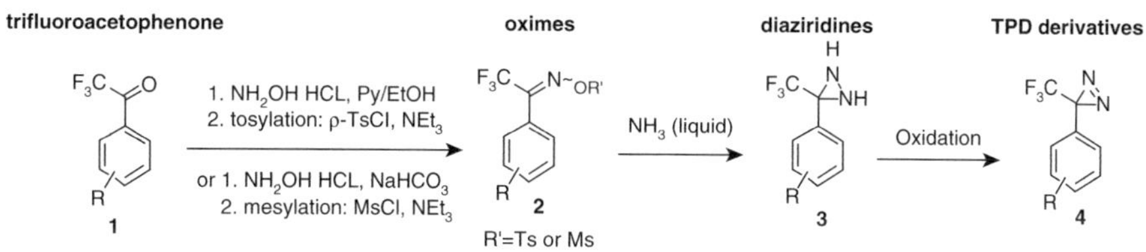

Fig. 6.1 Synthetic methods of TPD derivatives

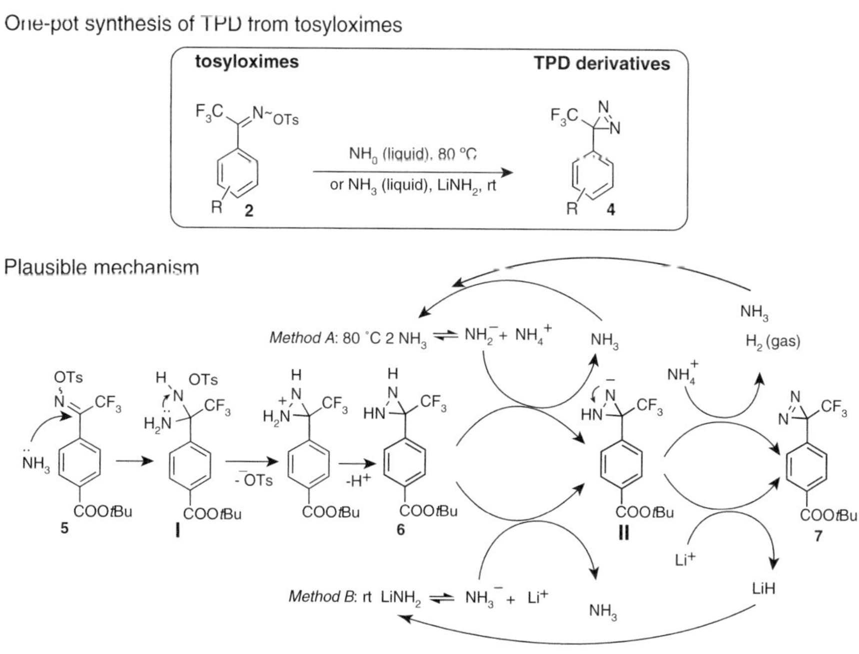

Fig. 6.2 One-pot synthesis of TPD from tosyloximes and mechanism study

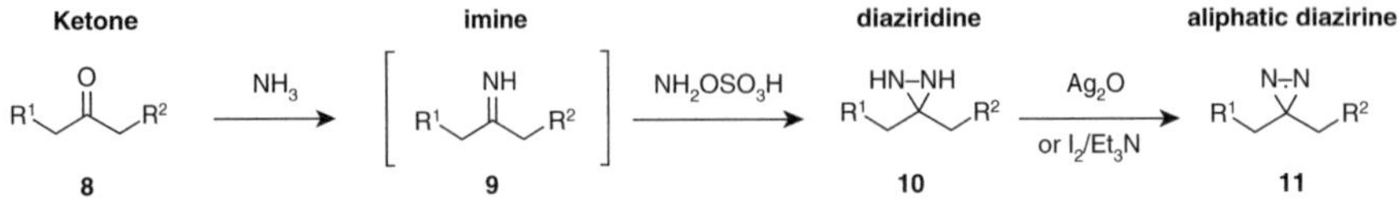

Fig. 6.3 Synthetic methods of aliphatic diazirine

obtain TPDs from corresponding tosyloximes without the isolation of diaziridines. Mechanism study indicated that the NH_2^- generated from liquid NH_3 at high temperature or $LiNH_2$ should be responsible for the formation of TPD ring from diaziridines (Wang et al. 2015).

6.2.2 Aliphatic Diazirines

Aliphatic diazirines (Suchanek et al. 2005) is another important member in diazirine family. Compared to TPD derivatives, the applications of aliphatic diazirines are limited due to the less generated carbene intermediate after photoreaction and the undesired intramolecular rearrangement reaction that decreases the efficiency. However, their small size and convenient synthetic procedure attract increasing attention in recent years.

Unlike the preparation of TPD derivatives, the ketone precursor **8** of aliphatic diazirines is initially converted to imine **9** under the treatment with methanolic ammonia or liquid ammonia under high pressure (Fig. 6.3). With the addition of hydroxylamine-*O*-sulfonic acid, diaziridine **10** is constructed. The obtained diaziridine is then subjected to oxidation by silver oxide or iodine with triethylamine to afford the aliphatic diazirine **11** (Al-Omari et al. 2006; Vila-Perello et al. 2007).

Details are summarized in Chap. 10 in the book.

6.3 Synthesis of Diazirinyl Bioactive Compounds

6.3.1 Optically Pure Diazirinylphenylalanines

Optically pure trifluoromethyldiazirinyl phenylalanine ((Tmd)Phe) has been used as important building block for investigation of peptides, proteins, and other biomacromolecule (Baldini et al. 1988; High et al. 1993; Ploug et al. 1998; Tippmann et al. 2007; Wang et al. 2013). Numerous methods are available to prepare (Tmd)Phe derivatives. Nassal (1984) reported a preparation of diazirinylphenylalanines from halogenaged TPD **12** (Fig. 6.4). Through ligation with diethyl acetamido-malonate in the presence of NaOEt, halogenaged TPD could be converted to TPD-based

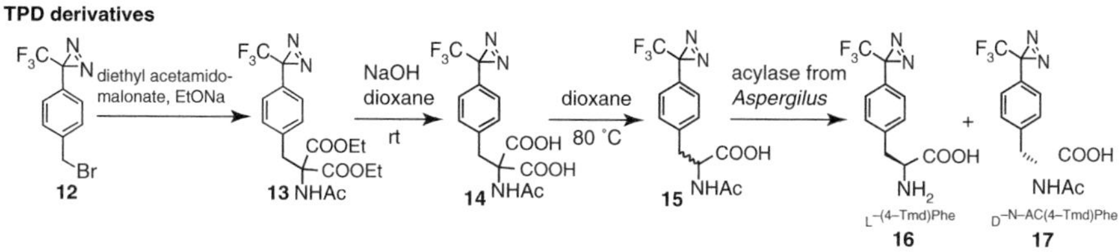

Fig. 6.4 Synthesis of (Tmd)Phe via reaction of halogenaged TPD and diethyl acetamidomalonate

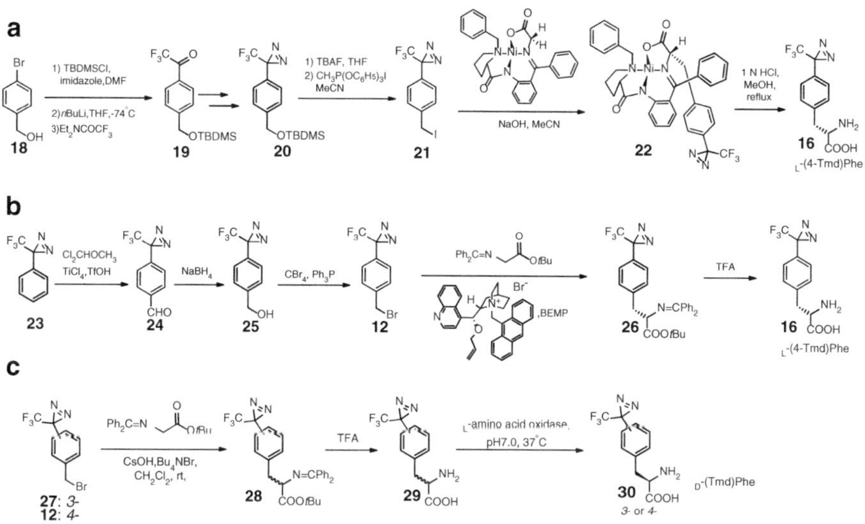

Fig. 6.5 Asymmetric syntheses of (Tmd)Phe

diester **13**. Following with further hydrolysis of ethyl ester groups under basic condition, and decarboxylation by heating at 80 °C, D,L-*N*-acetyl(Tmd)Phe **15** were obtained. The racemates were isolated with the enzymatic resolution of *N*-acyl amino acids. Under the treatment with acylase from *Aspergillus*, free L-(Tmd)Phe **16** and D-*N*-acetyl(Tmd)Phe **17** were constructed which were suitable for further separation.

Asymmetric syntheses were also suitable for the construction of (Tmd)Phe derivatives. Fishwick et al. (1994) reported an effective method to prepare L-(Tmd) Phe **16** through treatment of halogenated TPD **21** with the chiral Ni complex (Fig. 6.5a). Halogenated TPD was prepared from 4-bromobenzyl alcohol **18** via construction of ketone skeleton on benzene ring and the subsequent traditional procedure for TPD derivatives. The alkylation by Ni complex was the crucial step to afford the stereoselective intermediate **22**. Hydrolysis of the complex in refluxing

HCl/MeOH readily afforded to a desired product. Another work was presented by authors' group (Hashimoto et al. 2002) by using a similar strategy, many L-(Tmd) Phe-based derivatives were obtained successfully. Furthermore, it has been reported that through the reaction of halogenated TPD **12** and *tert*-butylglycinate benzophenone imine in the presence of cinchonidine-based asymmetric catalyst and 2-*tert*-butylimino-2-diethylamino-1,3-dimethylperhydro-1,3,2-diazaphosphorine (BEMP) as a base, L-(Tmd)Phe could be easily obtained (Nakashima et al. 2006) (Fig. 6.5b). In this work, they presented a novel synthetic strategy of aldehyde TPD **24** by treating TPD **23** with dichloromethyl methyl ether in the presence of $TiCl_4$ and trifluoromethanesulfonic acid (TfOH). Following further reduction and bromination, compound **12** was obtained. As an application, corresponding photoreactive calmodulin-binding peptide was readily prepared via conventional solid-phase peptide synthesis. In the absence of cinchonidine-based asymmetric catalyst and BEMP, the reaction of halogenated TPD and *tert*-butylglycinate benzophenone imine affords a racemate **28** (Fig. 6.5c). Under the treatment with L-amino acid oxidase, authors (Masuda et al. 2010) presented a convenient work to selectively afford D-(Tmd)Phe derivatives **30**.

Fillion et al. (2006) reported a stereoselective synthesis of Fmoc-L-(4-Tmd)Phe **41** with a 14% overall yield via 10 steps (Fig. 6.6). Initially, diiodobenzene **31** was converted into ketone **32** by reacting organolithium salt with ethyl α,α,α-trifluoroacetate. Corresponding ketone was subjected into the reported strategy for synthesis of diaziridine derivatives **35**. The preparation of chiral moiety was started from compound **36**. Through protection and iodination, iodate chiral moiety **39** was obtained. The key step next was a palladium-catalyzed coupling reaction of a *p*-diaziridinyliodoaryl derivative **35** with diprotected β-iodoalanine **39** in the presence of activated Zn* and tris(dibenzylideneacetone)dipalladium(0)/tri-*o*-tolylphosphine catalyst to generate diprotected phenylalanine diaziridine **40**. Following subsequent oxidation and deprotection, L-(Tmd)Phe derivative **41** was prepared, then incorporated into the C-terminal position of the human angiotensin II.

To develop a convenient strategy for construction of (Tmd)Phe derivatives, authors (Wang et al. 2015) reported a practical method by direct construction of trifluoromethyldiazirine on optically pure phenylalanines (Fig. 6.7). Iodophenylalanines **42** were used as the starting materials. Through the protection with *tert*-butyl group and the subsequent treatment with MeLi, *t*BuLi, and CF_3COOEt, trifluoroacetophenone derivatives **44** were constructed. Following the traditional procedures for synthesis of

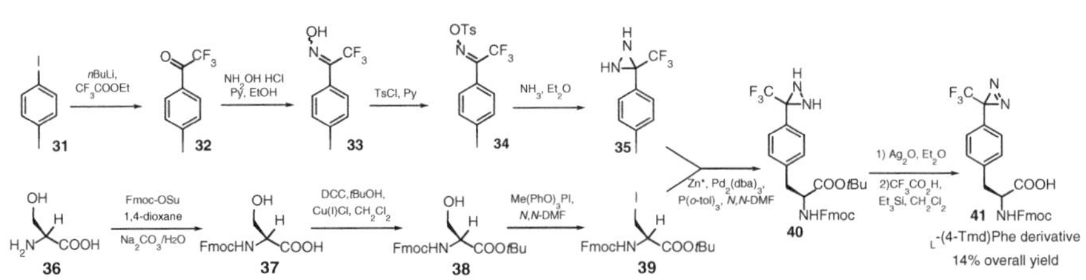

Fig. 6.6 Stereospecific synthesis of L-(Tmd)Phe derivative

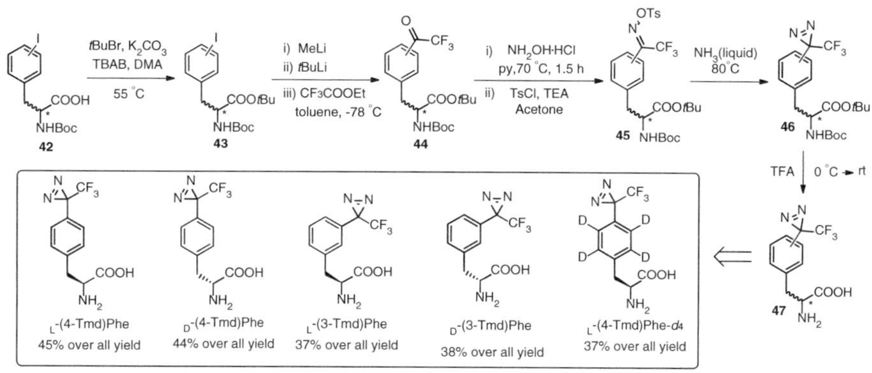

Fig. 6.7 Direct synthesis of (Tmd)Phe derivative from iodophenylalanines

TPD as well as our developed one-pot strategy (liquid NH_3 at 80 °C in sealed tube), we prepared L-(4-Tmd)Phe, D-(4-Tmd)Phe, L-(3-Tmd)Phe, D-(3-Tmd)Phe, and deuterated derivative L-(4-Tmd)Phe-d_4 (Wang et al. 2014) in good overall yields.

6.3.2 3-(Trifluoromethyl)diazirinyl Homo- and Bis-homo-phenylalanine

Homophenylalanine (hPhe) and bishomophenylalanine (bhPhe) elongate one or two methylene in a side chain of phenylalanine (Phe). Sometimes they show different biological activities when replacing Phe with h- or bh-Phe in bioactive peptides (Mosberg et al. 1990; Abiko and Sekino 1998). In addition, hPhe is well known as a skeleton of pharmaceutical products, which inhibit angiotensin converting enzyme, such as benazepril and enalapril (Chang and Yang 2003a, b). Synthesis of h- or bh-Phe has been reported using various methodologies, enzymatic resolution (Zhao et al. 2002), Suzuki-coupling (Barfoot et al. 2005), diastereoselective Michel addition (Yamada et al. 1998), and catalytic asymmetric hydrogenation (Xie et al. 2000). Additionally, it has been also reported that synthesis of hPhe using Friedel-Crafts reaction of Asp anhydrides with $AlCl_3$ under reflux for many hours (Xu et al. 2000). Although these methods are really useful for C–C bond coupling, diazirine moiety cannot tolerate under many of these reaction conditions. Recently, authors have overcome these difficulties by Friedel-Crafts acylation using TfOH. TfOH allowed the Friedel-Crafts reaction to proceed under mild conditions with TPD derivative **48** and side chain derivatives of Asp **49** or Glu **50** to afford products **51**, **52**. Subsequently, benzylcarbonyl group was subjected to hydrogenation with triethylsilane **53**, **54**, followed by removal of the protective groups to afford TPD containing hPhe **55** or bhPhe **56** without racemization of starting Asp or Glu (Murashige et al. 2009; Murai et al. 2009) (Fig. 6.8).

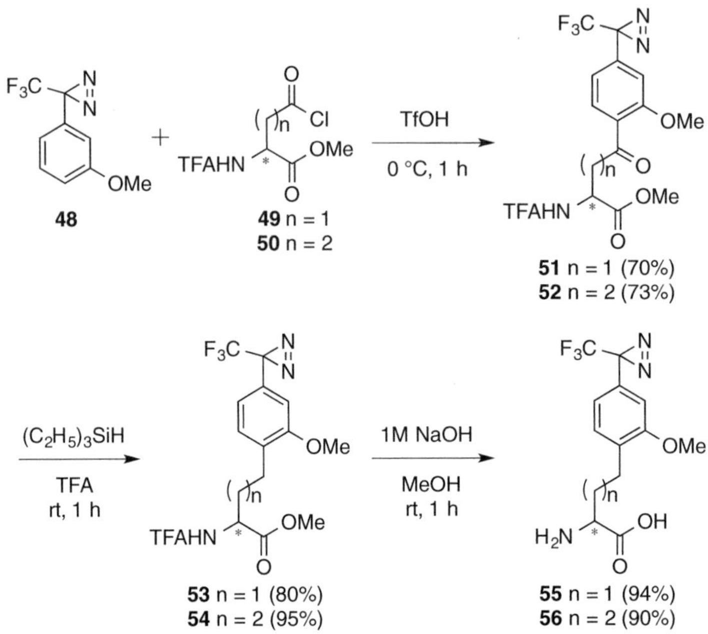

Fig. 6.8 Synthesis of diazirinyl hPhe **55** and bhPhe **56** from TPD (**48**) and optically pure Asp **49** or Glu **50** derivatives

6.3.3 Hydrogen/Deuterium Exchange of 3-(Trifluoromethyl) diazirinyl Phenylalanine

Photoaffinity reagents are pivotal straightforward tools for the rapid investigation of the interactions between ligand and its target proteins. Elucidation of photoaffinity labeled proteins is often achieved by MS and MS/MS analysis of enzymatic digestion production of its, however, the process is complicated by inevitable non-specific binding proteins. Deuterium is one of the simplest stable isotopes and can be utilized for elucidation of the different mass numbers derived from a mixture of deuterium-labeled and -unlabeled photoaffinity probe cross-linking target proteins. Recently authors (Murai et al. 2013) developed an effective H/D exchange method for natural aromatic α-amino acids and their containing bioactive peptides with trifluoromethanesulfonic acid-d (TfOD) at low temperature. Furthermore, authors succeeded H/D exchange of diazirinyl L-Phe at the same condition. Diazirinyl L-Phe **57** was subjected to H/D exchange with TfOD at 0 °C for 5 h. The H/D exchange proceeded in a time-dependent manner and the H/D exchange ratio was observed up to 35% (35% and 10% for ortho- and meta-position from diazirinyl group, respectively).

Diazirinyl *N*-Ac-L-Phe **58** was also subjected to H/D exchange at 0 °C, which promoted higher H/D exchange at both *ortho*- (up to 70%) and *meta*- (up to 35%) positions for 5 h (Wang et al. 2014) (Fig. 6.9).

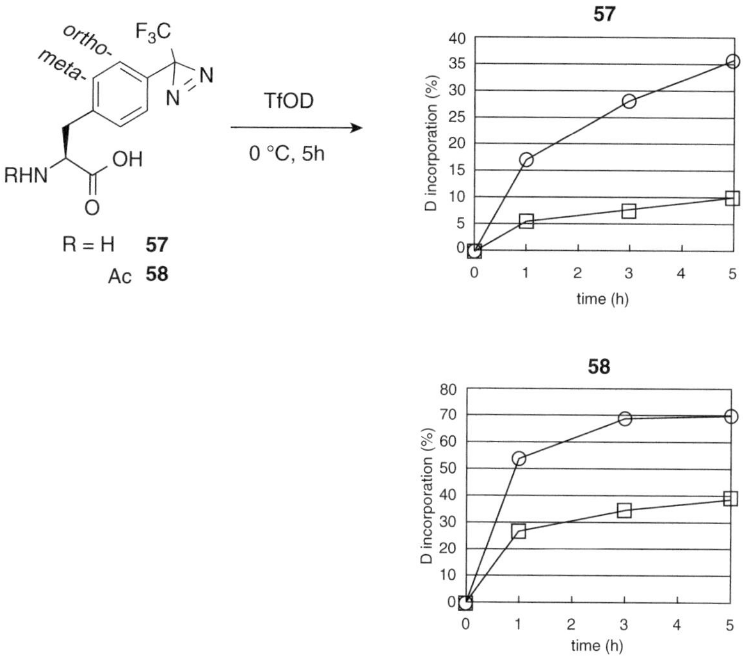

Fig. 6.9 Time-course H/D exchange ratio for *ortho*- and *meta*-position from diazirinyl group of compound **57** and **58** was presented by *open square* and *circle*, respectively

6.3.4 3-(Trifluoromethyl)diazirinyl Indole

A number of natural and synthetic products that contain indole moiety have variety of attractive biological activities such as pharmacology, toxicity, phytohormone, insecticides, so on. Therefore, the photoreactive indole promises the identification of the mechanism between biological target molecules of natural products. Photoreactive indoles containing an arylazide (Melhado and Leonard 1983), a benzophenone (Cherry et al. 2007) have been reported in the past, whereas there have been no reports to synthesize a 3-(trifluoromethyl)diazirinyl indole, is more stable before photo irradiation, and generates a higher active species, for use as a photoaffinity labeling reagent. Firstly, authors tried to synthesize a 3-(trifluoromethyl)diazirinyl indole from 3-(trifluoromethyl)phenyldiazirine by Fisher indole synthesis, Mori–Ban–Hegedus indole synthesis (Eto et al. 1989), and Larock heteroannulation (Ma et al. 2009), but 3-(trifluoromethyl)diazirine was decomposed during the reaction (Fig. 6.10).

In our recent study (Murai et al. 2012), authors have achieved the construction of a 3-(trifluoromethyl)diazirine on indole at the 5- or 6-position by directly treating the corresponding bromo indoles. Trifluoroacetylation of 5- or 6-bromoindole smoothly proceeded by treating with potassium hydride/*tert*-BuLi followed by the reaction with appropriate trifluoroacetylation reagents to obtain trifluoroacetyl indoles **60**. A trifluoroacetyl group was converted to 3-(trifluoromethyl)diazirine according to modified similarly methods because of the instability or a side reaction

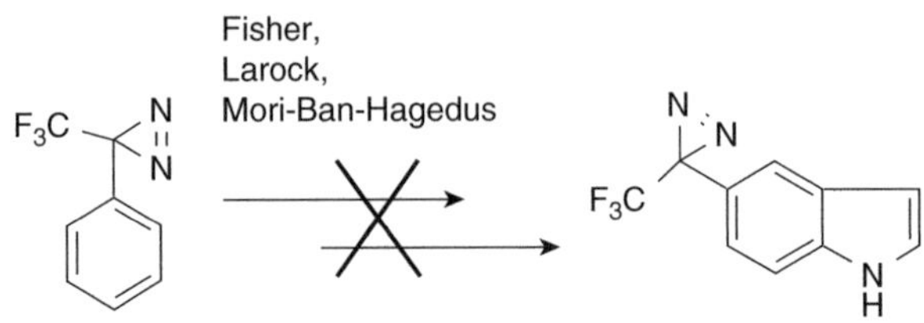

Fig. 6.10 Trials of 3-(trifluoromethyl)diazirinyl indole synthesis from 3-(trifluoromethyl) phenyldiazirine

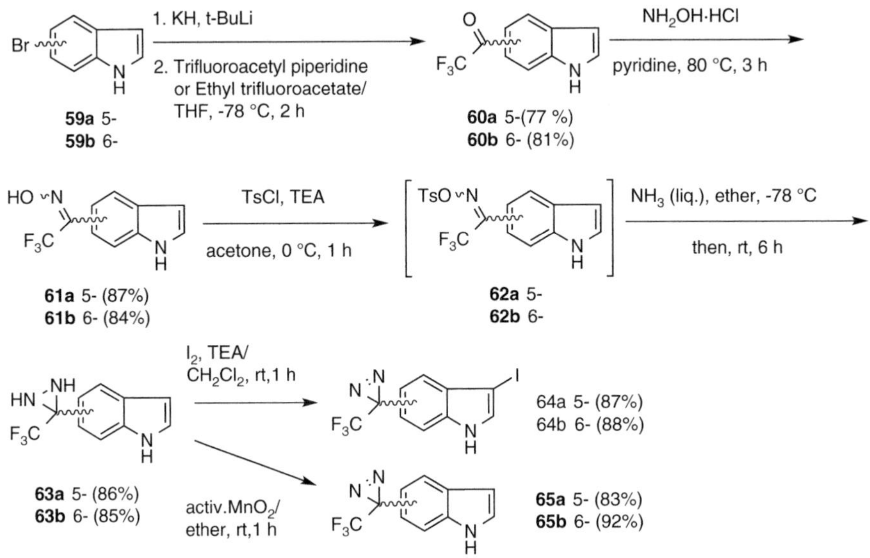

Fig. 6.11 Novel synthesis of 3-(trifluoromethyl)diazirinyl indole **65a** and **65b**

of intermediate products under previous reaction conditions. Tosylation of the oxime indole **61** proceeded with tosyl chloride in acetone and triethylamine at 0 °C. In addition, because the isolation yield of tosyloxime indoles **62** did decrease due to the hydrolysis of that in purification by silica gel chromatography, tosyloxime indoles **62** were not isolated and were directly subjected to conversion to diaziridine indoles **63** with liquid ammonia. Oxidation of diaziridine indoles to diazirines **65** used activated MnO_2 instead of I_2 to avoid a side reaction of iodination at the 3-position on indoles **64** (Fig. 6.11).

6.3.5 3-(Trifluoromethyl)diazirinyl Tryptophan

A derivatization of 3-(trifluoromethyl)diazirinyl indoles can introduce furthermore natural bioactive compounds, whereas it requires mild conditions to avoid the decomposition of 3-(trifluoromethyl)diazirin. Authors have also succeeded in the derivatizations of diazirinyl indoles **65**. Although tryptophan is one of the most significant amino acids and a precursor of various biological activities, it is difficult to apply previous synthetic methods to diazirinyl indoles due to harsh conditions. In the original paper (Snyder and MacDonald 1955), an active species was generated at high temperature in the presence of indole derivatives to construct a tryptophan moiety. The diazirinyl indoles were also decomposed in acetic acid under the reflux conditions. In authors' method, active species was firstly generated from L-serine in acetic anhydride and acetic acid under reflux, followed by addition of diazirinyl indoles at room temperature to get racemic *N*-acetyltryptophans **66** without decomposition of a diazirinyl ring. Optical pure diazirinyl tryptophans **67** were introduced by enzymatic resolution of **66** with L- or D-acylase. Furthermore, authors have also confirmed that D-diazirinyl tryptophan **67c** exhibited more sweetness activity than D-tryptophan. Immediately after publishing authors' paper, Lindel and coworkers (Wartmann and Lindel 2013) also reported the synthesis of a diazirinyl tryptophan with a similar approach. Gramine derivatives **68**, which play a defensive role in plants (Semenov and Granik 2004), were also constructed from diazirinyl indoles with CH_2=NMe_2I at room temperature for 24 h. Indole carbinol **70**, which has been reported to have anticarcinogenic, antioxidant, and antiatherogenic effects (Broadbent and Broadbent 1998), was constructed by the Vilsmeier-Haack reaction of diazirinyl indoles, followed by reduction of an aldehyde moiety **69** (Fig. 6.12).

Furthermore authors have synthesized diazirinyl indole-3-acetic acid (IAA), known as one of the phytohormones, auxin, that regulates cell growth and cell division (Zhao 2010). The diazirinyl indoles firstly reacted with oxalyl chloride, followed by methanolysis to afford methyl indole 3-oxoacetates **71**. The reduction of the α-keto moiety on methyl pyruvate with triethylsilane in TFA, which have no effect on the diazirinyl group (Hashimoto et al. 2003), occurred the side-reduction of alkene moiety between the 2 and 3-position of indoles **72** at the same time. Oxidation at the 2,3-position of **72** with MnO_2 can be smoothly applied to construct **74b**, whereas in **74a** was observed the decomposition of the diazirinyl ring due to its unstability. To construct 5-diazirinyl IAA, the α-carbonyl group of **71** was reduced with $NaBH_4$ to afford α-hydroxy esters **73**, followed by dehydration with P_2I_4 to get **74** without decomposition of diazirinyl group. This method was also able to apply to the 6-diazirinyl derivatives. Finally, by hydrolysis of the methyl ester under alkaline condition, 5- and 6-diazirinyl IAA **75** were obtained (Fig. 6.13).

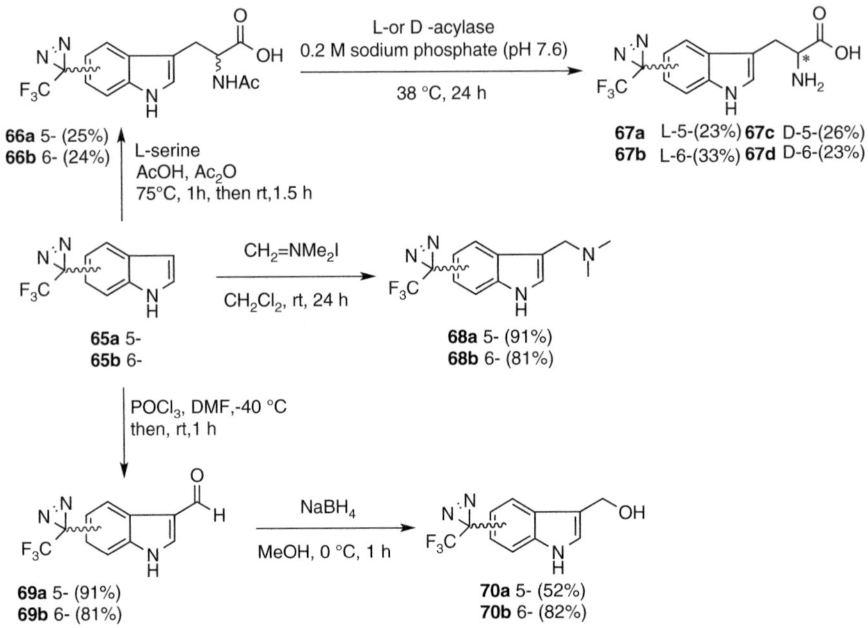

Fig. 6.12 Post-functional synthesis of diazirinyl tryptophan, indole carbinol, and gramine from 3-(trifluoromethyl)diazirinyl indole

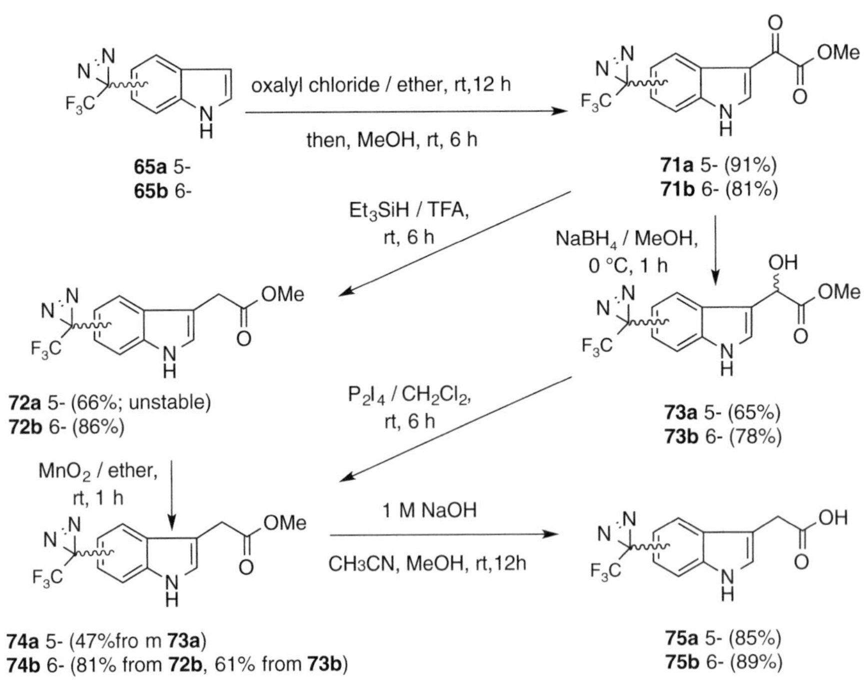

Fig. 6.13 Post-functional synthesis of diazirinyl indole-3-acetic acids from 3-(trifluoromethyl) diazirinyl indole

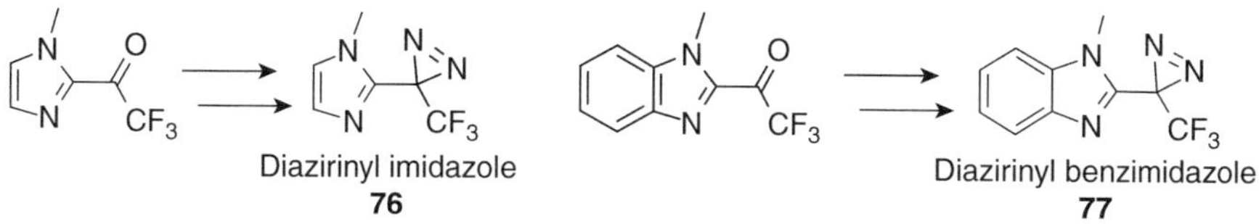

Fig. 6.14 Novel synthesis of 3-(trifluoromethyl)diazirinyl imidazole **76** and benzimidazole **77**

6.3.6 *3-(Trifluoromethyl)diazirinyl Imidazole and Benzimidazole*

Histidine, which is one of the essential amino acids, is a precursor of histamine and carnosine and contains an imidazole moiety. Although an imidazole also plays crucial roles such as chemical materials for medicines, insecticides, and industrial products, 3-(trifluoromethyl)diazirinyl imidazole has not been reported until recently. Lindel and coworkers (Raimer et al. 2014) have developed 3-(trifluoromethyl)diazirinyl imidazoles **76** and benzimidazoles **77** based on the imidazole or benzimidazole skeletons according to general methods (Fig. 6.14).

6.3.7 *3-(Trifluoromethyl)diazirinyl Pyrazole*

In pharmacology, pyrazole derivatives are widely used as analgesic, anti-inflammatory, antipyretic, tranquilizing, so on. Fipronil **78**, containing pyrazole moiety, is a major insecticide acting as a noncompetitive blocker of the γ-aminobutyric acid (GABA) receptor/chloride channel (Cole et al. 1993). Casida and coworkers (Sammelson and Casida 2003) synthesized a 3-(trifluoromethyl)diazirinyl fipronil derivative **82** to elucidate the noncompetitive blocker site of the GABA-gated chloride channel. Their synthetic strategy started from the iodination **79** of pyrazole, followed by coupling with 3,5-dichloro-4-fluorobenzotrifluoride **80**. Trifluoroacetylation of **80** proceeded by treating with *i*-PrMgCl at −78 °C, followed by the reaction with *N*-(trifluoroacetyl) piperidine at room temperature. Diazirinyl fipronil **82** was introduced from the ketone **81** according to general methods (Fig. 6.15).

6.3.8 *3-(Trifluoromethyl)diazirinyl (Benzo)thiophene*

Thiophene and benzothiophene have been used as ingredients for medicines, agrichemicals and coating materials. Oncodazole **83**, containing a thiophene skeleton, is a tubulin binding agent (Hoebeke et al. 1976; Lee et al. 1980; Head et al. 1985) that has potent biological activities for anthelmintic (Raeymaekers et al. 1978) and antifungal (Davidse and Flach 1978) activities. The study of binding mechanism between tubulin and inhibitors can elucidate the mechanism of the mitotic arrest and develop new drugs. Ladd et al. (1988) have developed

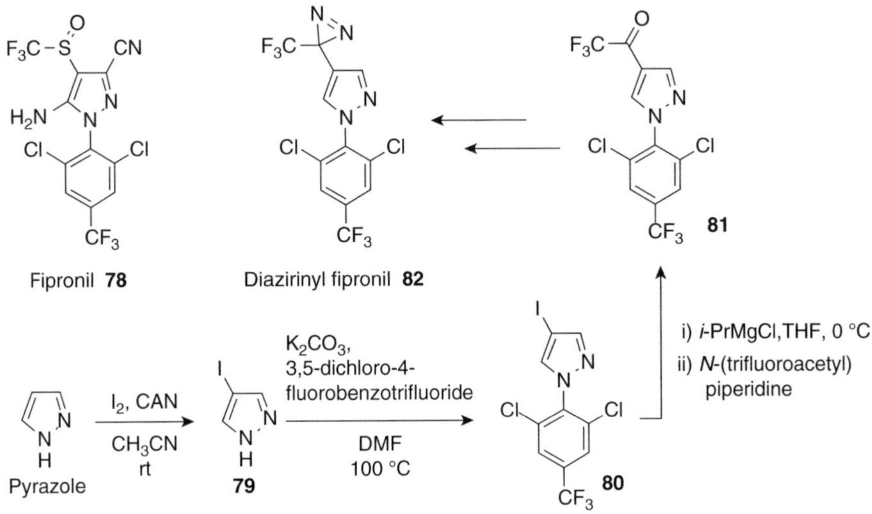

Fig. 6.15 Fipronil **78** and synthesis of 3-(trifluoromethyl)diazirinyl fipronil derivative **82**

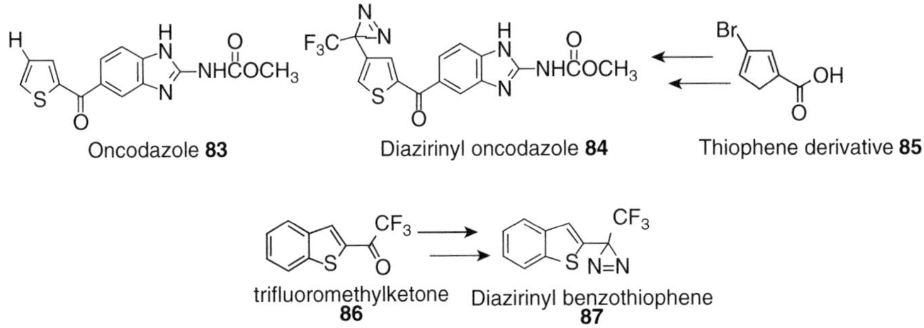

Fig. 6.16 Oncodazole **83** and synthesis of 3-(trifluoromethyl)diazirinyl oncodazole derivative **84** and -(trifluoromethyl)diazirinyl benzothiophene **87**

3-(trifluoromethyl)diazirinyl oncodazole derivative **85** from a thiophene derivative **84** to reveal the interaction of **83** with tubulin. 3-(Trifluoromethyl)diazirinyl benzothiophene **87** has also been derived from trifluoromethylketone **86** by Sheridan and coworkers (Wang and Sheridan 2007) (Fig. 6.16).

6.4 Conclusion

Owing to a great deal of studies for efficient synthesis of diazirinyl compounds, there have been increasing interests in the use of diazrine as a photophore in recent years. In addition to fast and easy preparation of diazirinyl α-aromatic amino acids, diazirinyl heterocyclic bioactive compounds would contribute the identification of their biological processes in the future. Furthermore, because deuterium can be utilized for elucidation of the different mass numbers derived from a mixture of deuterium-labeled and non-labeled photophores, easy preparation of deuterated diazirines can achieve the quick and simple identification of target molecules by MS analysis.

Acknowledgements This research was partially supported by Grants-in-Aid for Young Scientist B from the Japan Society for Promotion of Science (JSPS) (KAKENHI Grant No. 15K16552). The authors thank Prof. Hatanaka for giving us this kind of opportunity.

References

Abiko T, Sekino H (1998) Functional roles of phenylalinine of thymic humoral factor-γ2 in the impaired blastogenic response of uremic T-lymphocytes. Drug Dev Ind Pharm 24:569–572

Al-Omari M, Banert K, Hagedorn M (2006) Bi-3*H*-diazirin-3-yls as precursors of highly strained cycloalkynes. Angew Chem Int Ed 45:309–311

Baldini G, Martoglio B, Schachenmann A, Zugliani C, Brunner J (1988) Mischarging escherichia coli tRNAPhe with L-4′-[3-(trifluoromethyl)-3*H*-diazirin-3-yl]phenylalanine, a photoactivatable analog of phenylalanine. Biochemistry 27:7951–7959

Barfoot CW, Harvey JE, Kenworthy MN, Kilburn JP, Ahmed M, Taylor RJK (2005) Highly functionalised organolithium and organoboron reagents for the preparation of enantiomerically pure alpha-amino acids. Tetrahedron 61:3403–3417

Broadbent TA, Broadbent HS (1998) The chemistry and pharmacology of indole-3-carbinol (indole-3-methanol) and 3-(methoxymethyl)indole. Curr Med Chem 5:469–491

Brunner J, Senn H, Richards FM (1980) 3-Trifluoromethyl-3-phenyldiazirin. J Biol Chem 255:3313–3318

Burkard N, Bender T, Westmeier J, Nardmann C, Huss M, Wieczorek H, Grond S, von Zezschwitz P (2010) New fluorous photoaffinity labels (F-PAL) and their application in V-ATPase inhibition studies. Eur J Org Chem 2010(11):2176–2181

Carrera SS, Kerdelhué J-L, Langenwalter KJ, Brown N, Warmuth R (2005) Inner-phase reaction dynamics: the influence of hemicarcerand polarizability and shape on the potential energy surface of an inner-phase reaction. Eur J Org Chem 2005(11):2239–2249

Chang C-Y, Yang T-K (2003a) Synthesis of optically active α-aminobenzolactam via an oxidative–cyclization reaction. Tetrahedron Asymm 14:2081–2085

Chang C-Y, Yang T-K (2003b) Asymmetric synthesis of ACE inhibitor-benazepril HCl via a bioreductive reaction. Tetrahedron Asymm 14:2239–2245

Cherry K, Lebegue N, Leclerc V, Carato P, Yous S, Berthelot P (2007) Efficient synthesis of 5- and 6-tributylstannylindoles and their reactivity with acid chlorides in the Stille coupling reaction. Tetrahedron Lett 48:5751–5753

Cole LM, Nicholson RA, Casida JE (1993) Action of phenylpyrazole insecticides at the GABA-gated chloride channel. Pestic Biochem Physiol 46:47–54

Das J (2011) Aliphatic diazirines as photoaffinity probes for proteins: recent developments. Chem Rev 111:4405–4417

Davidse LC, Flach W (1978) Blue-fluorescent bovine α-crystallin. Biochim Biophys Acta 534:82–88

Delfino JM, Schreiber SL, Richards FM (1993) Design, synthesis, and properties of a photoactivatable membrane-spanning phospholipidic probe. J Am Chem Soc 115:3458–3474

Dubinsky L, Krom BP, Meijler MM (2013) Diazirine based photoaffinity labeling. Bioorg Med Chem 20:554–570

Eto H, Eguchi C, Kagawa T (1989) Production of L-tryptophan from γ-glutamaldehydic acid. Bull Chem Soc Jpn 62:961–963

Fillion D, Deraët M, Holleran BJ, Escher E (2006) Stereospecific synthesis of a carbene-generating angiotensin II analogue for comparative photoaffinity labeling: Improved incorporation and absence of methionine selectivity. J Med Chem 49:2200–2209

Fishwick CWG, Sanderson JM, Findlay JBC (1994) An efficient route to s-N-(9-fluorenylmethoxycarbonyl)-4-(1-azi-2,2,2-trifluoroethyl) phenylalanine. Tetrahedron Lett 35:4611–4614

Fleet GWJ, Porter RR, Knowles JR (1969) Affinity labelling of antibodies with aryl nitrene as reactive group. Nature 224:511–512

Galardy RE, Craig LC, Jamieson JD, Printz MP (1974) Photoaffinity labeling of peptide hormone binding sites. J Biol Chem 249:3510–3518

Hashimoto M, Hatanaka Y (2008) Recent progress in diazirine-based photoaffinity labeling. Eur J Org Chem 2008(15):2513–2523

Hashimoto M, Hatanaka Y, Sadakane Y, Nabeta K (2002) Synthesis of tag introducible (3-Trifluoromethyl)phenyldiazirine based photoreactive phenylalanine. Bioorg Med Chem Lett 12:2507–2510

Hashimoto M, Hatanaka Y, Nabeta K (2003) A facile synthesis of benzofuran derivatives: a useful synthon for preparation of trypsin inhibitor. Heterocycles 59:395–398

Hatanaka Y, Yoshida E, Nakayama H, Kanaoka Y (1989) Chromogenic diazirine: a new spectrophotometric approach for photoaffinity labeling. Bioorg Chem 17:482–485

Head J, Lee LL-Y, Field DJ, Lee JC (1985) Equilibrium and rapid kinetic studies on nocodazole-tubulin interaction. J Biol Chem 260:11060–11066

High S, Martoglio B, Gorlich D, Andersen SSL, Ashford AJ, Giner A, Hartmann E, Prehn S, Rapoport TA, Dobberstein B, Brunner J (1993) Site-specific photocross-linking reveals that Sec61p and TRAM contact different regions of a membrane-inserted signal sequence. J Biol Chem 268:26745–26751

Hiramatsu T, Guo Y, Hosoya T (2007) 3-Azidodifluoromethyl-3*H*-diazirin-3-yl group as an all-in-one functional group for radioisotope-free photoaffinity labeling. Org Biomol Chem 5:2916–2919

Hoebeke J, Van Nijen G, De Brabander M (1976) Interaction of oncodazole (R 17934), a new antitumoral drug, with rat brain tubulin. Biochem Biophys Res Commun 69:319–324

Kumar NS, Young RN (2009) Design and synthesis of an all-in-one 3-(1,1-difluoroprop-2-ynyl)-3*H*-diazirin-3-yl functional group for photo-affinity labeling. Bioorg Med Chem 17:5388–5395

Ladd DL, Harrsch PB, Kruse LI (1988) Synthesis and tubulin binding of 4′-(1-azi-2,2,2-trifluoroethyl)oncodazole, a photolabile analog of oncodazole. J Org Chem 53:417–420

Lee JC, Field DJ, Lee LLY (1980) Effects of nocodazole on structures of calf brain tubulin. Biochemistry 19:6209–6215

Ma J, Yin W, Zhou H, Liao X, Cook JM (2009) General approach to the total synthesis of 9-methoxy-substituted indole alkaloids: Synthesis of mitragynine, as well as 9-methoxygeissoschizol and 9-methoxy-N_b-methylgeissoschizol. J Org Chem 74:264–273

Masuda K, Koizumi A, Misaka T, Hatanaka Y, Abe K, Tanaka T, Ishiguro M, Hashimoto M (2010) Photoactive ligands probing the sweet taste receptor. Design and synthesis of highly potent diazirinyl D-phenylalanine derivatives. Bioorg Med Chem Lett 20:1081–1083

Melhado LL, Leonard NJ (1983) An efficient synthesis of azidoindoles and azidotryptophans. J Org Chem 48:5130–5133

Mosberg HI, Heyl DL, Haaseth RC, Omnaas JR, Medzihradsky F, Smith CB (1990) Cyclic dermorphin-like tetrapeptides with delta-opioid receptor selectivity. 3. Effect of residue 3 modification on in vitro opioid activity. Mol Pharm 38:924–928

Murai Y, Hatanaka Y, Kanaoka Y, Hashimoto M (2009) Effective synthesis of optically active 3-phenyl-3-(3-trifluoromethyl)diazirinyl bishomophenylalanine derivatives. Heterocycles 79:359–364

Murai Y, Masuda K, Sakihama Y, Hashidoko Y, Hatanaka Y, Hashimoto M (2012) Comprehensive synthesis of photoreactive (3-trifluoromethyl)diazirinyl indole derivatives from 5- and 6-trifluoroacetylindoles for photoaffinity labeling. J Org Chem 77:8581–8587

Murai Y, Wang L, Masuda K, Sakihama Y, Hashidoko Y, Hatanaka Y, Hashimoto M (2013) Rapid and controllable hydrogen/deuterium exchange on aromatic rings of α-amino acids and peptides. Eur J Org Chem 2013(23):5111–5116

Murashige R, Murai Y, Hatanaka Y, Hashimoto M (2009) Effective synthesis of optically active trifluoromethyldiazirinyl homophenylalanine and aroylalanine derivatives with the Friedel-Crafts reaction in triflic acid. Biosci Biotechnol Biochem 73:1377–1380

Nakashima H, Hashimoto M, Sadakane Y, Tomohiro T, Hatanaka Y (2006) Simple and versatile method for tagging phenyldiazirine photophores. J Am Chem Soc 128:15092–15093

Nassal M (1984) 4′-(1-Azi-2,2,2-trifluoroethyl)phenylalanine, a photolabile carbene-generating analog of phenylalanine. J Am Chem Soc 106:7540–7547

Ploug M, Østergaard S, Hansen LBL, Holm A, Danø K (1998) Photoaffinity labeling of the human receptor for Urokinase-type plasminogen activator using a decapeptide antagonist. Evidence for a composite ligand-binding site and a short interdomain separation. Biochemistry 37:3612–3622

Qiu ZH, Lu LH, Jian X, He C (2008) A diazirine-based nucleoside analogue for efficient DNA interstrand photocross-linking. J Am Chem Soc 130:14398–14399

Raeymaekers AHM, Van Gelder JLH, Roevens LFC, Janssen PAJ (1978) Synthesis and anthelminthic activity of alkyl-(5-acyl-1-H-benzimidazol-2-yl) carbamates. Arzneimittelforschung 28:586–594

Raimer B, Wartmann T, Jones PG, Lindel T (2014) Synthesis, stability, and photoreactivity of diazirinyl-substituted N-heterocycles based on indole, benzimidazole, and imidazole. Eur J Org Chem 2014(25):5509–5520

Sammelson RE, Casida JE (2003) Synthesis of a tritium-labeled, fipronil-based, highly potent, photoaffinity probe for the GABA receptor. J Org Chem 68:8075–8079

Semenov BB, Granik VG (2004) Chemistry of N-(1H-indol-3-ylmethyl)-N,N-dimethylamine (Gramine): a review. Pharm Chem J 38:287–310

Smith RAG, Knowles JR (1973) Aryldiazirines. Potential reagents for photolabeling of biological receptor sites. J Am Chem Soc 95:5072–5073

Snyder HR, MacDonald JA (1955) A synthesis of tryptophan and tryptophan analogs. J Am Chem Soc 77:1257–1259

Song Z, Zhang Q (2009) Fluorous aryldiazirine photoaffinity labeling reagents. Org Lett 11:4882–4885

Suchanek M, Radzikowska A, Thiele C (2005) Photo-leucine and photo-methionine allow identification of protein-protein interactions in living cells. Nat Methods 2:261–267

Tippmann EM, Liu WS, Summerer D, Mack AV, Schultz PG (2007) A genetically encoded diazirine photocrosslinker in Escherichia coli. ChemBioChem 8:2210–2214

Vedenyapina MD, Kuznetsov VV, Nizhnikovskii EA, Strel'tsova ED, Makhova NN, Struchkova MI, Vedenyapin AA (2006) Electrochemical synthesis of pentamethylenediazirine. Russ Chem Bull 55:2013–2015

Vila-Perello M, Pratt MR, Tulin F, Muir TW (2007) Covalent capture of phospho-dependent protein oligomerization by site-specific incorporation of a diazirine photo-cross-linker. J Am Chem Soc 129:8068–8069

Wang J, Sheridan RS (2007) A singlet aryl-CF3 carbene: 2-benzothienyl(trifluoromethyl)carbene and interconversion with a strained cyclic allene. Org Lett 9:3177–3180

Wang L, Hisano W, Murai Y, Sakurai M, Muto Y, Ikemoto H, Okamoto M, Murotani T, Isoda R, Kim D, Sakihama Y, Sitepu IR, Hashidoko Y, Hatanaka Y, Hashimoto M (2013) Distinct metabolites for photoreactive l-phenylalanine derivatives in Klebsiella sp. CK6 isolated from Rhizosphere of a wild dipterocarp sapling. Molecules 18:8393–8401

Wang L, Murai Y, Yoshida T, Okamoto M, Masuda K, Sakihama Y, Hashidoko Y, Hatanaka Y, Hashimoto M (2014) Hydrogen/deuterium exchange of cross-linkable α-amino acid derivatives in deuterated triflic acid. Biosci Biotechnol Biochem 78:1129–1134

Wang L, Murai Y, Yoshida T, Ishida A, Masuda K, Sakihama Y, Hashidoko Y, Hatanaka Y, Hashimoto M (2015) Alternative one-pot synthesis of (trifluoromethyl)phenyldiazirines from

tosyloxime derivatives: application for new synthesis of optically pure diazirinylphenylalanines for photoaffinity labeling. Org Lett 17:616–619
Wartmann T, Lindel T (2013) L-Phototryptophan. Eur J Org Chem 2013(9):1649–1652
Xie Y, Lou R, Li Z, Mi A, Jiang Y (2000) DPAMPP in catalytic asymmetric reactions: enantioselective synthesis of l-homophenylalanine. Tetrahedron Asymm 11:1487–1494
Xu Q, Wang G, Wang X, Wu T, Pan X, Chan ASC, Yang T (2000) The synthesis of L-(+)-homophenylalanine hydrochloride. Tetrahedron Asymm 11:2309–2314
Yamada M, Nagashima N, Hasegawa J, Takahashi S (1998) A highly efficient asymmetric synthesis of methoxyhomophenylalanine using michael addition of phenethylamine. Tetrahedron Lett 39:9019–9022
Zhao Y (2010) Auxin biosynthesis and its role in plant development. Annu Rev Plant Biol 61:49–64
Zhao H, Luo RG, Wei D, Malhotra SV (2002) Concise synthesis and enzymatic resolution of L-(+)-homophenylalanine hydrochloride. Enantiomer 7:1–3

Chapter 7
Photoreactive Biomacromolecules: Installation of Photoreactive Units and Applications for Analyzing Biological Interfaces

Yutaka Sadakane, Isao Ohtsuka, and Yasumaru Hatanaka

Abstract Photoaffinity labeling enables analysis of weak biological interactions through formation of a covalent bond between ligands and receptor proteins upon photolysis of photoreactive groups. The photoreactive group, 3-trifluoromethyl-3-phenyldiazirine, has many advantages over other photoreactive groups like phenylazide and benzophenone, but its use is limited due to the difficulty to incorporate this photophore within macromolecules. The chemoselective ligation of versatile photoreactive units to macromolecules is a reliable solution for this difficulty; this approach easily attaches the useful 3-trifluoromethyl-3-phenyldiazirine photophore to biomacromolecules. Photoreactive bromide unit bearing 3-trifluoromethyl-3-phenyldiazirine is chemoselectively attached to phosphorothioate linkages of modified oligonucleotides to prepare photoreactive DNA, which is then used in a photoaffinity electrophoretic mobility shift assay for analyzing the proteins bound to a specific DNA motif. This review focuses on the preparation of photoreactive biomacromolecules such as peptides, proteins, oligosaccharides, and DNA based on our photoreactive unit technique and describes their application as novel approaches for the photoaffinity labeling of biomacromolecules.

Keywords Agglutinin • Benzophenone • Calmodulin • Chemoselective reaction • Phosphorothioate • POU domain • 3-Trifluoromethyl-3-phenyldiazirine

---

Y. Sadakane (✉)
School of Pharmaceutical Sciences, Suzuka University of Medical Science, Suzuka 513-8670, Japan
e-mail: sadapon@suzuka-u.ac.jp

I. Ohtsuka
School of Pharmaceutical Sciences, Kyushu University of Health and Welfare, Nobeoka 882-8508, Japan

Y. Hatanaka
Graduate School of Medicine and Pharmaceutical Sciences, University of Toyama, Toyama 930-0194, Japan

Y. Hatanaka, M. Hashimoto (eds.), *Photoaffinity Labeling for Structural Probing Within Protein*, DOI 10.1007/978-4-431-56569-7_7

7.1 Introduction

Physiological affinity binding is composed of various weak interactions such as hydrogen bonds, ionic bonds, and Van der Waals interactions. Thus, direct analysis of biological interactions such a ligand and receptor protein interfaces requires mild conditions to avoid disturbing the native weak interactions. Photoaffinity labeling enables biological partners to crosslink to each other through a covalent bond. Using a ligand bearing the appropriate photoreactive group, biological complexes remain bound to each other even under denaturing conditions. The photoreactive group plays the role of a fishhook during fishing, that is, the photoreactive group catches the receptor proteins strongly, whereas native ligands catch the protein weakly, as fishing with a bait coiled up by a fishline (Fig. 7.1). Higher-order protein structure is absolutely necessary for maintaining biological complexes. However, crosslinking between ligands and receptors keeps the complexes intact even under conditions

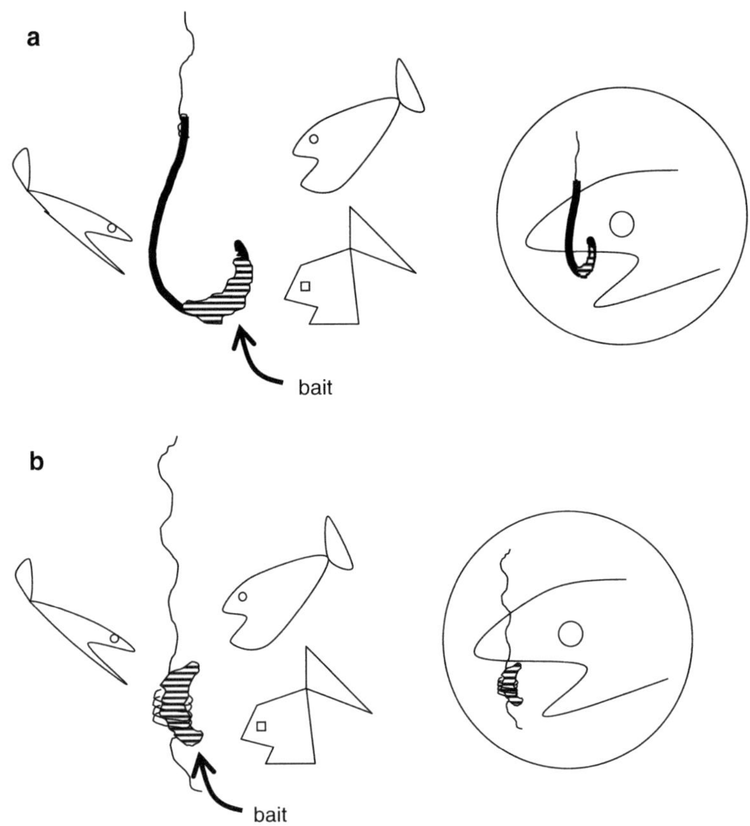

Fig. 7.1 Role of photoreactive groups. (**a**) The photoreactive group functions as a fishhook, onto which ligands are placed as fish bait. The fish does not detach from the bait. (**b**) Affinity selection with native ligands seems like fishing without a fishhook. It is difficult to catch the target fish

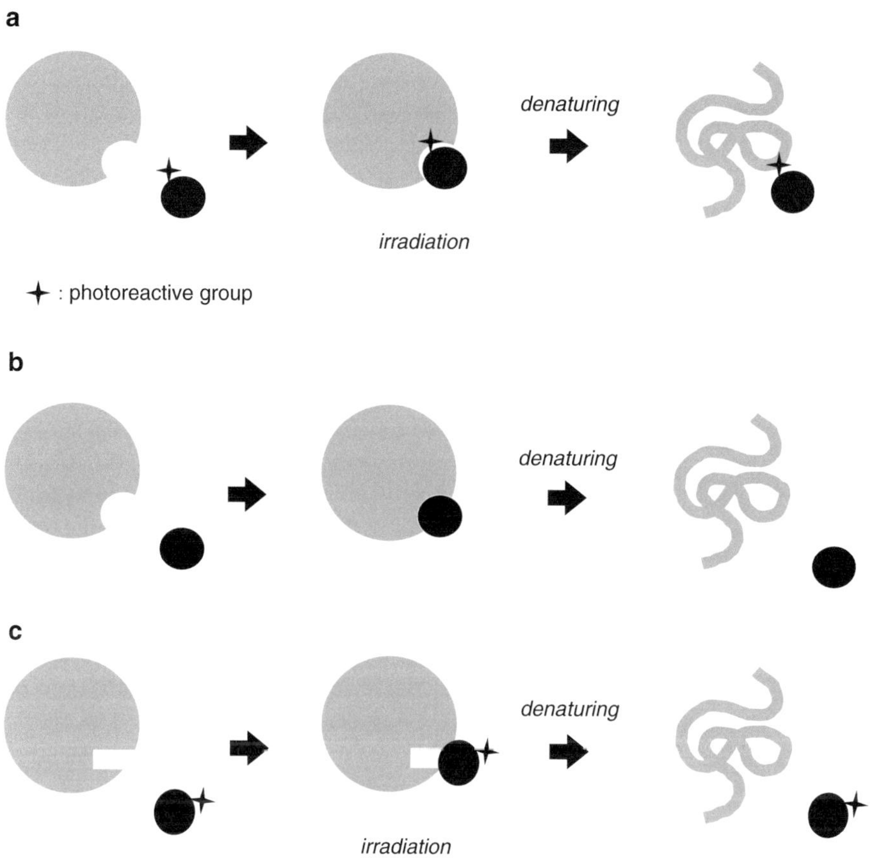

Fig. 7.2 Advantages of photoaffinity labeling. (**a**) A ligand bearing a photoreactive group is able to form a covalent bond with its receptor protein by UV irradiation; thus, the complex is maintained even after denaturation. (**b**) A native ligand is detached from the receptor protein after denaturation. (**c**) Nonspecifically bound ligand is detached from the protein after denaturation

that destroy the protein structure (Fig. 7.2). Thus, photoaffinity labeling makes it possible to analyze ligand-receptor protein interactions under denaturing conditions. Since analysis under denaturing conditions often provides high sensitivity and resolution due to low background, photoaffinity labeling technology expands the assay spectrum of biological functions and improves the analytical quality.

7.2 Photoreactive Groups

Azide, carbonyl, and diazirine are common photoreactive groups that have been used for analyzing biological interfaces, and choice of photoreactive group is an important factor in designing efficient photoaffinity labeling assays (Hatanaka and

Sadakane 2002; Sadakane and Hatanaka 2006; Hatanaka 2015). Phenylazide is the most widely used photoreactive group for analyzing biological interactions because of its easy synthesis and commercial availability. However, nitrene, generated by photolysis of azide, is less reactive than carbene, generated by photolysis of diazirine, and produces unstable nitrogen-heteroatom crosslinking (Maliarik and Goldstein 1988; Kemnitz et al. 1998). Benzophenone is a well-known compound that contains a photoreactive carbonyl group and is also widely used for studying biological interactions. The carbonyl group generates a reactive triplet state that initiates the crosslink, and the ground state can be regenerated from the excited state when a suitable C–H bond is absent near the carbonyl group (Galardy et al. 1973). Higher crosslinking efficiency is obtained by such repeated excitation in comparison with other photoreactive groups, which is a unique characteristic of carbonyl crosslinking. Wittelsberger et al. (2006) reported that methionine (Met) residues in proteins act as a magnet for photoaffinity labeling with benzophenone. Various amino acid residues in the receptor region of parathyroid hormone (PTH) receptor 1 (PTH1R) were replaced with Mets, and the crosslinking positions were determined. This experiment showed that the crosslinking position is shifted by the presence of Met regardless of the distance between benzophenone and amino acid target, that is, PTH bearing benzophenone reacted with Met residues through the range of almost 10 amino acids of PTH1R. The magnetic effect of Met seems to interfere with determination of the accurate binding site between a ligand and a receptor. Actually, some reports show that Met residues are a contact site for benzophenone crosslinking (O'Neil et al. 1989; Kage et al. 1996; Bremer et al. 2001; Sachon et al. 2003).

Diazirine generates a highly reactive carbene intermediate after photolysis, and the photoreactive group 3-trifluoromethyl-3-phenyldiazirine is commonly used for photoaffinity labeling to reduce undesirable side reactions (Brunner et al. 1980; Brunner 1993; Fleming 1995). The chemical advantages and limitations of 3-trifluoromethyl-3-phenyldiazirine have been reviewed previously (Platz 1995; Hatanaka et al. 1996; Hatanaka and Kanaoka 1998). Several practical examinations have shown that 3-trifluoromethyl-3-phenyldiazirine is an excellent photoreactive group for photoaffinity labeling of biomacromolecules. Photoreactive carbohydrates bearing 3-trifluoromethyl-3-phenyldiazirine successfully aided in analysis of glucose transporter distribution where photoreactive carbohydrates bearing phenylazide or benzophenone failed (Yang et al. 1992). The structure of RNA polymerase III (Pol III) was analyzed by attaching nucleotides with various photoreactive groups like phenylazide, benzophenone, and 3-trifluoromethyl-3-phenyldiazirine, and contacts of Pol III to DNA could be analyzed by the DNA bearing 3-trifluoromethyl-3-phenyldiazirine but not by the DNA bearing other photoreactive groups (Tate et al. 1998). The photoreactive pentapeptide thymopentin bearing 3-trifluoromethyl-3-phenyldiazirine could be crosslinked to target proteins by a shorter period of UV irradiation compared to thymopentin bearing benzophenone (Weber and Beck-Sickinger 1997). These reports show that the use of 3-trifluoromethyl-3-phenyldiazirine as a photoreactive group seems to be promising for achieving efficient photoaffinity labeling. Dubinsky et al. (2012) reviewed various applications of 3-trifluoromethyl-3-phenyldiazirine for analyzing biological functions.

7.3 Photoreactive Units

In photoaffinity labeling experiments, the use of 3-trifluoromethyl-3-phenyldiazirine is limited for analyzing biological functions compared to other photoreactive groups like phenylazide and benzophenone. Difficult synthesis and commercial unavailability are its main limitations although 3-trifluoromethyl-3-phenyldiazirine also has several advantages for photoaffinity labeling. We developed the "photoreactive unit technique," which is a solution for the difficult preparation of photoreactive biomacromolecules integrated with 3-trifluoromethyl-3-phenyldiazirine (Sadakane 2007; Sadakane and Hatanaka 2008; Hatanaka 2015). The photoreactive units can convert peptides, proteins, and DNA to photoreactive ligands by a chemoselective reaction (Fig. 7.3). Biomacromolecules possess specific reactive groups, which are used as target groups to introduce functional molecular tags. In proteins, thiol groups of cysteine (Cys) residues and amino groups of lysine residues are widely used as chemoselective targets for introducing fluorescent compounds or biotin tags. The aldehyde groups at reducing ends of carbohydrates are also used for modifying functional compounds. Additionally, artificial target groups can be easily incorporated into specific positions of biomacromolecules during their synthesis because biomacromolecules are constructed of building blocks such as amino acids and nucleotides. For example, specific phosphodiester bonds of oligonucleotides can be replaced with phosphorothioate groups, which can be used for introducing functional compounds. Chemoselective ligations described above are the foundations of designing photoreactive units to prepare photoreactive biomacromolecules.

A photoreactive unit bearing methane thiosulfonate **1** can be easily introduced into Cys residues of proteins and peptides by formation of a disulfide bond. Treatment with appropriate reducing agents separates the ligands from 3-trifluoromethyl-3-phenyldiazirine. As a result, a small unit of photolyzed compound remains on the crosslinked molecules, and then the crosslinked position can be determined by mass spectrometry. Photoreactive amino acids protected by fluorenylmethoxycarbonyl (Fmoc) groups **2** have been developed as building blocks for solid-phase synthesis by an automated peptide synthesizer (Nakashima et al. 2006). The unprotected form of the three-membered diazirine ring is stable enough to withstand various peptide synthesis reactions. Thus, an automated peptide synthesizer can easily prepare μmol amounts of photoreactive peptides. Photoreactive units bearing bromide **3** can be integrated into modified oligonucleotides where one phosphodiester linkage is replaced with a phosphorothioate linkage, and thus photoreactive DNA bearing 3-trifluoromethyl-3-phenyldiazirine is easily prepared

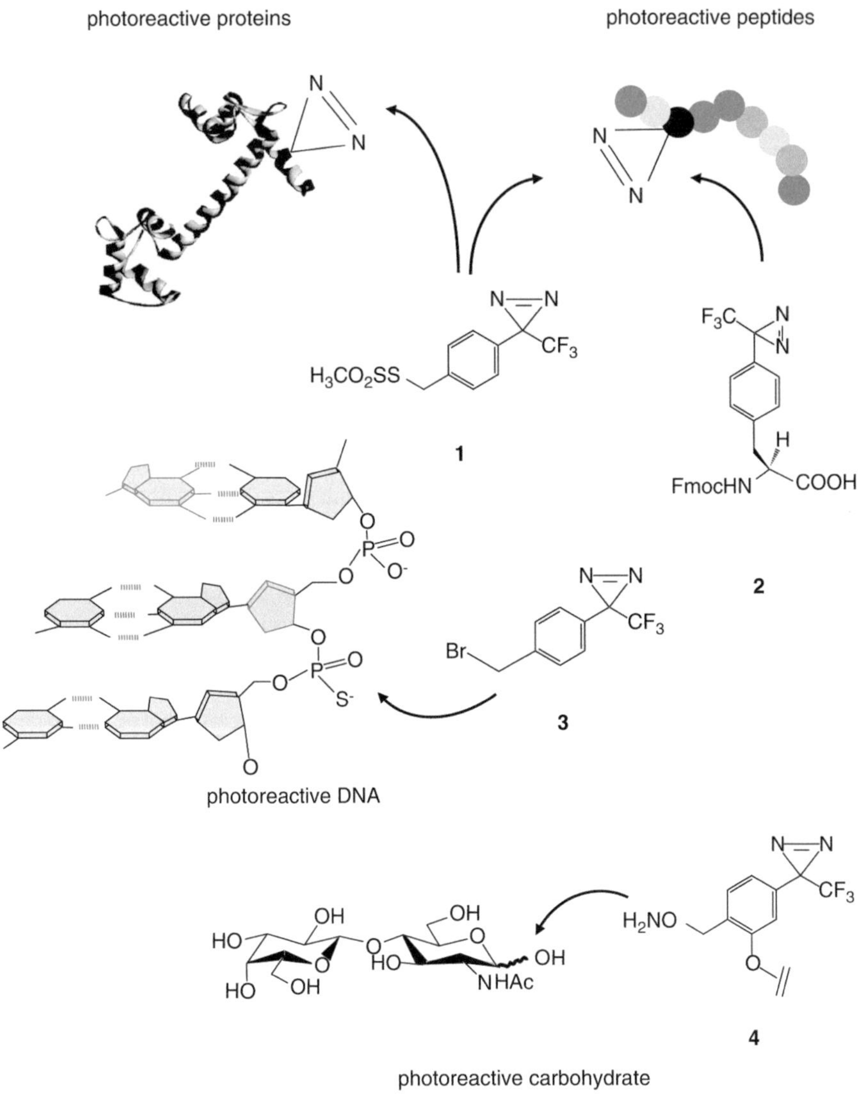

Fig. 7.3 Photoreactive units for preparing photoreactive proteins, peptides, and DNA and multifunctional photoreactive tools for carbohydrates. Photoreactive proteins are prepared with photoreactive unit ***1***, *S*-[4-(3-trifluoromethyl-*3H*-diazirin-3-yl)-benzyl]-methane thiosulfonate, and photoreactive peptides are prepared with ***1*** or ***2***, *N*-(9-fluorenylmethoxycarbonyl)-4-[3-(trifluoromethyl)-*3H*-diazirin-3-yl]-L-phenylalanine (Fmoc-TmdPhe). Photoreactive DNA is prepared with ***3***, 4-[3-(trifluoromethyl)-*3H*-diazirin-3-yl]benzyl bromide using DNA containing phosphorothioate linkages. Photoreactive carbohydrates are prepared with the multifunctional photoreactive tool ***4***, 3-trifluoromethyl-3-phenyldiazirine bearing an aminooxy group (commercially available, named AffiLight-CHO), which can bind to reducing ends of natural carbohydrates

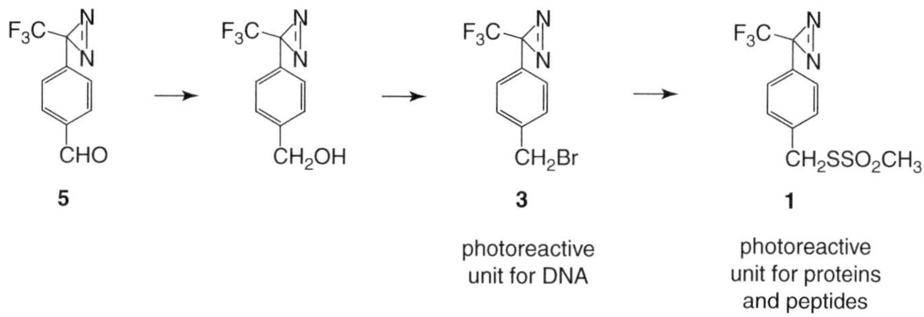

Fig. 7.4 Synthesis of photoreactive units ***1*** and ***3***. 4-[3-(trifluoromethyl)-*3H*-diazirin-3-yl] benzaldehyde ***5*** is the starting compound

(Sadakane and Hatanaka 2016). The photoreactive unit technique provides an easy way for preparing photoreactive peptides, proteins, and DNA. Biochemists who use photoaffinity labeling can prepare desired photoreactive biomacromolecules using their physiological macromolecules of interest. The synthesis of photoreactive units has also been developed. 3-Trifluoromethyl-3-phenyldiazirine is synthesized from commercially available 2,2,2-trifluoro-1-phenyl-ethanone on a large scale (Brunner et al. 1980). However, this compound does not have any chemical tether for connection to appropriate ligands. Nakashima et al. (2006) developed a synthesis method for integrating an aldehyde tether by a Friedel-Crafts reaction using an appropriate Lewis acid. The aldehyde compound **5** can be easily converted to a bromide compound **3** (Nassal 1984) or a thiosulfonate compound **1** (Kaneda et al. 2003), which are photoreactive units for preparing photoreactive DNA and peptides/proteins, respectively (Fig. 7.4).

The multifunctional photoreactive tool **4** possesses three functional groups: an aminooxy group for chemoselective ligation onto natural carbohydrates, a biotin tag for detection or identification of crosslinked proteins, and 3-trifluoromethyl-3-phenyldiazirine for photoaffinity labeling (Hatanaka et al. 2000). The conversion reaction takes advantage of the oxime formation between the aminooxy group and the aldehyde group on the reducing end of a carbohydrate. This multifunctional tool **4** makes it possible to prepare photoreactive carbohydrates by mixing with natural carbohydrates in mild conditions. One of the few disadvantages for our multifunctional tool **4** is that the reducing end of modified carbohydrates has been changed to an open-ring form, whose structure cannot often be recognized by carbohydrate-binding proteins. Especially in the case of short oligosaccharides such as disaccharides, this disadvantage of our tool **4** is a critical drawback for analyzing carbohydrate-binding interfaces. We synthesized a disaccharide attached to an aldehyde linker to solve this critical problem (Ohtsuka et al. 2011). Lactose bearing an aldehyde linker **6** can be synthesized from lactose bearing allyl groups after treatment with O_3 gas, which enables preparation of the multifunctional photoreactive probe attached to closed-ring lactose (Fig. 7.5). We compared performances between multifunctional photoreactive probes that were attached to closed-ring **7** or open-ring lactose **8** using various carbohydrate-binding proteins like soybean agglutinin (SBA), *Erythrina crista-galli* agglutinin (ECA), and peanut agglutinin (PNA). The multifunctional carbohydrate probe was mixed with agglutinins and then irradiated for 30 min on ice with long-wavelength UV light to promote the formation of cross-links. The probe-agglutinin complexes were analyzed by SDS-PAGE. After transferring to polyvinylidene difluoride membranes, the complexes were visualized by chemiluminescence of a luminol reaction using a streptavidin-horseradish peroxidase (HRP) conjugate. Photoaffinity labeling using PNA is shown in Fig. 7.5. The photoreactive probe attached to closed-ring lactose **8** bound to PNA more strongly

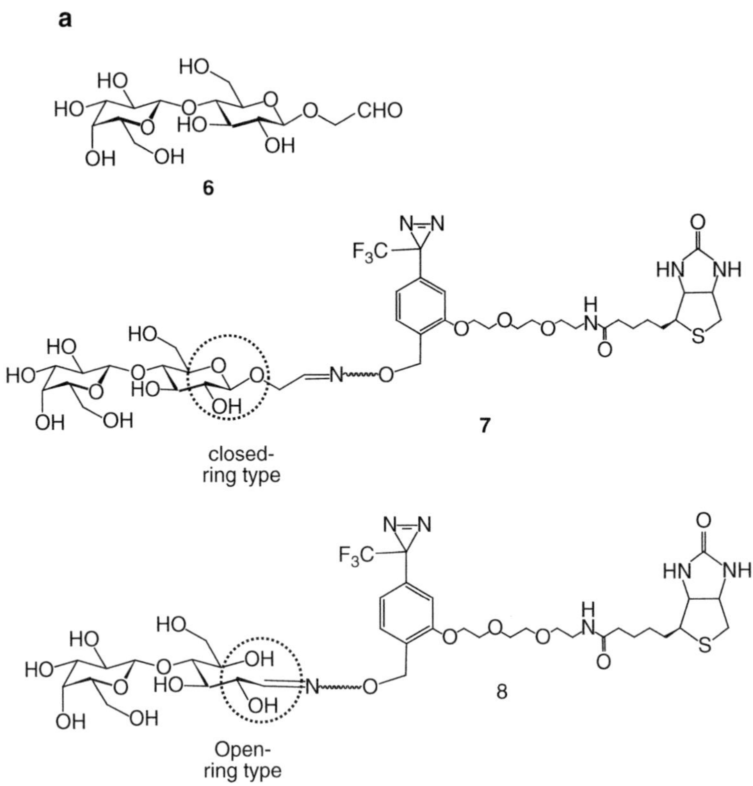

Fig. 7.5 Photoaffinity labeling of carbohydrate-binding proteins using multifunctional tool ***4***. (**a**) The D-lactose analog ***6*** (2-oxyoethyl β-D-garactopyranosyl-(1 → 4)-β-D-glucopyranoside) is used for preparing the photoreactive probe of closed-ring lactose ***7***. The open-ring lactose probe ***8*** was prepared from natural D-lactose. (**b**) Photoaffinity labeling of peanut agglutinin with photoreactive probes bearing open-ring lactose ***8*** (lanes 1 and 2) and closed-ring lactose ***7*** (lanes 3 and 4). Competitive inhibitor experiments were performed with excess amount of lactose (lanes 2 and 4). The carbohydrate-agglutinin complex was visualized by chemiluminescence using a streptavidin-HRP conjugate (indicated by *arrow*). M1: molecular weight markers, M2: biotinylated molecular weight markers

Fig. 7.5 (continued)

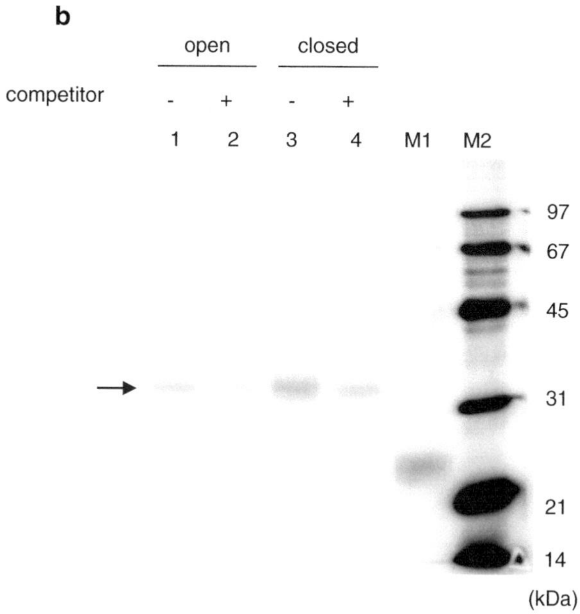

than open-ring lactose **7** did. The complex formation of closed-ring lactose **8**, but not open-ring lactose **7**, was inhibited by addition of excess D-lactose. Similar results were obtained from the experiments using SBA and ECA. Since the agglutinins SBA, ECA, and PNA are reported to bind specifically to D-galactose and D-galactosamine, these results correspond with the binding properties of the agglutinins.

We successively synthesized three types of D-lactosamine, the β-bonds of which consisted of 1 → 3, 1 → 4, and 1 → 6 (Fig. 7.6), bearing aldehyde linkers to prepare the photoreactive probe attached to closed-ring carbohydrates (Ohtsuka et al. 2014). The synthetic D-lactosamines with different bonds, **9**, **10**, and **11** were attached to the multifunctional photoreactive tool **4** by a one-step chemoselective ligation, and binding efficiency to ECA was examined. The β(1 → 4) form of D-lactosamine **10** was alone strongly bound to the ECA, and a competition assay using excess D-lactosamine inhibited complex formation. The other types of D-lactosamine, **9** and **11**, were faintly bound to ECA, and no significant differences in binding intensity were observed in the presence or absence of excess competitor. The multifunctional photoreactive probe attached to the closed-ring carbohydrate makes it possible to analyze the binding efficiency of different glycosidic bonds. Photoreactive compounds bearing an aminooxy group and 3-trifluoromethyl-3-phenyldiazirine are also used for preparing photoreactive solid matrix immobilized carbohydrates to screen for carbohydrate-binding proteins (Sadakane and Hatanaka 2004). We prepared carbohydrate-immobilized photoreactive matrices and examined the photoaffinity capture of carbohydrate-binding proteins on the surface of the matrices.

Fig. 7.6 Three types of disaccharide derivatives of D-galactose and D-glucosamine used to prepare the closed-ring type carbohydrates photoreactive probes. β(1 → 3) ***9***, 2-oxyoethyl β-D-galactopyranosyl-(1 → 3)-2-deoxy-2-*N*-acetyl-β-D-glucopyranoside; β(1 → 4) ***10***, 2-oxyoethyl β-D-galactopyranosyl-(1 → 4)-2-deoxy-2-*N*-acetyl-β-D-glucopyranoside; and β(1 → 6) ***11***: 2-oxyoethyl β-D-galactopyranosyl-(1 → 6)-2-deoxy-2-*N*-acetyl-β-D-glucopyranoside

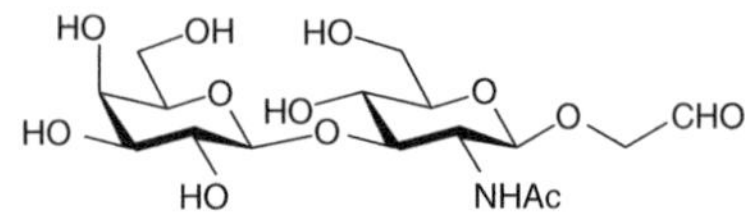

β(1→3) **9**

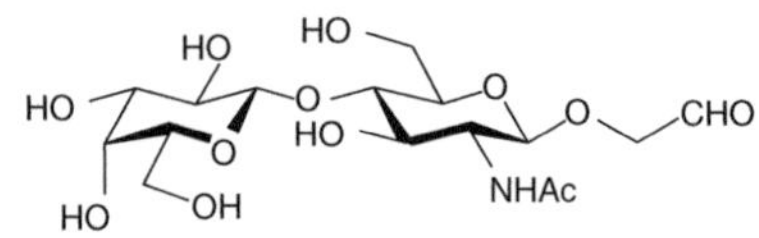

β(1→4) **10**

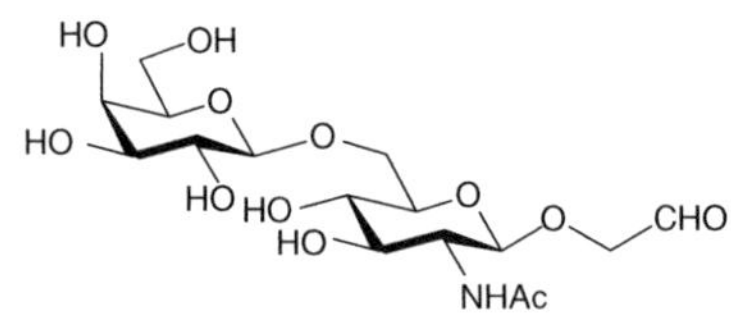

β(1→6) **11**

Photoaffinity labeling using photoreactive carbohydrates has increasingly become one of the major techniques for studying the biology of carbohydrate interfaces (Lauc et al. 2000; Yu et al. 2012; Sakurai et al. 2015).

7.4 Photoreactive Peptides and Proteins

Since peptides are easily built by automated solid-phase peptide synthesizer systems, various types of photoreactive amino acids have been developed to prepare photoreactive peptides, including the phenylazide form (Schwyzer and Caviezel 1971; Escher and Schwyzer 1974), benzophenone form (Kauer et al. 1986; Wilson et al. 1997), and the 3-trifluoromethyl-3-phenyldiazirine form (Nassal 1984; Shih and Bayley 1985; Nakashima et al. 2006). One of the key considerations for photoreactive amino acids is their stability during standard solid-phase synthesis due to the easy preparation of photoreactive peptides. Phenylazide is not stable enough for solid-phase synthesis; thus, multistep procedures are required to utilize phenylazide-type amino acids as building blocks (Escher et al. 1978). Benzophenone-bearing amino acids have been utilized as building blocks for solid-phase synthesis (Kauer et al. 1986; Egnaczyk et al. 2001). *p*-Benzoyl-L-phenylalanine was substituted for the 3rd residue (tryptophan (Trp)) in the 17-residue calmodulin (CaM)-binding peptide (BP), the sequence of which is KLWKKLLKLLKKLLKLG. This Trp residue is important for the binding interface between CaM and CaM-BP (O'Neil et al. 1987; Meador et al. 1992). CaM-BP bearing benzophenone successfully aided in determination of the crosslinked amino acids in the CaM protein by repetitive HPLC manipulation to isolate crosslinked peptides (O'Neil et al. 1989). *p*-Benzoyl-L-phenylalanine is now widely used for preparing photoreactive peptides because of its commercial availability and good stability for solid-phase synthesis. Photoreactive Alzheimer β-amyloid peptide, in which residue 4 (phenylalanine (Phe)) was replaced with *p*-benzoyl-L-phenylalanine, has been synthesized and used to analyze the inter-strand contact region (Egnaczyk et al. 2001), and photoreactive vasoactive intestinal peptide has been synthesized to analyze the interaction with its receptor protein (Tan et al. 2003) in this manner.

Nakashima et al. (2006) reported the use of amino acids bearing 3-trifluoromethyl-3-phenyldiazirine as building blocks for solid-phase synthesis. The synthetic strategy for the photoreactive amino acid L-4-(3-trifluoromethyl-3*H*-diazirin-3-yl)-Phe (TmdPhe) was developed and applied to prepare the 17-residue photoreactive CaM-BP. We have taken an alternative approach for preparing photoreactive peptides using a photoreactive unit bearing methane thiosulfonate **1** (Sadakane 2007). The 17-residue CaM-BP was used for the photoaffinity labeling of CaM protein as a model experiment (Fig. 7.7). The Trp residue at the third position of CaM-BP was replaced with a Cys, and then the photoreactive unit **1** was mixed with the protein at neutral pH. The peptide integrated with photoreactive unit **1** via a disulfide bond and then was purified and used as the photoreactive CaM-BP. The crosslinking efficiency between CaM and CaM-BP was affected by the presence or absence of calcium ions, and trifluoperazine, a specific inhibitor of CaM, inhibited complex formation in a concentration-dependent manner. The crosslinking efficiency of photoaffinity labeling reached its maximum within 2 min when CaM-BP was bearing 3-trifluoromethyl-3-phenyldiazirine, but it took over 45 min when bearing benzophenone. This shows that benzophenone requires a longer period of UV irradiation than 3-trifluoromethyl-3-phenyldiazirine. Since long periods of UV exposure may cause damage to biomacromolecules, 3-trifluoromethyl-3-phenyldiazirine is superior for photoaffinity labeling of CaM.

For preparing photoreactive proteins, a technique for site-specific incorporation of the photoreactive compound into proteins is required. The sulfhydryl group of Cys residues within proteins is often used for integrating bromide reagents or thiosulfonate reagents (Kenyon and Bruice 1977). Such chemoselective reactions are used for preparing photoreactive proteins. Hixson and Hixson (1975) demonstrated the integration of *p*-azidophenacyl bromide into the sulfhydryl group of specific Cys residues of D-glyceroaldehyde-3-phosphate dehydrogenase (GAPDH). A catalytic free sulfhydryl group in the active site of GAPDH was the target for modification with the photoreactive agent. Kaneda et al. (2003) incorporated the photoreactive unit **1** at the same site in GAPDH as *p*-azidophenacyl bromide and developed a photoaffinity capture system for efficient discovery of small ligands. Since Cys residues can be artificially integrated by site-directed mutagenesis, the photoreactive compounds can be attached at the desired position of target proteins. Pendergrast et al. (1992) used photoaffinity labeling to study protein-DNA interactions by artificially integrating Cys residues into proteins. Catabolite activator protein (CAP), a helix-turn-helix motif sequence-specific DNA-binding protein, has a unique free

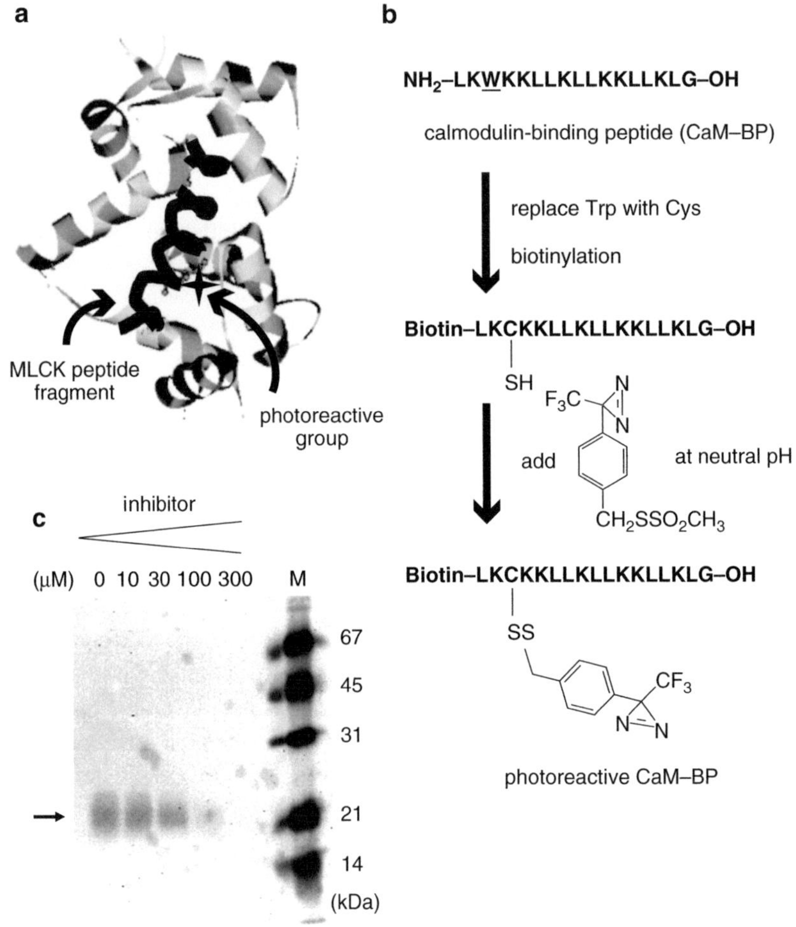

Fig. 7.7 Photoaffinity labeling of CaM protein. (**a**) The crystal structure of CaM protein bound to a peptide fragment of the CaM-binding region from myosin light chain kinase (MLCK). The position integrated with photoreactive unit ***1*** corresponds to the Trp residue on the peptide fragment of MLCK. (**b**) Preparative strategy of photoreactive CaM-BP. The Trp residue of CaM-BP was replaced with a Cys and biotin was attached to the N-terminus of the peptide. Photoreactive unit ***1*** was mixed with the peptide at a neutral pH and the photoreactive CaM-BP was purified by HPLC. (**c**) Photoaffinity labeling of CaM using the photoreactive peptide bearing unit ***1***. Competitive inhibition with trifluoperazine, a specific inhibitor of CaM, was performed. Various concentrations of trifluoperazine (indicated above the gel image) were mixed with CaM, and the photoaffinity labeling experiment was performed by irradiation at 0 °C for 2 min with 30-W, 360-nm UV light. Specimens were separated by SDS-PAGE and the crosslinked CaM protein was visualized by chemiluminescence using a streptavidin-HRP conjugate

sulfhydryl group on a Cys at residue 10. To study details of the interface, recombinant CAP, where residues 2 and 10 were replaced with Cys and Ser residues, respectively, was prepared. *p*-Azidophenacyl bromide was integrated into either of these CAP positions for preparing two photoreactive CAPs, and the orientation of the DNA-binding motif in the complex of CAP and DNA was examined. Cleary et al.

(1997) studied protein-DNA interaction using modified proteins artificially integrated with phenylazide. The agent, *p*-azidophenacyl bromide, was integrated into the Cys residues of the octamer-binding transcription factor 1 (Oct-1) protein where various amino acids had been replaced with Cys residues. Studies of photoreactive Oct-1 protein revealed its conformational flexibility during binding to specific DNA sequences.

The other strategy for site-specific incorporation of a photoreactive amino acid into proteins is using a cell-free translation system with suppressor tRNAs chemically charged with a photoreactive amino acid. The cell-free translation system is supplemented with an amber suppressor tRNA charged with photoreactive amino acids bearing 3-trifluoromethyl-3-phenyldiazirine, TmdPhe (High et al. 1993). To examine the details of protein translocation machinery, TmdPhe was incorporated into nascent chains upon recognition of the UAG stop codon in mRNA. The role of a novel factor, YidC, was found to be an essential protein for both Sec-dependent and Sec-independent insertion of inner membrane proteins (Samuelson et al. 2000; Scotti et al. 2000; Driessen et al. 2001). The photoreactive proteins integrated with photoreactive groups at specific sites promote the structural analysis of various biological interfaces as well as the discovery of medicinal leads.

7.5 Photoreactive DNA

Photoreactive DNAs are useful tools for analyzing the interaction between nucleotides and proteins. Two strategies have been reported for preparing photoreactive DNA: one is the use of photoreactive nucleobases as substitutes for the corresponding normal bases, and the other is the integration of a photoreactive unit onto a specific site of the DNA where one or more phosphodiester linkages are replaced by phosphorothioate linkages. In the former strategy, various types of photoreactive nucleotide analogues have been utilized. Photoreactive cCTP analogs bearing phenylazide, 4-[*N*-(*p*-azidobenzoyl)-2-aminoethyl]-dCTP, were synthesized (Lannutti et al. 1996) and applied to analyze the chromatin remodeling complex SWI/SNF, whose molecular mass is over 2000 kDa (Sengupta et al. 2001). Photoreactive dUTP analogs bearing 3-trifluoromethyl-3-phenyldiazirine were synthesized and used for photoaffinity labeling human immunodeficiency virus 1 reverse transcriptase (Yamaguchi and Saneyoshi 1996) and DNA polymerase β (Yamaguchi et al. 1997). Zofall and Bartholomew (2000) reported two novel photoreactive nucleotides (Fig. 7.8), in which phenylazide or 3-trifluoromethyl-3-phenyldiazirine was attached to the N-6 amino group of adenine by a short tether, because dATP bearing a biotin tag at the N-6 amino group did not interfere with its enzymatic incorporation into DNA. Adenine has other potential attaching positions, C-2 and C-8. However, photoreactive nucleotides conjugated with phenylazide analogs at C-2 and C-8 positions were relatively difficult to incorporate into DNA because these modifications interfered with base pairing (Moore et al. 1996) and resulted in unsuitable conformations (Costas et al. 2000), respectively. The photoreactive group 3-trifluoromethyl-3-phenyldiazirine is widely used for preparation of

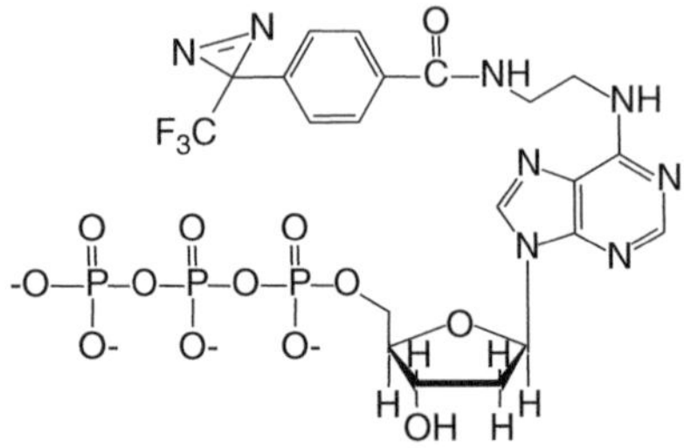

12

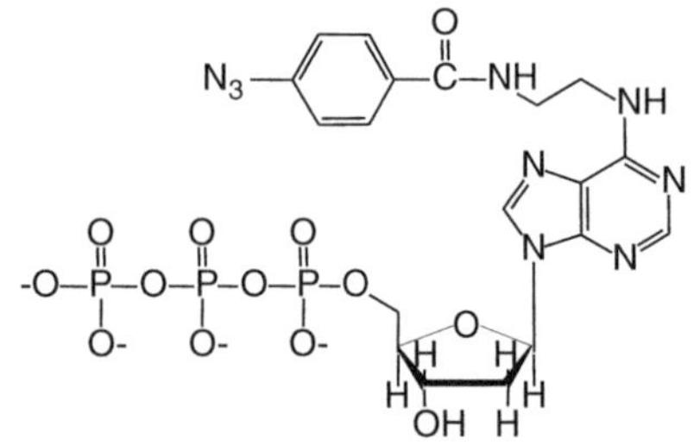

13

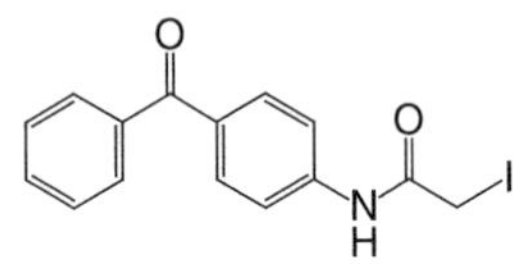

14

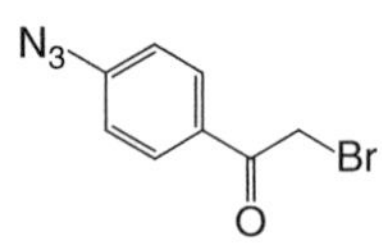

15

Fig. 7.8 Photoreactive compounds for preparing photoreactive DNA. dATP analogs bearing phenyldiazirine ***12***, N^6-[4-[3-(trifluoromethyl)-diazirin-3-yl]benzoyl-(2-aminoethyl)]-2′-deoxyadenosine-5′-triphosphate or phenylazide ***13***, and N^6-[4-azidobenzoyl-(2-aminoethyl)]-2′-deoxyadenosine-5′-triphosphate. These compounds are used as substitutes for native dATP at specific positions during DNA synthesis. Photoreactive compounds bearing benzophenone ***14***, 4-iodoacetamide-benzophenone or phenylazide ***15***, 4-azidophenacyl bromide are the photoreactive units integrated into specific sites of oligonucleotides where the phosphodiester bond has been replaced with a phosphorothioate bond

photoreactive DNA because this group is stable enough to withstand automated DNA synthesis. Compound design and the synthetic scheme for nucleotides bearing 3-trifluoromethyl-3-phenyldiazirine were developed for efficient photoaffinity labeling (Shigdel et al. 2008; Qiu et al. 2008). The biological functions of DNA repair enzymes such as formamidopyrimidine-DNA glycosylase were revealed by photoaffinity labeling experiments using 3-trifluoromethyl-3-phenyldiazirine (Taranenko et al. 2003; Winnacker et al. 2009).

In the latter strategy, commercially available photoreactive reagents such as 4-azidophenacyl bromide **15** and 4-(iodoacetamide)-benzophenone **14** are widely used for analyzing the biological interface between DNA and its binding proteins (Fig. 7.8). Nucleophilic attack by sulfur displaces the bromine or iodine of the photoreactive reagents, and then the reagent is coupled to the DNA phosphate backbone. An oligonucleotide with one or more phosphodiester linkages replaced with phosphorothioate linkages can be obtained commercially by custom synthesis; thus, any complicated synthetic operations are not required for the preparation of photoreactive DNA. In addition, photoreactive units can be attached to any position of DNA because the unit is incorporated into phosphate backbone, and DNA bearing photoreactive units in the phosphate backbone can detect proteins interacting in both the minor and major groove. Yang and Hash (1994) showed that photoreactive DNA with phenylazide conjugated to the phosphate backbone is a useful tool for analyzing protein-DNA interactions. Oligonucleotides bearing phenylazide on the

Fig. 7.9 Photoreactive units used in our studies. 4-[3-(Trifluoromethyl)-*3H*-diazirin-3-yl]benzyl bromide **3**, 4-[3-(trifluoromethyl)-*3H*-diazirin-3-yl] benzyl iodoacetamide ***16***, and 4-bromomethyl-benzophenone ***17***

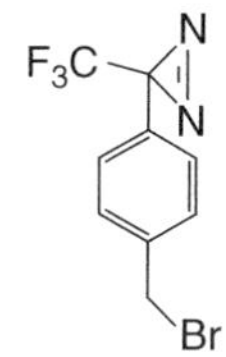

3

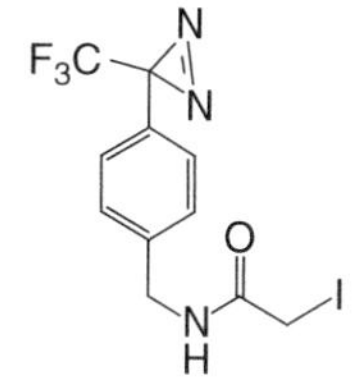

16

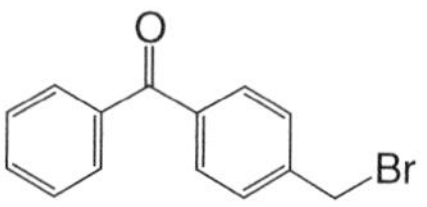

17

phosphate backbone were used to identify crosslinked amino acids to examine the contact point between target DNA and integration host factor (IHF), an accessory protein for integrative recombination by phage DNA. These photoaffinity labeling experiments succeeded in identifying the amino acid residues crosslinked to IHF. Phenylazide-bearing photoreactive DNA has also been used to identify amino acid residues near the active site of *Taq* I restriction endonuclease (Mayer and Barany 1995) and to determine the orientation and positional mapping of transcription factor subunits relative to the enhancer sequence of major histocompatibility complex class II (Westerheide and Boss 1999). Photoreactive RNA prepared with benzophenone **14** was used to analyze the interface between the tRNA and tRNA synthetase (Musier-Forsyth and Schimmel 1994).

We have developed a method for integrating photoreactive 3-trifluoromethyl-3-phenyldiazirine units **3** into oligonucleotides bearing phosphorothioate (Sadakane and Hatanaka 2016). The integration efficiencies of units **3** and **16** into 4-mer oligonucleotides bearing phosphorothioate linkages (5′-AGsCT-3′ in which “s” indicates phosphorothioate) were determined (Fig. 7.9). Both units were specifically integrated into the phosphorothioate sulfhydryl groups by a chemoselective reaction with the same efficiency. We determined the optimum conditions for integrating bromide units **3** into phosphorothioate linkages based on previous reports describing the integration of phenylazide or benzophenone (Fig. 7.10). HPLC profiles revealed that photoreactive oligonucleotides bearing 3-trifluoromethyl-3-phenyldiazirine (peak **b** in Fig. 7.11) were also detected by measuring absorption at 360 nm, which corresponds to the n to π^* absorption of diazirine (λ_{max} 360 nm). In previous reports, 4-azidophenacyl bromide **15** was reacted in 20 mM sodium bicarbonate (pH 9.0) with 45% dimethyl sulfoxide for 3 h at room temperature (Musier-Forsyth and Schimmel 1994) or 50 mM phosphate buffer (pH 7.0) with 50% methanol for 1 h at room temperature (Mayer and Barany 1995), and 4-(iodoacetamide)-benzophenone **14** was reacted in the appropriate buffer (pH 6–7) with 60% ethanol for 24 h at 50 °C (Westerheide and Boss 1999). However, the integration of 3-trifluoromethyl-3-phenyldiazirine **3** into 5′-AGsCT-3′ was considerably inhibited by the presence of H_2O; thus, coupling reactions with unit **3** were performed in the absence of H_2O.

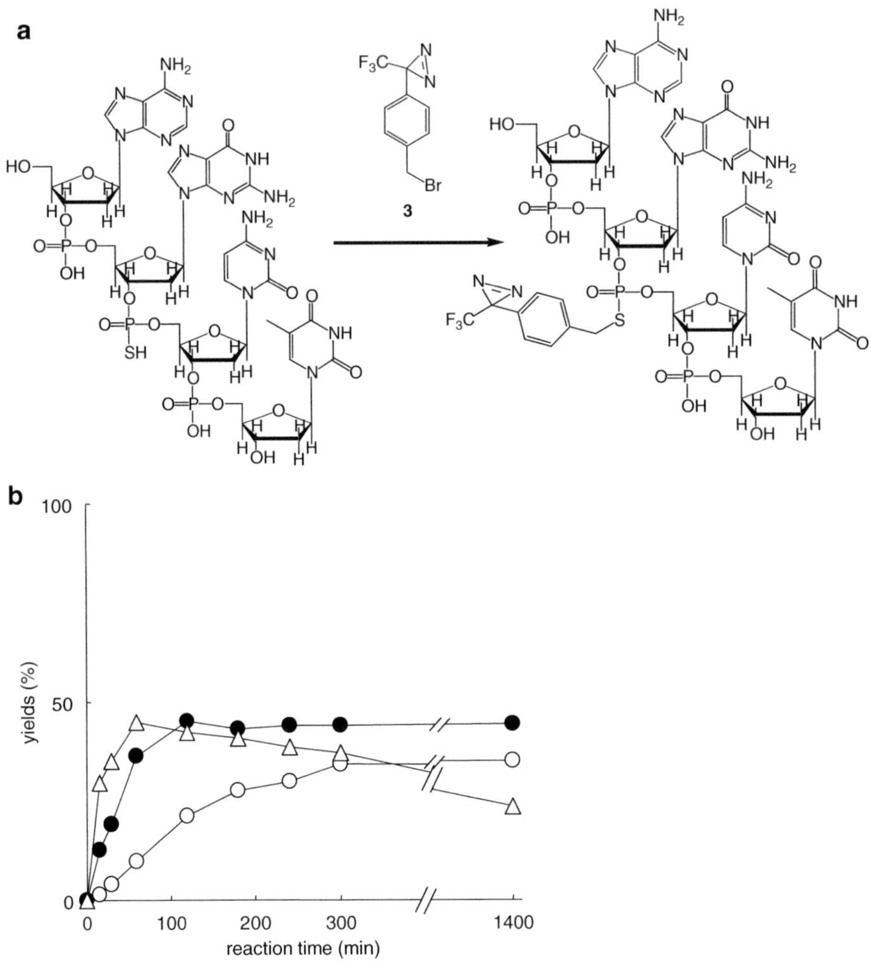

Fig. 7.10 Integration of the photoreactive unit ***3*** into a 4-mer oligonucleotide bearing a phosphorothioate linkage. (**a**) Reaction scheme for integrating unit ***3*** into 5′-AGsCT-3′ (**b**) Time course of unit ***3*** integration into a phosphorothioate-bearing 4-mer oligonucleotide, 5′-AGsCT-3′. The reaction yields at various incubation temperatures are shown: 20 °C (*open circles*), 37 °C (*closed circles*), and 56 °C (*open triangles*) as determined by HPLC analysis

7.6 Photoaffinity EMSA

Electrophoretic mobility shift assays (EMSAs) were initially developed to analyze protein binding to specific DNA sequences, such as the interaction of *lac* repressor-operator and the mapping of transcriptional complexes (Chelm and Geiduschek 1979; Fried and Crothers 1981; Garner and Revzin 1981), and these assays are

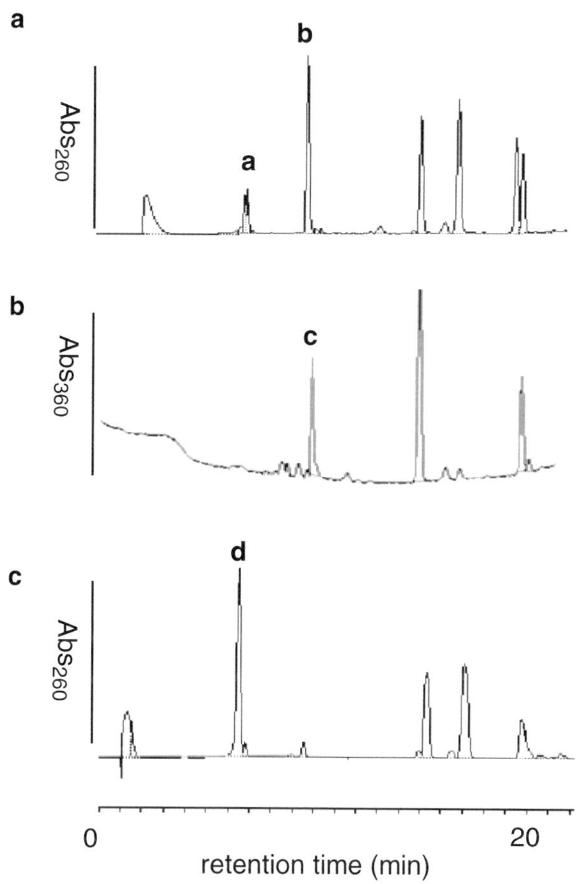

Fig. 7.11 HPLC profiles of reaction mixtures with unit ***3***. (**a**) The oligonucleotide bearing a phosphorothioate linkage, 5′-AGsCT-3′, was mixed with unit ***3*** for 2 h at 37 °C and analyzed by HPLC with monitoring at 260 nm absorption. Peak ***a***, the oligonucleotide 5′-AGsCT-3′; peak ***b***, 5′-AGsCT-3′ attached with unit ***3***. (**b**) The same reaction mixture used in (**a**) was analyzed by HPLC with monitoring at 360 nm absorption, which corresponds to the absorption of diazirine. The retention time of peak ***c*** is identical to that of peak ***b***. (**c**) The oligonucleotide 5′-AGCT-3′ was mixed with unit ***3*** for 2 h at 37 °C and analyzed by HPLC with monitoring at 260 nm absorption. Peak ***d***: the oligonucleotide 5′-AGCT-3′. All HPLC profiles were obtained by a linear acetonitrile gradient of 0–80% containing 10 mM triethylamine-acetate (pH 7.1) over 20 min at a flow rate of 1 mL/min

now widely used as a major molecular biological methodology. The EMSA is based on the decrease in DNA mobility when bound to proteins; thus, the DNA-protein complex should be maintained during the electrophoretic assay. EMSAs are often performed on non-denaturing polyacrylamide or agarose gels in order to maintain DNA-protein complexes. In non-denaturing polyacrylamide gel electrophoresis (PAGE), protein mobility is considerably influenced by charge and conformation, which results in low resolution of protein separation because of broadened protein bands. Additionally, the molecular weight of the proteins cannot be determined. We developed a photoaffinity EMSA using photoreactive units **3** for analyzing the DNA-protein interaction with sodium dodecyl sulfate (SDS)-PAGE in denaturing conditions, which offers high-resolution analysis and information about protein molecular weight (Sadakane and Hatanaka 2016). The outline of this photoaffinity EMSA is summarized in Fig. 7.12; photoreactive DNA bearing 3-trifluoromethyl-3-phenyldiazirine is prepared with a biotinylated oligonucleotide, and then the target DNA-binding protein is crosslinked with the photoreactive DNA by UV irradiation and separated by SDS-PAGE under denaturing conditions. The oligonucleotide-protein complexes are visualized by chemiluminescence using a luminol reaction with a streptavidin-HRP conjugate, which is tightly bound to an oligonucleotide biotin tag. We developed a photoaffinity EMSA using a model complex of a photoreactive DNA bearing octamer motif and its binding protein (Fig. 7.13). The proteins binding to the octamer motif, ATGCAAAT, were identified as mammalian transcription factors, named Oct-1 and Oct-2 (Sturm et al. 1988). The Oct-1 protein contains a POU domain, which is the region responsible for binding to the octamer motif, and is named from the initial letter of three genes, pit-1, Oct-1/Oct-2, and *unc*-86. The POU domain consists of two helix-turn-helix subdomains, a POU-specific domain, and a POU-homeodomain, and both subdomains bound to opposite faces of the octamer DNA (Ko et al. 1988; Herr et al. 1988; Klemm et al. 1994). The biotinylated oligonucleotide 5′-biotin-TGTATGsCAAATAAGG-3′, containing the octamer motif and phosphorothioate linkage, was used to prepare the photoreactive DNA octamer.

The photoaffinity EMSA was developed by using recombinant POU domain protein from Oct-1 in an *E. coli* glutathione-*S*-transferase (GST) fusion expression system. The affinity-purified recombinant fusion proteins contained three distinct proteins with molecular weights of 34, 36, and 45 kDa. Matrix-assisted laser desorption/ionization time-of-flight mass spectrometry revealed that the 45 kDa protein contained both subdomains of the POU domain, but the 34 and 36 kDa proteins contained the POU-specific domain alone (Fig. 7.14b). The individual binding properties of the three proteins were analyzed simultaneously on a single lane by photoaffinity EMSA, whereas a conventional EMSA using DNA octamers could not distinguish the binding ability of the three proteins. Competition experiments also functioned well in the photoaffinity EMSA; that is, excess competitor inhibited crosslinking of the 45 kDa fusion protein, but only partially inhibited crosslinking of the 34 and 36 kDa fusion proteins, and mutated competitor did not inhibit the crosslinking to any protein. The photoaffinity labeling efficiency was compared between DNA bearing 3-trifluoromethyl-3-phenyldiazirine and benzophenone. Crosslinking efficiency of 3-trifluoromethyl-3-phenyldiazirine was superior to benzophenone (Fig. 7.14a) because prolonged UV irradiation causes damage to biomacromolecules such as

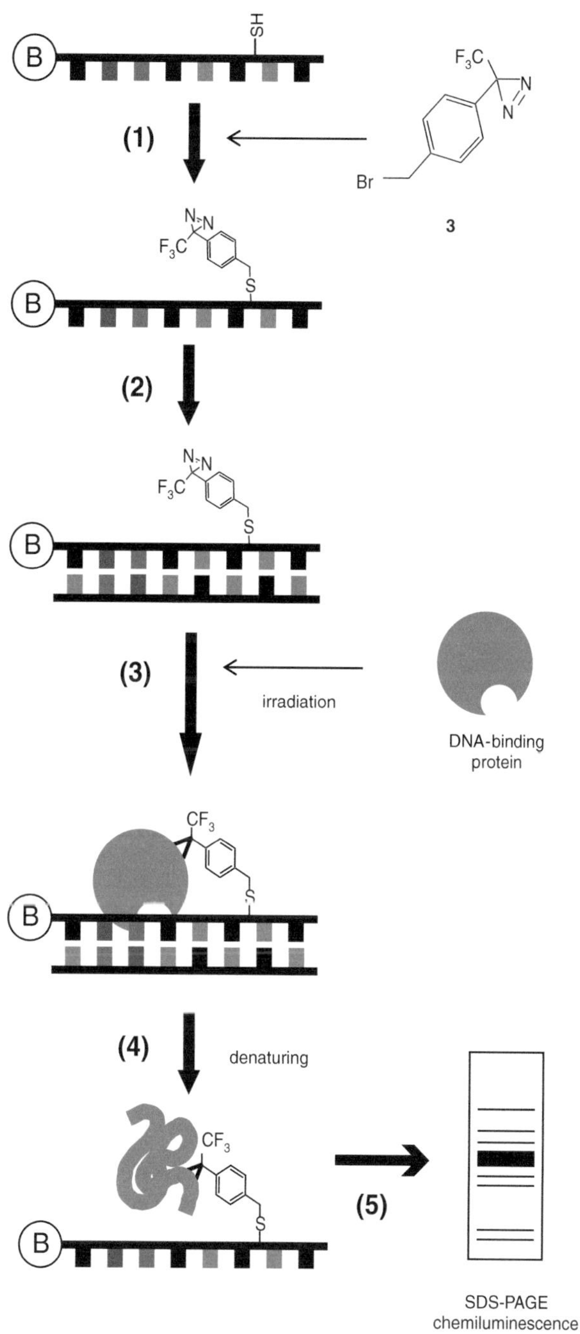

Fig. 7.12 Outline of the photoaffinity electrophoretic mobility shift assay (EMSA). (*1*) Photoreactive unit ***3*** was chemoselectively integrated into the sulfhydryl group of phosphorothioate on a biotinylated oligonucleotide. (*2*) Double-stranded photoreactive DNA was then prepared by annealing with its complimentary oligonucleotide. (*3*) Photoreactive DNA was then mixed with the sample containing the DNA-binding protein and irradiated with 360-nm UV light. (*4*) A covalent bond was formed between the photoreactive DNA and its binding protein. The oligonucleotide-protein complex was stable even under denaturing conditions. (*5*) The complex was separated by SDS-PAGE and visualized by chemiluminescence using a streptavidin-HRP conjugate

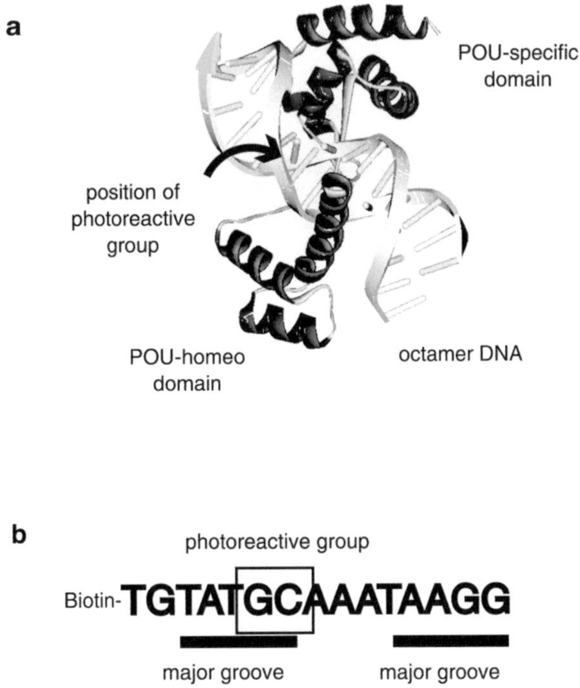

Fig. 7.13 Model complex for photoaffinity EMSA. (**a**) Complex of the Oct-1 POU domain and the DNA octamer. The position integrated with photoreactive unit ***3*** is an interface between POU-specific domain and POU-homeodomain. (**b**) Sequence of biotinylated 15-mer oligonucleotides containing the octamer sequence ATGCAAAT. The position integrated with photoreactive unit ***3*** is surrounded with a square

DNA and protein. We further applied the photoaffinity EMSA to identify octamer-DNA-motif-binding proteins from the nuclear extract of HeLa cells (Fig. 7.15). The photoreactive DNA bearing 3-trifluoromethyl-3-phenyldiazirine crosslinked with several HeLa cell proteins, whereas DNA bearing benzophenone crosslinked with fewer proteins. Photoaffinity EMSA revealed that at least five proteins from the nuclear extract of HeLa cells could bind to the octamer motif whereas conventional EMSA showed only few band. Various proteins such as Oct-1 (Sturm et al. 1988), Oct-2 (Ko et al. 1988), Oct-3 (Scholer et al. 1989), Oct-6 (Meijer et al. 1990), and NonO (Yang et al. 1997) were reported to recognize the octamer motif (Andersen and Rosenfeld 2001). Since the photoaffinity EMSA characterized multiple proteins that could bind to this specific DNA motif in a single assay, the all-inclusive nature of the photoaffinity EMSA is suitable for proteomic studies of DNA-binding proteins.

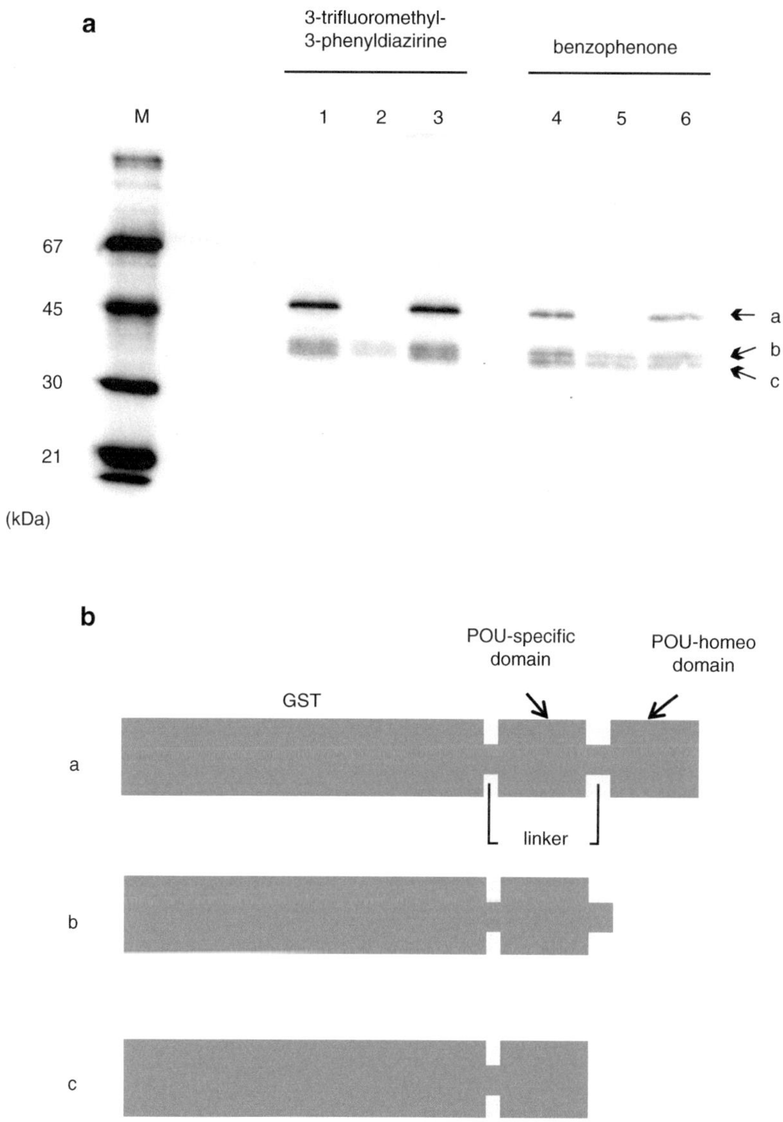

Fig. 7.14 Photoaffinity EMSA of recombinant POU-GST fusion proteins. (**a**) Recombinant POU-GST fusion protein was mixed with a photoreactive DNA octamer bearing 3-trifluoromethyl-3-phenyldiazirine (lanes 1–3) or benzophenone (lanes 4–6) and then irradiated with 360-nm UV light for 5 and 20 min, respectively. Competition experiments were performed with excess specific competitor (lanes 2 and 5, 5′-TGTATGCAAATAAGG-3′) or mutated competitor (lanes 3 and 6, 5′-TGTCTGCAAATAAGG-3′). The crosslinked proteins were separated by SDS-PAGE and detected by chemiluminescence with the streptavidin-HRP conjugate. Biotinylated molecular weight markers are shown in left side of the gel (lane M). (**b**) Illustration of the recombinant POU-GST fusion protein. The purified fusion protein contained three proteins whose molecular weights were 45 (***a***), 36 (***b***), and 34 (***c***) kDa, which corresponded to ***a***, ***b***, and ***c*** in (**a**). The POU protein consists of a POU-specific domain and a POU-homeodomain. Mass spectrometry revealed that the fusion proteins of 36 (***b***) and 34 (***c***) kDa lacked the POU-homeodomain

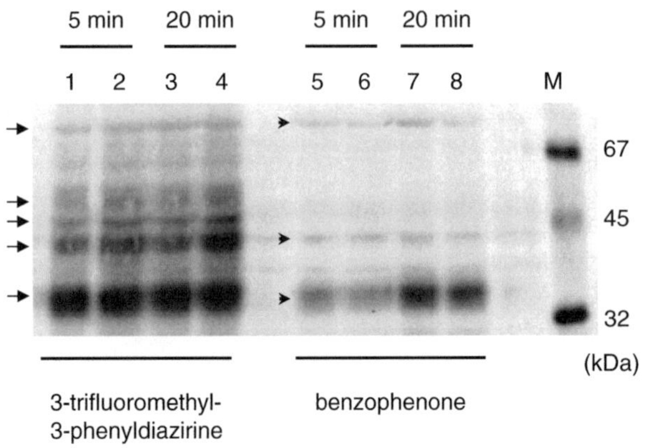

Fig. 7.15 Photoaffinity EMSA of HeLa cell nuclear extracts using photoreactive DNA bearing 3-trifluoromethyl-3-phenyldiazirine (lanes 1–4) or benzophenone (lanes 5–8). HeLa cell nuclear extract was incubated with the photoreactive DNAs, and the mixture was irradiated with 360-nm UV light for 5 min (lanes 1, 2, 5, and 6) or for 20 min (lanes 3, 4, 7, and 8). The crosslinked proteins were separated by SDS-PAGE and detected by chemiluminescence with a streptavidin-HRP conjugate. Duplicate experiments were performed for each sample. Arrows and arrowheads indicate the crosslinked proteins via 3-trifluoromethyl-3-phenyldiazirine and benzophenone, respectively. Biotinylated molecular weight markers are shown on the right side of the gel (lane M)

7.7 Novel Approaches for Photoaffinity Labeling

Li et al. (2013) reported DNA-programmed photoaffinity labeling, which is classified as a DNA-templated chemistry, as a unique system for labeling receptor proteins and small ligand molecules (Fig. 7.16). This system consists of two oligonucleotide tools: one is the small ligand molecule bearing an oligonucleotide in the 5′ region, and the other is the complimentary oligonucleotide bearing a 3′ alkyl diazirine and a 5′ tag. DNA-programmed photoaffinity labeling was applied for analysis of 12-kDa FK506-binding protein (FKBP12) using AP1497, an inhibitor of FKBP12, bearing an oligonucleotide. Using the complementary oligonucleotide bearing an alkyl diazirine and fluorescein tag, FKBP12 was identified in Jurkat cell lysate by SDS-PAGE analysis. DNA-programmed photoaffinity labeling has also been utilized to analyze the interaction between transcription factors and DNA by using an oligonucleotide bearing the double-stranded binding sequence of p50 in its 5′ region (Liu et al. 2015). The crosslinking efficiency of DNA-programmed photoaffinity labeling has been examined using various photoreactive groups such as phenylazide, 3-trifluoromethyl-3-phenyldiazirine, and benzophenone (Li et al. 2014).

Chin et al. (2002a) reported an elaborate technique for preparation of photoreactive proteins in vivo (Fig. 7.17). Wang et al. (2001) created both an orthogonal suppressor tRNA that can derive photoreactive amino acids and an expanded aminoacyl-tRNA synthetase that incorporates the photoreactive amino acids but not any of the endogenous 20 amino acids and observed site-specific incorporation of

Fig. 7.16 DNA-programmed photoaffinity labeling. (*1*) An oligonucleotide attached to a ligand molecule at its 5′ region is mixed with target receptor protein. (*2*) A complementary oligonucleotide attached with a photoreactive group at its 3′ region and a detection tag at its 5′ region is added to the mixture. (*3*) Both oligonucleotides are hybridized followed by UV irradiation to form a crosslink between the receptor protein and the complementary oligonucleotide. (*4*) SDS-PAGE analysis is performed under denaturing conditions to separate crosslinked proteins

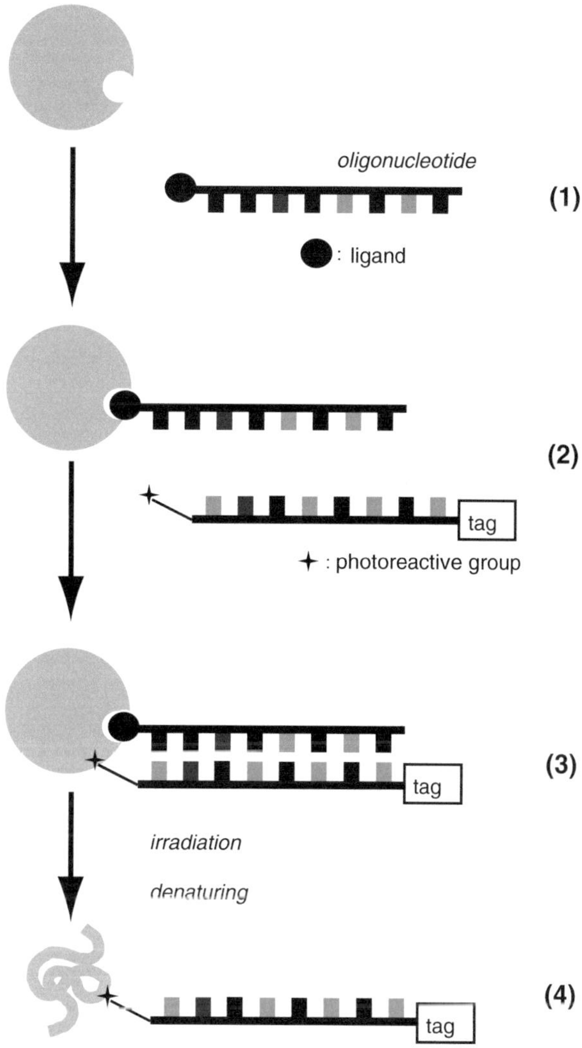

photoreactive amino acids with high translational fidelity into proteins in vivo. *p*-Benzoyl-L-phenylalanine, an amino acid bearing benzophenone, was site specifically incorporated into GST at position 52 (phenylalanine) or 198 (tyrosine) by the engineered aminoacyl-tRNA synthetase/tRNA pair (Chin et al. 2002a). The GST protein is a dimer of two identical subunits: position 52 is located at the dimer interface, but position 198 is located at the opposite side of the interface. The photoreactive GST protein substituted at position 52 resulted in a covalently linked homodimer after UV-irradiation, but GST substituted at position 198 was left in monomeric form. The photoreactive GST protein, where amino acids 52 or 198 were replaced with *p*-azide-L-phenylalanine, also showed the same result as GST

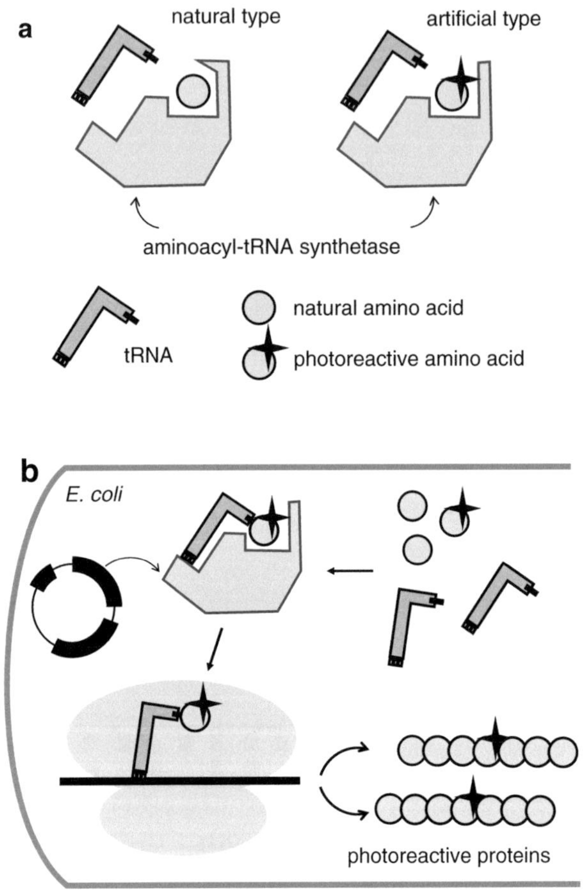

Fig. 7.17 Preparation of photoreactive proteins by an artificial *Escherichia coli* (*E. coli*) system in which tRNA and aminoacyl-tRNA synthetases are engineered for site-specific incorporation of unnatural amino acids into proteins. (**a**) The system to deliver unnatural amino acids site specifically into proteins in vivo. The system consists of two artificial units, an orthogonal suppressor tRNA for delivery of an unnatural amino acid that is not acylated by any endogenous *E. coli* aminoacyl-tRNA synthetase and an aminoacyl-tRNA synthetase that charges the orthogonal suppressor tRNA that was produced by multiple rounds of DNA shuffling. (**b**) An artificial *E. coli* that generates the photoreactive proteins. The engineered aminoacyl-tRNA synthetase/tRNA pair facilitated the incorporation of photoreactive amino acids into proteins in vivo

bearing *p*-benzoyl-L-phenylalanine (Chin et al. 2002b). The site-specific substitution of photoreactive amino acids is an excellent tool for examining protein-protein interactions. The protocol for preparing photoreactive proteins genetically incorporated with *p*-benzoyl-L-phenylalanine is summarized elsewhere (Farrell et al. 2005).

Suchanek et al. (2005) synthesized two novel photoreactive amino acids, photo-leucine (Leu) and photo-methionine (Met), to establish a strategy for in vivo incorporation of photoreactive amino acids that does not require any modification

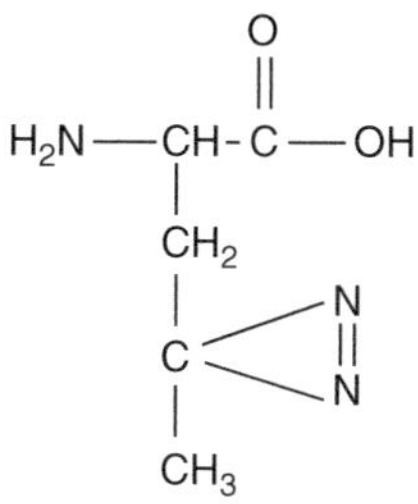
18

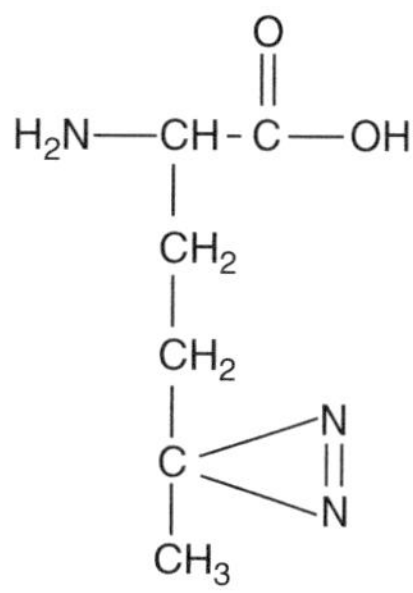
19

Fig. 7.18 Photoreactive amino acids incorporated into proteins without any modification of the translation machinery. The structures of photo-leucine (Leu) ***18***, L-2-amino-4,4′-azi-pentanoic acid, and photo-methionine (Met) ***19***, L-2-amino-5,5′-azi-hexanoic acid, which closely resemble natural Leu and Met, respectively

of the translation machinery (Fig. 7.18). These photoreactive amino acids closely resemble natural amino acids; thus, mammalian cells can uptake these compounds from culture medium and use them like a normal amino acid during protein synthesis. This technology has been applied to studying protein-protein interactions in a complex of membrane proteins involved in regulation of cellular lipid homeostasis.

In conclusion, photoreactive biomacromolecules have been applied for the analysis of various biological interfaces, and various skillful strategies have been developed for preparing photoreactive biomacromolecules such as peptides, proteins, and DNAs. Novel methodologies for photoaffinity labeling using photoreactive biomacromolecules will continue to aid in analysis of biological functions with high sensitivity and high resolution.

Acknowledgments This work was supported by JSPS KAKENHI Grants Nos. 23510286 and 16K08207 for Y. Sadakane and 24790033 for I. Ohtsuka.

References

Andersen B, Rosenfeld MG (2001) POU domain factors in the neuroendocrine system: lessons from developmental biology provide insights into human disease. Endocr Rev 22:2–35

Bremer AA, Leeman SE, Boyd ND (2001) Evidence for spatial proximity of two distinct receptor regions in the substance P (SP)-neurokinin-1 receptor (NK-1R) complex obtained by photolabeling the NK-1R with *p*-benzoylphenylalanine[3]-SP. J Biol Chem 276:22857–22861

Brunner J (1993) New photolabeling and crosslinking methods. Annu Rev Biochem 62:483–514

Brunner J, Senn H, Richards FM (1980) 3-Trifluoromethyl-3-phenyldiazirine. A new carbene generating group for photolabeling reagents. J Biol Chem 255:3313–3318

Chelm BK, Geiduschek EP (1979) Gel electrophoretic separation of transcription complexes: an assay for RNA polymerase selectivity and a method for promoter mapping. Nucleic Acids Res 7:1851–1867

Chin JW, Martin AB, King DS, Wang L, Schultz PG (2002a) Addition of a photocrosslinking amino acid to the genetic code of *Escherichia coli*. Proc Natl Acad Sci U S A 99:11020–11024

Chin JW, Santoro SW, Martin AB, King DS, Wang L, Schultz PG (2002b) Addition of *p*-azido-L-phenylalanine to the genetic code of *Escherichia coli*. J Am Chem Soc 124:9026–9027

Cleary MA, Pendergrast PS, Herr W (1997) Structural flexibility in transcription complex formation revealed by protein-DNA photocrosslinking. Proc Natl Acad Sci U S A 94:8450–8455

Costas C, Yuriev E, Meyer KL, Guion TS, Hanna MM (2000) RNA-protein crosslinking to AMP residues at internal positions in RNA with a new photocrosslinking ATP analog. Nucleic Acids Res 28:1849–1858

Driessen AJ, Manting EH, van der Does C (2001) The structural basis of protein targeting and translocation in bacteria. Nat Struct Biol 8:492–498

Dubinsky L, Krom BP, Meijler MM (2012) Diazirine based photoaffinity labeling. Bioorg Med Chem 20:554–570

Egnaczyk GF, Greis KD, Stimson ER, Maggio JE (2001) Photoaffinity cross-linking of Alzheimer's disease amyloid fibrils reveals interstrand contact regions between assembled beta-amyloid peptide subunits. Biochemistry 40:11706–11714

Escher E, Schwyzer R (1974) *p*-Nitrophenylalanine, *p*-azidophenylalanine, *m*-azidophenylalanine, and *o*-nitro-*p*-azido-phenylalanine as photoaffinity labels. FEBS Lett 46:347–350

Escher EH, Nguyen TM, Robert H, St-Pierre SA, Regoli DC (1978) Photoaffinity labeling of the angiotensin II receptor. 1. Synthesis and biological activities of the labeling peptides. J Med Chem 21:860–864

Farrell IS, Toroney R, Hazen JL, Mehl RA, Chin JW (2005) Photo-cross-linking interacting proteins with a genetically encoded benzophenone. Nat Methods 2:377–384

Fleming SA (1995) Chemical reagents in photoaffinity labeling. Tetrahedorn 51:12479–12520

Fried M, Crothers DM (1981) Equilibria and kinetics of *lac* repressor-operator interactions by polyacrylamide gel electrophoresis. Nucleic Acids Res 9:6505–6525

Galardy RE, Craig LC, Printz MP (1973) Benzophenone triplet: a new photochemical probe of biological ligand-receptor interactions. Nat New Biol 242:127–128

Garner MM, Revzin A (1981) A gel electrophoresis method for quantifying the binding of proteins to specific DNA regions: application to components of the *Escherichia coli* lactose operon regulatory system. Nucleic Acids Res 9:3047–3060

Hatanaka Y (2015) Development and leading-edge application of innovative Photoaffinity labeling. Chem Pharm Bull 63:1–12

Hatanaka Y, Kanaoka Y (1998) Biotinyl diazirine photophore: an approach to high-resolution photoaffinity labeling for probing receptor-ligand interface. Heterocycles 47:625–632

Hatanaka Y, Sadakane Y (2002) Photoaffinity labeling in drug discovery and developments: chemical gateway for entering proteomic frontier. Curr Top Med Chem 2:271–288

Hatanaka Y, Nakayama H, Kanaoka Y (1996) Diazirine-based photoaffinity labeling: chemical approach to biological interfaces. Rev Heteroatom Chem 14:213–243

Hatanaka Y, Kempin U, Park JJ (2000) One-step synthesis of biotinyl photoprobes from unprotected carbohydrates. J Org Chem 65:5639–5643

Herr W, Sturm RA, Clerc RG, Corcoran LM, Baltimore D, Sharp PA, Ingraham HA, Rosenfeld MG, Finney M, Ruvkun G et al (1988) The POU domain: a large conserved region in the mammalian pit-1, oct-1, oct-2, and *Caenorhabditis elegans* unc-86 gene products. Genes Dev 2:1513–1516

High S, Martoglio B, Görlich D, Andersen SS, Ashford AJ, Giner A, Hartmann E, Prehn S, Rapoport TA, Dobberstein B, Brunner J (1993) Site-specific photocross-linking reveals that Sec61p and TRAM contact different regions of a membrane-inserted signal sequence. J Biol Chem 268:26745–26751

Hixson SH, Hixson SS (1975) *p*-Azidophenacyl bromide, a versatile photolabile bifunctional reagent. Reaction with glyceraldehyde-3-phosphate dehydrogenase. Biochemistry 14:4251–4254

Kage R, Leeman SE, Krause JE, Costello CE, Boyd ND (1996) Identification of methionine as the site of covalent attachment of a *p*-benzoyl-phenylalanine-containing analogue of substance P on the substance P (NK-1) receptor. J Biol Chem 271:25797–25800

Kaneda M, Sadakane Y, Hatanaka Y (2003) A novel approach for affinity-based screening of target specific ligands: application of photoreactive D-glyceraldehyde-3-phosphate dehydrogenase. Bioconjug Chem 14:849–852

Kauer JC, Erickson-Viitanen S, Wolfe HR Jr, DeGrado WF (1986) *p*-Benzoyl-L-phenylalanine, a new photoreactive amino acid. Photolabeling of calmodulin with a synthetic calmodulin-binding peptide. J Biol Chem 261:10695–10700

Kemnitz CR, Karney WL, Borden WT (1998) Why are nitrenes more stable than carbenes? An ab initio study. J Am Chem Soc 120:3499–3503

Kenyon GL, Bruice TW (1977) Novel sulfhydryl reagents. Methods Enzymol 47:407–430

Klemm JD, Rould MA, Aurora R, Herr W, Pabo CO (1994) Crystal structure of the Oct-1 POU domain bound to an octamer site: DNA recognition with tethered DNA-binding modules. Cell 77:21–32

Ko HS, Fast P, McBride W, Staudt LM (1988) A human protein specific for the immunoglobulin octamer DNA motif contains a functional homeobox domain. Cell 55:135–144

Lannutti BJ, Persinger J, Bartholomew B (1996) Probing the protein-DNA contacts of a yeast RNA polymerase III transcription complex in a crude extract: solid phase synthesis of DNA photoaffinity probes containing a novel photoreactive deoxycytidine analog. Biochemistry 35:9821–9831

Lauc G, Lee RT, Dumic J, Lee TC (2000) Photoaffinity glycoprobes—a new tool for the identification of lectins. Glycobiology 10:357–364

Li G, Liu Y, Liu Y, Chen L, Wu S, Liu Y, Li X (2013) Photoaffinity labeling of small-molecule-binding proteins by DNA-templated chemistry. Angew Chem Int Ed Eng 52:9544–9549

Li G, Liu Y, Yu X, Li X (2014) Multivalent photoaffinity probe for labeling small molecule binding proteins. Bioconjug Chem 25:1172–1180

Liu Y, Zheng W, Zhang W, Chen N, Liu Y, Chen L, Zhou X, Chen X, Zhenga H, Li X (2015) Photoaffinity labeling of transcription factors by DNA-templated crosslinking. Chem Sci 6:745–751

Maliarik MJ, Goldstein IJ (1988) Photoaffinity labeling of the adenine binding site of the lectins from lima bean, *Phaseolus lunatus*, and the kidney bean, *Phaseolus vulgaris*. J Biol Chem 263:11274–11279

Mayer AN, Barany F (1995) Photoaffinity cross-linking of *Taq*I restriction endonuclease using an aryl azide linked to the phosphate backbone. Gene 153:1–8

Meador WE, Means AR, Quiocho FA (1992) Target enzyme recognition by calmodulin: 2.4 angstrom structure of a calmodulin-peptide complex. Science 257:1251–1255

Meijer D, Graus A, Kraay R, Langeveld A, Mulder MP, Grosveld G (1990) The octamer binding factor Oct6: cDNA cloning and expression in early embryonic cells. Nucleic Acids Res 18:7357–7365

Moore BM 2nd, Jalluri RK, Doughty MB (1996) DNA polymerase photoprobe 2-[(4-azidophenacyl) thio]-2′-deoxyadenosine 5′-triphosphate labels an *Escherichia coli* DNA polymerase I Klenow fragment substrate binding site. Biochemistry 35:11642–11651

Musier-Forsyth K, Schimmel P (1994) Acceptor helix interactions in a class II tRNA synthetase: Photoaffinity cross-linking of an RNA miniduplex substrate. Biochemistry 33:773–779

Nakashima H, Hashimoto M, Sadakane Y, Tomohiro T, Hatanaka Y (2006) Simple and versatile method for tagging phenyldiazirine photophores. J Am Chem Soc 128:15092–15093

Nassal M (1984) 4′-(1-Azi-2,2,2-trifluoroethyl)phenylalanine, a photolabile carbene-generating analog of phenylalanine. J Am Chem Soc 106:7540–7545

O'Neil KT, Wolfe HR Jr, Erickson-Viitanen S, DeGrado WF (1987) Fluorescence properties of calmodulin-binding peptides reflect alpha-helical periodicity. Science 236:1454–1456

O'Neil KT, Erickson-Viitanen S, DeGrado WF (1989) Photolabeling of calmodulin with basic, amphiphilic alpha-helical peptides containing *p*-benzoylphenylalanine. J Biol Chem 264:14571–14578

Ohtsuka I, Sadakane Y, Higuchi M, Hada N, Hada J, Kakiuchi N, Sakushima A (2011) The effect of structural differences in the reducing terminus of sugars on the binding affinity of carbohydrates and proteins analyzed using photoaffinity labeling. Bioorg Med Chem 19:894–899

Ohtsuka I, Sadakane Y, Hada N, Higuchi M, Atsumi T, Kakiuchi N (2014) The development of new molecular tools containing a chemically synthesized carbohydrate ligand for the elucidation of carbohydrate roles via photoaffinity labeling: carbohydrate-protein interactions are affected by the structures of the glycosidic bonds and the reducing-end sugar. Bioorg Med Chem 22:3829–3837

Pendergrast PS, Chen Y, Ebright YW, Ebright RH (1992) Determination of the orientation of a DNA binding motif in a protein-DNA complex by photocrosslinking. Proc Natl Acad Sci U S A 89:10287–10291

Platz MS (1995) Comparison of phenylcarbene and phenylnitrene. Acc Chem Res 28:487–492

Qiu Z, Lu L, Jian X, He C (2008) A diazirine-based nucleoside analogue for efficient DNA interstrand photocross-linking. J Am Chem Soc 130:14398–14399

Sachon E, Bolbach G, Lavielle S, Karoyan P, Sagan S (2003) Met174 side chain is the site of photoinsertion of a substance P competitive peptide antagonist photoreactive in position 8. FEBS Lett 544:45–49

Sadakane Y (2007) Endeavors to make the photophore, diazirine easy to use. Yakugaku Zasshi 127:1693–1699

Sadakane Y, Hatanaka Y (2004) Multifunctional photoprobes for rapid protein identification. In: Darvas F, Guttman A, Dorman G (eds) Chemical genomics. Marcel Dekker Inc, New York, pp 199–214

Sadakane Y, Hatanaka Y (2006) Photochemical fishing approaches for identifying target proteins and elucidating the structure of a ligand-binding region using carbene-generating photoreactive probes. Anal Sci 22:209–218

Sadakane Y, Hatanaka Y (2008) Development of new method for molecular biology using the photophore, diazirine. Yakugaku Zasshi 128:615–1622

Sadakane Y, Hatanaka Y (2016) Photoaffinity electrophoretic mobility shift assay using photoreactive DNA bearing 3-trifluoromethyl-3-phenyldiazirine in its phosphate backbone. Anal Biochem 506:1–7. doi:10.1016/j.ab.2016.04.016

Sakurai K, Ozawa S, Yamaguchi T (2015) Photoaffinity labeling studies of the carbohydrate-binding proteins with different affinities. Bioorg Med Chem 23:5319–5325

Samuelson JC, Chen M, Jiang F, Möller I, Wiedmann M, Kuhn A, Phillips GJ, Dalbey RE (2000) YidC mediates membrane protein insertion in bacteria. Nature 406:637–641

Scholer HR, Balling R, Hatzopoulos AK, Suzuki N, Gruss P (1989) Octamer binding proteins confer transcriptional activity in early mouse embryogenesis. EMBO J 8:2551–2557

Schwyzer R, Caviezel M (1971) *p*-Azido-L-phenylalanine: a photo-affinity probe related to tyrosine. Helv Chim Acta 54:1395–1400

Scotti PA, Urbanus ML, Brunner J, de Gier JW, von Heijne G, van der Does C, Driessen AJ, Oudega B, Luirink J (2000) YidC, the *Escherichia coli* homologue of mitochondrial Oxa1p, is a component of the sec translocase. EMBO J 19:542–549

Sengupta SM, Van Kanegan M, Persinger J, Logie C, Cairns BR, Peterson CL, Bartholomew B (2001) The interactions of yeast SWI/SNF and RSC with the nucleosome before and after chromatin remodeling. J Biol Chem 276:12636–12644

Shigdel UK, Zhang J, He C (2008) Diazirine-based DNA photo-cross-linking probes for the study of protein-DNA interactions. Angew Chem Int Ed Eng 47:90–93

Shih LB, Bayley H (1985) A carbene-yielding amino acid for incorporation into peptide photoaffinity reagents. Anal Biochem 144:132–141

Sturm RA, Das G, Herr W (1988) The ubiquitous octamer-binding protein Oct-1 contains a POU domain with a homeo box subdomain. Genes Dev 2:1582–1599

Suchanek M, Radzikowska A, Thiele C (2005) Photo-leucine and photo-methionine allow identification of protein-protein interactions in living cells. Nat Methods 2:261–267

Tan YV, Couvineau A, Van Rampelbergh J, Laburthe M (2003) Photoaffinity labeling demonstrates physical contact between vasoactive intestinal peptide and the N-terminal ectodomain of the human VPAC1 receptor. J Biol Chem 278:36531–36536

Taranenko M, Rykhlevskaya A, Mtchedlidze M, Laval J, Kuznetsova S (2003) Photochemical cross-linking of *Escherichia coli* Fpg protein to DNA duplexes containing phenyl(trifluoromethyl) diazirine groups. Eur J Biochem 270:2945–2949

Tate JJ, Persinger J, Bartholomew B (1998) Survey of four different photoreactive moieties for DNA photoaffinity labeling of yeast RNA polymerase III transcription complexes. Nucleic Acids Res 26:1421–1426

Wang L, Brock A, Herberich B, Schultz PG (2001) Expanding the genetic code of *Escherichia coli*. Science 292:498–500

Weber PJ, Beck-Sickinger AG (1997) Comparison of the photochemical behavior of four different photoactivatable probes. J Pept Res 49:375–383

Westerheide SD, Boss JM (1999) Orientation and positional mapping of the subunits of the multicomponent transcription factors RFX and X2BP to the major histocompatibility complex class II transcriptional enhancer. Nucleic Acids Res 27:1635–1641

Wilson CJ, Husain SS, Stimson ER, Dangott LJ, Miller KW, Maggio JE (1997) *p*-(4-Hydroxybenzoyl)phenylalanine: a photoreactive amino acid analog amenable to radioiodination for elucidation of peptide-protein interaction. Application to substance P receptor. Biochemistry 36:4542–4551

Winnacker M, Breeger S, Strasser R, Carell T (2009) Novel diazirine-containing DNA photoaffinity probes for the investigation of DNA-protein-interactions. Chembiochem 10:109–118

Wittelsberger A, Thomas BE, Mierke DF, Rosenblatt M (2006) Methionine acts as a “magnet” in photoaffinity crosslinking experiments. FEBS Lett 580:1872–1876

Yamaguchi T, Saneyoshi M (1996) A photolabile 2′,3′-dideoxyuridylate analog bearing an aryl(trifluoromethyl)diazirine moiety: photoaffinity labeling of HIV-1 reverse transcriptase. Nucleic Acids Res 24:3364–3369

Yamaguchi T, Suyama K, Narita K, Kohgo S, Tomikawa A, Saneyoshi M (1997) Synthesis and evaluation of oligodeoxyribonucleotides containing an aryl(trifluoromethyl)diazirine moiety as the cross-linking probe: Photoaffinity labeling of mammalian DNA polymerase β. Nucleic Acids Res 25:2352–2358

Yang SW, Hash HA (1994) Specific photocrosslinking of DNA-protein complexes: identification of contacts between integration host factor and its target DNA. Proc Natl Acad Sci U S A 91:12183–12187

Yang J, Clark AE, Kozka IJ, Cushman SW, Holman GD (1992) Development of an intracellular pool of glucose transporters in 3T3-L1 cells. J Biol Chem 267:10393–10399

Yang YS, Yang MC, Tucker PW, Capra JD (1997) NonO enhances the association of many DNA-binding proteins to their targets. Nucleic Acids Res 25:2284–2292

Yu SH, Wands AM, Kohler JJ (2012) Photoaffinity probes for studying carbohydrate biology. J Carbohydr Chem 31:325–352

Zofall M, Bartholomew B (2000) Two novel dATP analogs for DNA photoaffinity labeling. Nucleic Acids Res 28:4382–4390

Chapter 8
Covalently Capturing Protein Interactions in Living Cells by Site-Specific Incorporation of Photo-Cross-Linkable Amino Acids

Nobumasa Hino and Kensaku Sakamoto

Abstract Protein functions in living cells are strictly regulated to precisely execute cellular processes, which are accomplished by the interaction of proteins in a weak and transient manner. Protein photo-cross-linking is a powerful way to stabilize such interactions by covalently bonding the interacting proteins to each other. However, the in vivo use of low molecular weight cross-linkers has been very limited owing to the difficulty of creating specific modifications of the desired proteins for many cellular proteins. Recent developments in genetic code expansion technology have overcome such difficulties to create a variety of photoreactive moieties that can be site-specifically incorporated into the desired protein as unnatural amino acids in living cells. In this chapter, we present an overview of in vivo protein photo-cross-linking with genetically encoded photo-cross-linkable amino acids (PAAs) and then discuss its application for covalently capturing extremely weak and transient interactions, identification of direct binding partners of a target protein, and precise mapping of the binding interface between interacting proteins under physiologically relevant conditions.

Keywords Protein–protein interaction • Protein complex • Cross-link • Unnatural amino acid • Genetic code expansion

---

N. Hino (✉)
Laboratory of Molecular Medicine, Graduate School of Pharmaceutical Sciences, Osaka University, 1-6 Yamadaoka, Suita, Osaka 565-0871, Japan
e-mail: hino@phs.osaka-u.ac.jp

K. Sakamoto
Division of Structural and Synthetic Biology, RIKEN Center for Life Science Technologies, 1-7-22 Suehiro-cho, Tsurumi, Yokohama 230-0045, Japan
e-mail: kensaku.sakamoto@riken.jp

Y. Hatanaka, M. Hashimoto (eds.), *Photoaffinity Labeling for Structural Probing Within Protein*, DOI 10.1007/978-4-431-56569-7_8

8.1 Introduction

Protein–protein interactions are fundamental to all cellular processes, including proliferation, differentiation, and programmed cell death. To investigate what proteins are interacting with each other in living cells, the method by which protein complexes are isolated from the cells is an important consideration. An affinity-based method, called immunoprecipitation, has been widely employed for this purpose. Recent developments in highly sensitive mass spectrometry and growing genome databases have enabled the identification of a large number of proteins obtained from the immunoprecipitated complexes and have facilitated the elucidation of protein–protein interactions occurring in living cells (Markham et al. 2007; Dunham et al. 2012). However, the results do not often reflect the interactions in the native state because weakly bound protein complexes often dissociate during the immunoprecipitation procedure and some complexes that are not bound in the native state may form during or after the extraction procedure for the immunoprecipitation.

Photoaffinity protein cross-linking provides an effective way to overcome this problem (Vasilescu et al. 2004; Bruce 2012). Photo-cross-linkers introduced on protein surfaces in living cells can be activated by photo-irradiation to form covalent bonds with neighboring proteins to identify protein interactions. The cross-linked complexes show negligible dissociation during the purification steps, and the identified protein–protein interactions accurately reflect what is occurring in the cells. However, conventional chemical photo-cross-linkers, which usually equip amine or thiol-reactive groups in addition to the photoaffinity moiety, are sometimes inappropriate for in vivo modification of proteins owing to their low cell-membrane permeability and the difficulties in specific modification of lysine or cysteine residues of the desired protein among many other proteins. Instead, recent research efforts have focused on direct incorporation of photoaffinity probes as amino acid analogs into proteins during their translation (Tanaka et al. 2008).

There are two methods to incorporate the photo-cross-linkable amino acids (PAAs) into proteins: residue-specific and site-specific incorporation. Residue-specific incorporation employs amino acid analogs that structurally resemble natural amino acids. Suchanek et al. designed photo-leucine and photo-methionine PAAs that have similarities to natural amino acids L-leucine and L-methionine, respectively (Suchanek et al. 2005). The leucyl- and methionyl-tRNA synthetases can aminoacylate the PAAs to their cognate tRNAs, and the aminoacylated tRNAs are employed to translate leucine and methionine codons. Therefore, by incubating the cells in a medium containing photo-leucine and depleted L-leucine, photo-leucine is incorporated into proteins in place of leucine, which results in the expression of photo-cross-linkable proteins. In a similar way, photo-lysine has recently been designed and incorporated into proteins in place of lysine (Yang et al. 2015). Additionally, recent developments in genetic code expansion technology have enabled site-specific incorporation of a variety of unnatural amino acids, including PAAs, into proteins in living cells (Davis and Chin 2012). This technology relies on the expression of "orthogonal" pairs of tRNA and aminoacyl-tRNA synthetase (aaRS), which do not cross-react with any aaRSs and tRNAs of the host cells (Wang et al. 2001; Chin et al. 2003; Sakamoto et al. 2002). By using tRNA mutant that can read a nonsense codon, together with aaRS variants engineered to recognize PAAs as a substrate, PAAs can be genetically encoded and assigned at a position defined by an in-frame nonsense codon on the target gene.

In this chapter, we will discuss various types of genetically encoded PAAs, which can be introduced at the desired position of proteins in living cells, and their applications.

8.2 In Vivo Protein Photo-Cross-Linking with Genetically Encoded PAAs

8.2.1 An Overview

A representative scheme of site-specific incorporation of a genetically encoded PAA into a target protein and the subsequent photo-cross-linking analysis of interacting proteins in living cells is shown in Fig. 8.1. An orthogonal pair of UAG

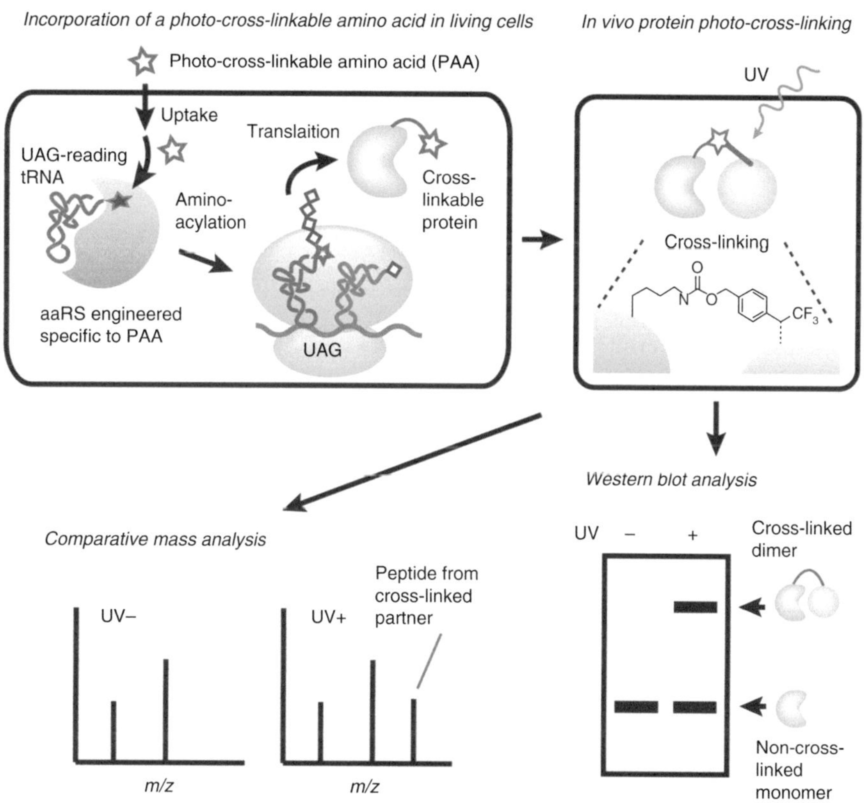

Fig. 8.1 Outline of in vivo photo-cross-linking analysis with genetically encoded PAAs. A PAA supplemented in culture medium is taken up by the cell and incorporated at the UAG codon-directed site of a target protein by the pair of UAG-reading tRNA and aaRS specific to the PAA. By UV exposure to the cells, the PAA-containing protein covalently bonds ("cross-links") to its interacting partners. The cross-linked complex is purified and subjected to analysis such as Western blotting or mass spectrometry to identify what proteins are interacting with the target protein in living cells

codon-reading tRNA and aaRS specific to a PAA is expressed in cells, together with a target gene with an in-frame TAG triplet sequence at the desired site. The intended PAA is supplemented in the cell culture medium to be transported into the cells and added to the 3′-end of the tRNA by the PAA-specific aaRS. The aminoacyl-tRNA is transferred to the ribosome and binds to the UAG codon of transcripts of the target gene, and thus a PAA is site-specifically incorporated into the target protein. The incorporation of the PAA can simply be ascertained by examining the expression of full-length protein because the UAG codon stops protein translation if it is not read through by the PAA-charged tRNA. In this regard, C-terminal tagging of the target protein is useful to detect and purify the PAA-incorporated protein and its complex. By photo-irradiating the cells, the PAA-containing protein is activated and covalently bound to a neighboring protein via the PAA-incorporated surface of the target protein. Then, the cross-linked complexes are purified from the cell extract and analyzed by Western blotting or mass spectrometry. In the immunoblot of the antibody reaction of the target protein, the cross-linked product is detected as a shifted band, in which the molecular mass of the product corresponds to the sum of the masses of target protein and interacting partner protein. If the partner is known or can be predicted from other experiments, it can be verified by using an antibody of the protein in a similar way. Identification of the partners is also performed by mass analysis of the components of cross-linked products, followed by peptide mass fingerprint database searching. The analysis may be individually performed for each of the cross-linked products visualized individually by silver staining (or similar methods after electrophoresis) or collectively by a comparative and quantitative approach. The latter approach relies on the fact that the cross-linked complex is stable and does not dissociate during the purification procedure; the interacting partners should be more abundant in the purified fraction from a cross-linked sample than from a non-cross-linked control.

8.2.2 Photoreactive Groups of Genetically Encoded PAAs

PAAs that have been encoded by genetic code expansion technology can be classified into two types by their structural analogy with natural amino acids and further divided into four subclasses based on the different photoreactive groups (Table 8.1). PAAs usually contain one of four types of photoreactive groups, benzophenone, aryl azide, aryl diazirine, or alkyl diazirine, which have widely been used as chemical cross-linkers (Bayley and Knowles 1977; Kauer et al. 1986; Brunner 1993; Dormán and Prestwich 1994). Benzophenone becomes a highly reactive biradical intermediate by excitation with long-wavelength (350–365 nm) UV light and reacts with neighboring C–H bonds to form a new covalent C–C bond. The intermediate can be relaxed back to the ground state if an accessible C–H bond is absent and then activated again with UV light. Therefore, prolonging irradiation time of the UV

Table 8.1 Genetically encoded photo-cross-linkable amino acids (PAAs)

	Amino acids	Symbols (synonyms)	Chemical structures	Host cells (references)
Phenylalanine and its analogs	L-Phenylalanine[a]	Phe	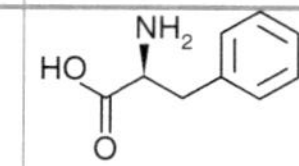	
	p-Benzoyl-L-phenylalanine	*p*Bpa (Bpa, BzF, BzPhe)	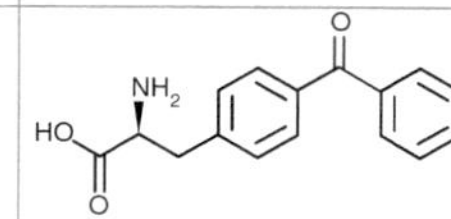	Bacteria (Chin et al. 2002a); yeast (Chin et al. 2003); mammal (Hino et al. 2005)
	p-Azido-L-phenylalanine	*p*Azpa (AzF, Azi)	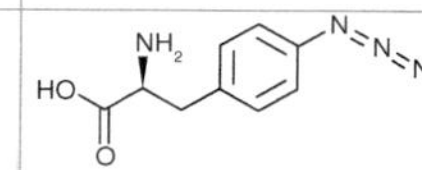	Bacteria (Chin et al. 2002b); yeast (Chin et al. 2003); mammal (Liu et al. 2007)
	4′-[3-(Trifluoromethyl)-3*H*-diazirin-3-yl]-L-phenylalanine	tmdPhe (TfmdPhe)	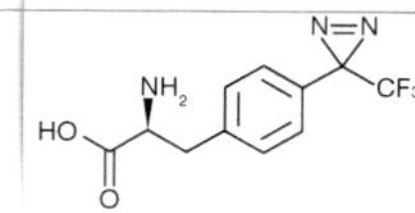	Bacteria (Tippmann et al. 2007); mammal (Hino et al. 2011)

(continued)

Table 8.1 (continued)

	Amino acids	Symbols (synonyms)	Chemical structures	Host cells (references)
Pyrrolysine and its analogs	L-Pyrrolysine[a]	Pyl	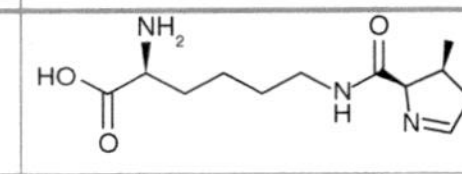	–
	3′-Azibutyl-*N*-carbamoyl-lysine	AbK	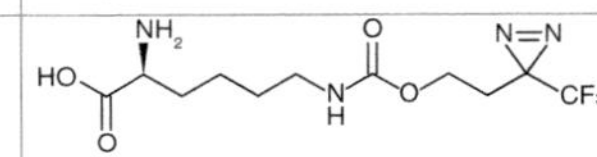	Bacteria (Hancock et al. 2010); yeast (Chou et al. 2010); mammal (Ai et al. 2011)
	(3-(3-Methyl-3*H*-diazirine-3-yl)-propaminocarbonyl-N^{ε}-L-lysine	DiZPK	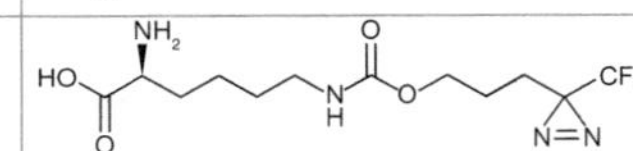	Bacteria (Lin et al. 2011)
	N^{ε}-Benzyloxycarbonyl-L-lysine[a]	ZLys	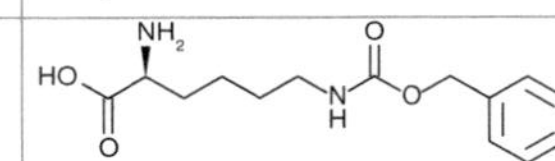	Bacteria (Yanagisawa et al. 2008); mammal (Yanagisawa et al. 2008)
	N^{ε}-[((4-(3-(Trifluoromethyl)-3*H*-diazirin-3-yl)-benzyl)oxy) carbonyl]-L-lysine	pTmdZLys	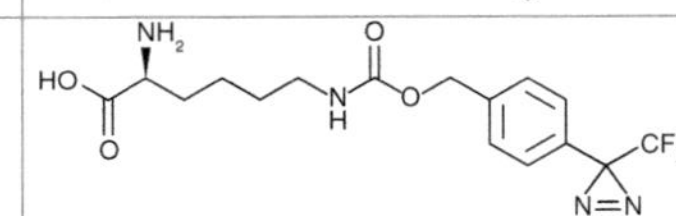	Bacteria (Yanagisawa et al. 2012); mammal (Yanagisawa et al. 2012)
	N^{ε}-[((3-(3-(Trifluoromethyl)-3*H*-diazirin-3-yl)-benzyl)oxy) carbonyl]-L-lysine	mTmdZLys	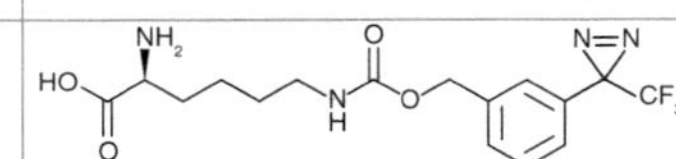	Mammal (Kita et al. 2016)
	N^{ε}-(3-Amino-5-azido-benzyloxycarbonyl)-L-lysine	AmAzZLys	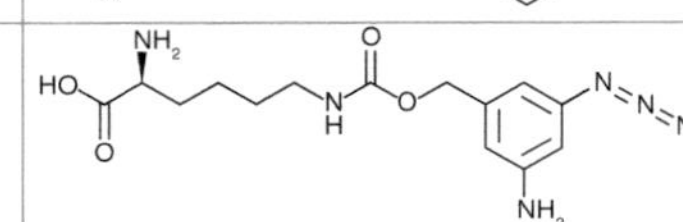	Bacteria (Yamaguchi et al. 2016)
	N^{ε}-(3-(3-Methyl-3*H*-diazirine-3-yl)-propaminocarbonyl-γ-seleno-L-lysine	DiZSeK	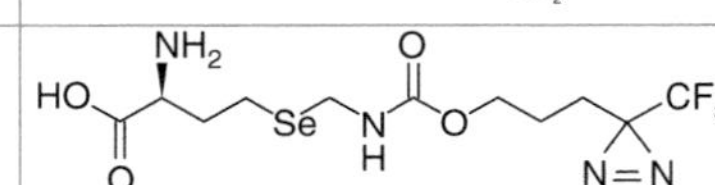	Bacteria (Lin et al. 2014)
	Se-(*N*-(3-(3-methyl-3*H*-diazirin-3-yl) propyl)propanamide)-3-yl-homoselenocysteine)	DiZHSeC	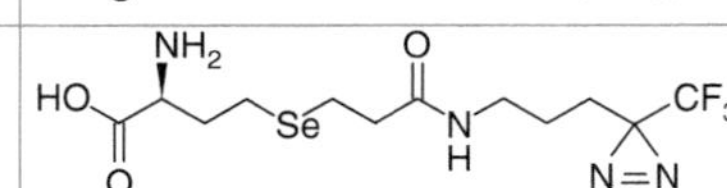	Bacteria (Yang et al. 2016); Mammal (Yang et al. 2016)

[a]These natural and unnatural amino acids do not act as a cross-linker but are shown for a comparison

light can increase the number of cross-linked products. However, the relative bulky size and the distinctive reactivity influenced by geometric constraints can limit the suitable positions to be incorporated as PAAs. Aryl azide and aryl diazirine have relatively small sizes compared with benzophenone. Therefore, the PAAs containing these photoreactive groups are not likely to interfere with the interaction of proteins, even if it is incorporated into or near the binding interface. Aryl azide and aryl diazirine can be photolyzed by 350–365 nm light, although it has been believed that aryl azide requires shorter-wavelength (250–300 nm) UV light for its activation, which causes severe damage to cellular proteins and nucleic acids (Takimoto et al. 2009). After the photolysis, these groups produce nitrene and carbene intermediates, respectively, and react with neighboring C–H and heteroatom–H bonds, forming C–C and C–heteroatom bonds. Their broad reactivity allows them to capture neighboring targets without discrimination. However, an absence of targets results in loss of the functional group by reacting with water molecules. Alkyl diazirine has the smallest size in the photoreactive group and has a reactivity similar to that of aryl diazirine. Therefore, PAAs containing this group can be designed relatively free from structural restrictions to be recognized by aaRS (Suchanek et al. 2005; Yang et al. 2015).

8.2.3 Genetic Encoding of PAAs as Phenylalanine Derivatives

PAAs have usually been incorporated into proteins as a phenylalanine or lysine analogs by using an orthogonal pair of aaRS and tRNA originally dedicated to tyrosine or pyrrolysine, respectively. A pioneering study from the Schultz lab achieved site-specific incorporation of *p*-benzoyl-L-phenylalanine (pBpa), a benzophenone-containing phenylalanine analogous PAA, into proteins in *E. coli* cells (Chin et al. 2002a). For genetic encoding of pBpa, they used a UAG-reading tyrosine tRNA mutant and tyrosyl-tRNA synthetase (TyrRS) from *Methanococcus jannaschii*, the orthogonality of this pair in *E. coli* cells having been established in their previous work (Wang et al. 2001). First, they prepared a *M. jannaschii* TyrRS library of mutants possessing five randomized residues surrounding the tyrosine-binding pocket and screened the library to evolve a TyrRS mutant recognizing pBpa but none of the canonical 20 amino acids. Then, they incorporated pBpa into glutathione *S*-transferase (GST) at position 52, which is in the dimer interface, and at position 198, which is on the opposite surface from the dimer interface. Upon exposure of the proteins to light in vitro, only the GST mutant containing pBpa at position 52 formed a cross-linked dimer, showing the applicability of this technology for probing protein–protein interactions. They also demonstrated that the cross-linking of the GSTs could be induced in living cells by direct photo-irradiation of the cells (Chin and Schultz 2002). Other phenylalanine analogous PAAs, *p*-azido-L-phenylalanine (pAzpa) and *p*-trifluoromethyl-diazirinyl-L-phenylalanine (tmdPhe), were genetically encoded in a similar way and site-specifically incorporated into proteins in *E. coli* cells (Chin et al. 2002b; Tippmann et al. 2007).

To extend this system to use in eukaryotic cells, it is necessary to employ another TyrRS/tRNA pair because the *M. jannaschii* TyrRS/tRNA pair is incompatible with eukaryotic cells. Chin et al. demonstrated that *E. coli* TyrRS and tyrosine tRNA satisfied the orthogonality with yeast pairs of aaRS and tRNA and successfully encoded unnatural amino acids, including pBpa and pAzpa, in yeast proteins by developing *E. coli* TyrRS variants specific to these PAAs (Chin et al. 2003). Independently, we found that a UAG-reading tyrosine tRNA mutant derived from *Bacillus stearothermophilus* can be functionally expressed in mammalian cells and used to encode an unnatural amino acid, 3-Iodo-L-tyrosine, by being paired with an *E. coli* TyrRS variant (Sakamoto et al. 2002; Kiga et al. 2002). In accordance with these observations, we surmised that pBpa can be encoded in mammalian cells by the combination of the pBpa-specific *E. coli* TyrRS and the *B. stearothermophilus* tRNA and generated the site-specific incorporation of the PAA into the cell signaling protein GRB2 in Chinese hamster ovary (CHO) cells (Hino et al. 2005). The GRB2 variant with pBpa at its ligand-binding surface, in the Src homology 2 (SH2) domain, allowed successful cross-linking with its binding partners, including the epidermal growth factor receptor (EGFR) and its family protein ERBB2. In a similar way, Takimoto et al. achieved the encoding of pAzpa in mammalian cells by using *E. coli*-derived UAG-reading tyrosine tRNA paired with *E. coli* pAzpa-specific TyrRS variant, which was evolved by screening in yeast cells (Takimoto et al. 2009). For the genetic encoding of tmdPhe in mammalian cells, we took another approach appropriating bacterial TyrRS variants that have been established to encode other phenylalanine analogs instead of the de novo development of a tmdPhe-specific enzyme (Hino et al. 2011). Since tmdPhe belongs to the group of L-phenylalanine derivatives with a bulky substituent at position 4 of the phenyl ring, we hypothesized that some of the reported *E. coli* TyrRS variants specific for such amino acids could recognize tmdPhe (Chin et al. 2003; Liu et al. 2007). Actually, we found that the variant for *para*-acetyl-L-phenylalanine efficiently recognized tmdPhe, demonstrating the applicability of the panels of established TyrRS variants, which eliminates the time-consuming screening process for the development of a new enzyme.

8.2.4 Genetic Encoding of PAAs as Lysine Derivatives

The aforementioned phenylalanine-derivative PAAs are defined as "short-range" cross-linkers, for which distances between the alpha carbon (Ca) and the photoreactive center are 6–8 Å. These short linkers are beneficial for precise mapping of the protein–protein interaction interface. However, they require steady efforts to find appropriate incorporation positions for efficient cross-linking. To address this inconvenience and achieve more efficient protein cross-linking, we and others have developed "long-range" PAAs that can be incorporated into proteins as pyrrolysine analogs. Pyrrolysine is the 22nd amino acid found in certain methanogenic archaea, including *Methanosarcina barkeri* and *Methanosarcina mazei* (Srinivasan et al.

2002; Blight et al. 2004). This amino acid has the longest side chain of naturally occurring amino acids used in biosynthesis of proteins. In 2002, Krzycki and coworkers discovered that pyrrolysine is encoded by a specific aaRS/tRNA pair and translationally incorporated at in-frame UAG codons (Srinivasan et al. 2002). Furthermore, the pair of pyrrolysine tRNA and pyrrolysyl-tRNA synthetase (PylRS) was shown to be orthogonal to prokaryotic and eukaryotic aaRS/tRNA pairs (Polycarpo et al. 2006; Ambrogelly et al. 2007; Mukai et al. 2008; Hancock et al. 2010). This means that the PylRS variant established using an *E. coli* screening system can readily be used in yeast or mammalian cells.

The first genetically encoded pyrrolysine-based PAA was 3′-azibutyl-*N*-carbamoyl lysine (AbK), which is an alkyl diazirine-containing lysine derivative (Table 8.1) (Hancock et al. 2010). Owing to its structural similarity with pyrrolysine, AbK is recognized by the wild-type *M. barkeri* or *M. mazei* PylRS/tRNA^{Pyl} pair and incorporated into proteins in *E. coli*, yeast, and mammalian cells (Hancock et al. 2010; Chou et al. 2010; Ai et al. 2011). Lin et al. encoded (3-(3-methyl-3*H*-diazirine-3-yl)-propaminocarbonyl-N^{ε}-L-lysine (DiZPK), which is longer by one methylene unit than AbK, by developing a *M. barkeri* PylRS mutant specific to this amino acid (Lin et al. 2011). More recently, we achieved genetic encoding of aryl diazirine-containing lysine derivatives N^{ε}-[((4-(3-(trifluoromethyl)-3*H*-diazirin-3-yl)-benzyl)oxy)carbonyl]-L-lysine (pTmdZLys) and N^{ε}-[((3-(3-(trifluoromethyl)-3*H*-diazirin-3-yl)-benzyl)oxy) carbonyl]-L-lysine (mTmdZLys), by using *M. mazei* PylRS that was engineered to recognize another aryl group-containing lysine derivative, N^{ε}-benzyloxycarbonyl-L-lysine (Yanagisawa et al. 2012, 2008; Kita et al. 2016). pTmdZLys is the longest PAA that has been encoded, for which the distance between the reactive center and the Cα is 15 Å. mTmdZLys shows high incorporation efficiency into proteins, although it has a slightly shorter length than pTmdZLys (Kita et al. 2016). The rest of the PAAs listed in Table 8.1 are N^{ε}-(3-amino-5-azido-benzyloxycarbonyl)-L-lysine (AmAzZLys), N^{ε}-(3-(3-methyl-3*H*-diazirine-3-yl)-propaminocarbonyl-γ-seleno-L-lysine (DiZSeK), and Se-(*N*-(3-(3-methyl-3*H*-diazirin-3-yl)propyl)propanamide)-3-yl-homoselenocysteine) (DiZHSeC), which have other functional groups providing a chemical handle for further modification and a cleavable site, which can support specific identification of the cross-linked products and the cross-linking sites (Yamaguchi et al. 2016; Lin et al. 2014; Yang et al. 2016). Interestingly, recently developed polyspecific PylRS variants were shown to recognize not only pyrrolysine analogs but also phenylalanine derivatives, including pBpa and pAzpa (Lacey et al. 2013; Wang et al. 2013; Guo et al. 2014).

8.2.5 *Side-by-Side Comparison Between PAAs: pBpa, tmdPhe, and pTmdZLys*

Side-by-side comparison of the cross-linking properties between pBpa, tmdPhe, and pTmdZLys clearly explains their unique benefits for spatial and temporal aspects in protein–protein interaction analysis (Hino et al. 2005, 2011; Yanagisawa et al. 2012). We incorporated these PAAs at positions 103–112 of GRB2. These

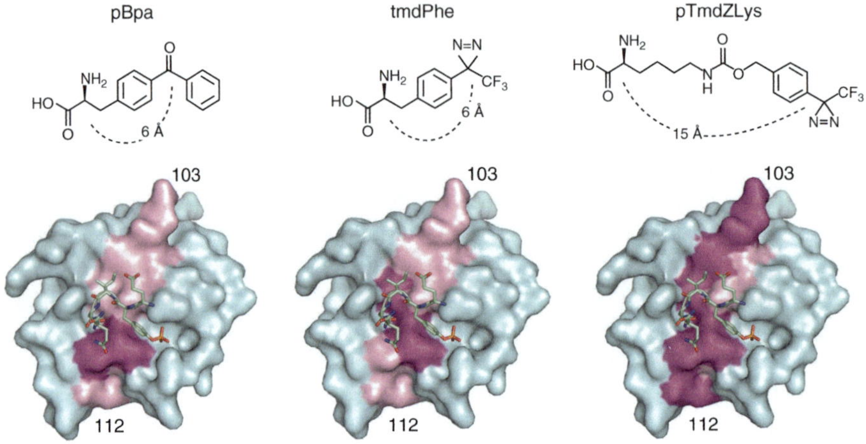

Fig. 8.2 Side-by-side comparison between PAAs: pBpa, tmdPhe, and pTmdZLys. (*Upper panels*) The structures of pBpa, tmdPhe, and pTmdZLys and the distance between the α-carbon and photo-reactive center developing after the photolysis. (*Lower panels*) The structure of the SH2 domain of GRB2 (*cyan*) complexed with a peptide ligand derived from EGFR (shown as *green sticks*) (Rahuel et al. 1998). The GRB2 residues replaced by each of the PAAs and efficient cross-linking with EGFR are shown in *pink* and *magenta*, respectively (Hino et al. 2005, 2011; Yanagisawa et al. 2012)

positions are located in the SH2 domain and are in proximity to a bound EGFR peptide with phosphor-tyrosine residue in the reported crystal structure (Fig. 8.2) (Rahuel et al. 1998). The cross-linking analysis between the GRB2 mutants and EGFR shows that pBpa at positions 109 and 111 and tmdPhe at 108 and 109 formed cross-links most efficiently. Among the tested positions, position 108 is the most proximal to the peptide ligand in the crystal structure and allowed cross-linking by tmdPhe, showing less steric hindrance than pBpa. On the other hand, the cross-linking pattern with pTmdZLys showed that positions 104, 106, 108, 109, and 111 allowed efficient cross-linking with EGFR, revealing that the cross-linkable range with pTmdZLys is significantly extended. Moreover, this comparison clearly shows that pTmdZLys can act as a "wide-range" cross-linker, cross-linking interacting proteins over a long distance as well as covering shorter ranges. These properties of pTmdZLys increase the chances of cross-linking and help to find the binding interfaces between proteins, while the short-reaching pBpa and tmdPhe are still useful to map protein–protein interfaces with high resolution.

The time course for photo-cross-linking reaction of the GRB2 mutant to EGFR using tmdPhe and pBpa was also compared (Hino et al. 2011). The cross-linked product with tmdPhe was detected as early as 1 min after the initiation of photo-irradiation, and the yield of the product was nearly maximal within 5 min. In contrast, cross-linking with pBpa reached its maximal level more slowly, within 10–15 min, but the product yield was more abundant than with tmdPhe at the time point of 15 min. This observation indicates that diazirine-based PAA is useful to interrogate the changes in protein interactions with high temporal resolution, while benzophenone-based PAA is superior to investigate the interactions of low abundant partners.

8.3 Applications of In Vivo Protein Photo-Cross-Linking

As described above, genetic encoding of PAAs enables site-specific incorporation of a variety of photoreactive moieties onto a desired surface of proteins in living cells. The benefits of this technology have especially been useful in the identification of direct binding partners of target proteins in living cells, trapping extremely weak and transient interactions, and precise mapping of binding interface between interacting proteins under physiologically relevant conditions.

8.3.1 Identification of Direct Binding Partners of a Target Protein in Living Cells

We applied this strategy to analyze cell signaling protein–protein interactions occurring in living cells (Hino et al. 2005). GRB2 is a signaling adaptor protein consisting of one SH2 domain and two terminal SH3 domains; the SH2 domain interacts with various tyrosine-phosphorylated proteins, including EGFR. From the reported tertiary structure of the SH2 domain of GRB2 complexed with a ligand peptide, we chose the leucine residue at position 111 to be replaced by pBpa because it is located in the neighborhood of the bound ligand but is not involved in the binding (Fig. 8.2). The pBpa-containing GRB2 mutant was functionally expressed in CHO cells and cross-linked to concurrently expressed EGFR by photo-exposure. This cross-link formation was dependent on the EGF treatment of the cells, which induces activation and phosphorylation of EGFR and association of GRB2 with the receptor. Furthermore, we found that the GRB2-pBpa mutant was also cross-linked to the EGF receptor family member protein ERBB2, which is endogenously expressed in CHO cells. These results demonstrate that the genetic encoding of PAAs is useful to study physiological protein–protein interactions occurring in living cells. In a subsequent study, we conducted comprehensive identification of GRB2-SH2 binding proteins by using in vivo photo-cross-linking and mass analysis in combination (Hino et al. 2011). We incorporated another PAA tmdPhe at position 109 of the SH2 domain of GRB2 and showed that the GRB2-tmdPhe mutant was cross-linked to not only EGFR but also SHC, which is known to bind to GRB2-SH2 after phosphorylation of tyrosine residues. This observation suggested that this position is suitable for the tmdPhe-mediated cross-linking of a variety of proteins directly binding to the SH2 domain of GRB2. To identify the proteins cross-linked with the PAA variant in human embryonic kidney (HEK) 293 c18 cells, we applied a quantitative and comparative mass spectrometric strategy based on stable isotope labeling of amino acids in cell culture (SILAC), which enables comparisons of protein abundance between samples (Fig. 8.3) (Ong et al. 2002). Cells were divided into two lines, with one being grown in a medium with normal or "light" lysine and arginine and the other being grown in a medium with lysine and arginine containing isotopic or "heavy" atoms. GRB2-tmdPhe was expressed in both cell lines, and only

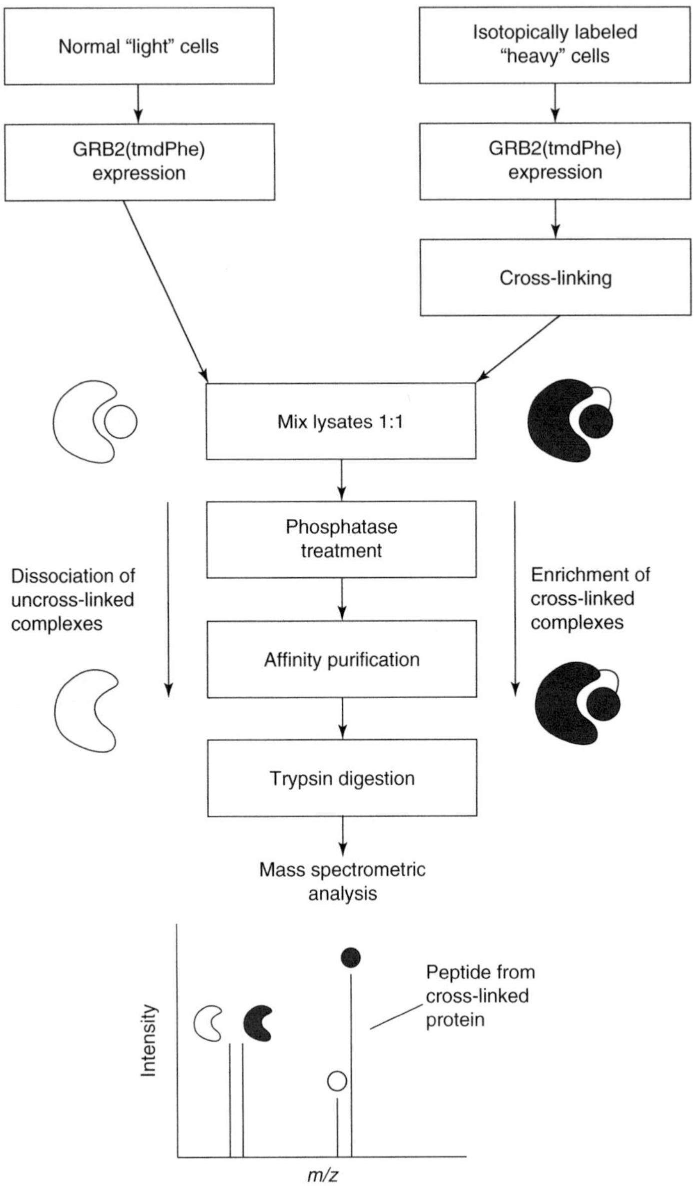

Fig. 8.3 Strategy for identifying the direct binding partners of GRB2-SH2 by using in vivo photo-cross-linking and quantitative mass spectrometric approaches in combination. HEK293 c18 cell cultures are divided into two populations and differentially labeled with "light" and "heavy" arginine and lysine. GRB2(tmdPhe) is expressed in both of the cells, while only the "heavy" cells are exposed to UV to induce a cross-linking reaction. Lysates from the two cultures are mixed at a 1:1 ratio and then treated with alkaline phosphatase, followed by immunoprecipitation. During this process, non-cross-linked complexes are dissociated, while cross-linked complexes are enriched in the "heavy" fraction. Therefore, cross-linked partners of GRB2(tmdPhe) should be abundant in the "heavy" fraction. Each peptide generated from the precipitated proteins has an arginine or lysine with different masses, enabling comparison of the abundance of two identical proteins included in the non-cross-linked or cross-linked populations by mass spectrometric analysis

the "heavy" cells were exposed to light to allow cross-link formation. Lysates of the "heavy" and "light" cells were mixed together and treated with alkaline phosphatase to dissociate non-covalently bound proteins from GRB2 interacting via phosphotyrosine. Cross-linked proteins of GRB2-SH2 were expected to be concentrated in the "heavy" fraction after immunoprecipitation of GRB2 complexes. We found that the overproduced EGFR and the endogenous SHP2, which are known to be direct targets of GRB2-SH2, were enriched in the "heavy" fraction as compared with the "light" fraction with ratios of 6.0 and 1.6, respectively. On the other hand, known targets of the SH3 domain of GRB2, such as SOS, dynamin, and synaptojanin, were equally detected in both fractions, indicating that the site-specifically incorporated PAA exclusively captured direct binders of the SH2 domain of GRB2. Other proteins showing a ratio >1.5 included two signaling proteins (AF6 and GIT1) and five RNA-binding proteins (hnRNPs F, H1, H2 and K, and p54nrb) and were considered to be good candidates as direct binding partners of GRB2-SH2. Subsequent photo-cross-linking analysis of GRB2-tmdPhe co-expressed with each of these candidate proteins clearly demonstrated their direct binding.

In a similar method, Okada et al. incorporated pBpa into the SH3 domains of nine representative Rho family guanosine triphosphatase-activating proteins (GAPs) to cross-link with their binding partners in HEK293 cells and identified a total of 860 candidates by mass spectrometric analysis (Okada et al. 2011). The following in silico validation and in vitro binding assays elucidated 54 SH3-mediated interactions, including 51 novel interactions. Notably, the pBpa-modified SH3 domains retain the ligand preference of the wild-type domains, indicating that pBpa does not impair the ligand recognition of the domain if it is not incorporated at the site directly involved in the interaction. Furthermore, they functionally characterized one of the identified interactions, between the neuronal proteins gephyrin and SRGAP or the SRGAP homologous protein WRP, and found that WRP facilitates the postsynaptic association between gephyrin and $GABA_A$ receptors.

Zhang et al. utilized this technology to probe protein interactions occurring under severe conditions (Zhang et al. 2011). HdeA, an acid chaperone of enteric bacteria, prevents acid-induced aggregation of periplasmic proteins when the bacteria pass through the mammalian stomach to reach the intestine. Zhang et al. first incorporated DiZPK into the dimer interface of the bacterial acid chaperone protein HdeA in *E. coli* cells and found that the protein exclusively cross-linked with itself under neutral pH conditions, but cross-linked with many other proteins at extremely acidic pH conditions. This result is consistent with previous observations showing that HdeA forms a stable homodimer at neutral pH and transforms into a plastic monomeric conformation for exerting its chaperone activity (Hong et al. 2005; Tapley et al. 2009). Next, they conducted mass spectrometric analysis of the captured proteins under acidic conditions and identified a total of 32 proteins including SurA and DegP, both of which are periplasmic factors associated with protein quality control (Sklar et al. 2007). Further functional analysis of the identified interaction revealed that HdeA prevents acid-induced aggregation of SurA and DegP and these proteins assist HdeA-mediated substrate refolding.

8.3.2 Covalently Capturing Weak and Transient Interactions

Photo-cross-linking becomes a "killer application" in aim to capture extremely weak and transient protein interactions, especially in substrate recognition of chaperones or protein trafficking (Schlieker et al. 2004; Weibezahn et al. 2004; Haslberger et al. 2007; Freinkman et al. 2011; Tamura et al. 2009; Shiota et al. 2011, 2015). Schlieker et al. clearly demonstrated the usability of site-specific photo-cross-linking technique to reveal the mechanism of substrate recognition by the AAA+ chaperone protein ClpB (Schlieker et al. 2004). ClpB is a major heat shock protein of *E. coli* and cooperates with the DnaK chaperone system to solubilize and refold aggregated proteins (Mogk et al. 1999; Motohashi et al. 1999). The substrate recognition by ClpB has been shown to be directly mediated by either or both of the two ATPase domains called AAA domains (Weibezahn et al. 2003). However, it was unclear where the substrate-binding site is located. Through several biochemical analyses and developing a hexameric structure model of ClpB, the authors estimated the central pore of the first AAA domain to be a potential substrate-binding site, which is formed by the conserved residues Tyr251, Glu254, and Glu257. To demonstrate direct interactions between the ClpB pore site and substrates, the Tyr251 residue was replaced by pBpa. A cross-linking product was observed when the reconstituted complex of the ClpB mutant and a peptide substrate was exposed to UV, indicating that the substrate-binding site is located near the central pore of the first AAA domain. This cross-linking was observed only in the presence of ATPγS, consistent with the previous finding that the association of ClpB with substrates is ATP-dependent (Srinivasan et al. 2002). An identical result was also obtained when a protein substrate was used in place of the peptide.

Mitochondrial proteins are mainly synthesized in the cytosol and then transported and sorted into subcompartments of the mitochondria. These processes are mediated by the mitochondrial membrane–protein complexes called translocators. Endo and coworkers have elucidated the mechanism of protein transport by dynamic cooperation of the translocators in the mitochondrial outer and inner membranes by taking snapshots of interactions between the translocators and their client proteins using site-specific photo-cross-linking analysis in yeast cells and in isolated mitochondria (Tamura et al. 2009; Shiota et al. 2011, 2015). To investigate how the precursor proteins are transported from the cytosol into the mitochondria, they first incorporated pBpa into various sites of the Tom22 protein, which is one of the subunits of the outer membrane translocator TOM40 complex (Shiota et al. 2011). Tom22 spans the outer membrane by its central transmembrane segment, with its N-terminal and C-terminal domains exposed to the cytosol and intermembrane space, respectively. They overexpressed Tom22 mutants containing pBpa at various positions in the cytosolic domain and found that the mutants with pBpa at positions 44–52 were cross-linked to the Tom20 protein, another component of TOM40 complex. In addition, in the presence of a substrate precursor protein, the Tom22 mutant with pBpa at position 48 was cross-linked to the substrate, while the amount of cross-linked product between the Tom22 mutant and Tom20 decreased. This result suggests that the clamp of the cytosolic receptor domains of Tom22 and Tom20 opens to accept the pre-sequence of the precursor protein for targeting signal recognition. On the other hand, a Tom22 mutant having pBpa at position 136 in the intermembrane space domain was cross-linked to the Tom50 protein, an essential subunit of the inner-membrane translocator TIM23 complex. In the presence of substrate proteins, the Tom22-Tom50 cross-linking was also decreased, showing that Tom22 and Tom50 dissociate from each other upon precursor protein transfer from the TOM40 complex to the TIM23 complex. They concluded that such dynamic interactions between Tom22 and Tom50 may facilitate efficient pre-sequence transfer from the TOM40 complex to the downstream TIM23 complex. Then, they approached the next question: how are precursor proteins passed through the outer membrane of mitochondria (Shiota et al. 2015)? The channel-forming β-barrel protein Tom40 is the core subunit of the outer membrane translocator TOM40 complex and plays a major role in protein entry into mitochondria. However, it had been controversial whether proteins transit through the lumen of the Tom40 β-barrel or via the interstitial space between multiple β-barrels. To resolve this, they separately introduced pBpa at 108 different positions in the 387-residue Tom40 protein in yeast cells and analyzed cross-linking to precursor proteins. In consideration with the modeled structure of Tom40, all of the cross-linked positions were on the interior of the channel, indicating that proteins in transit are located inside the β-barrel pore of Tom40. Furthermore, through detailed mapping of cross-linked sites of Tom40 to different precursor proteins, they found that positively charged pre-sequences pass the Tom 40 channel along the acidic pathway of the inner wall, while other proteins without such pre-sequences pass through the channel along the hydrophobic residues.

8.3.3 *Validation of Protein Complex Structures Under Physiologically Relevant Conditions*

Although X-ray crystallography is a well-established tool for structural determination of protein complexes, native proteins are unable to be crystallized in many cases. Therefore, many protein structures are based on a truncated or fragmented form of the protein, which has been heterogeneously expressed in *E. coli* cells, purified, mixed to regenerate a complex, and then crystalized under unnatural conditions. This requires additional analysis to ascertain whether the solved structure reflects the native state of the protein complex, such as testing the effects of amino acid mutations in the binding interface on its affinity (Cunningham and Wells 1989; Weiss et al. 2000). However, such experimental settings are sometimes difficult because most single-site mutations produce only minor effects on protein interaction and multisite mutations can disrupt local or overall conformation of the protein. In contrast, site-specific incorporation of PAAs into the determined interface gives obvious and detailed information about the contacting points of the proteins in

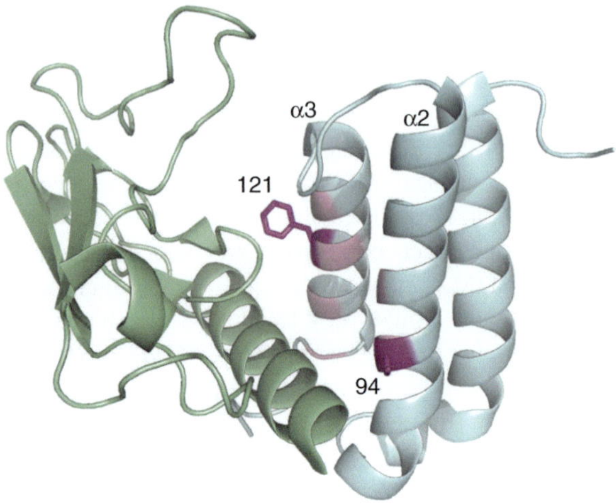

Fig. 8.4 Site-specific cross-linking between membrane proteins RAMP2 and CRLR to validate the determined co-crystalized structures. The extracellular domains of RAMP2 and CRLR are shown in *cyan* and *green*, respectively. The residues replaced by pBpa on the α2 and α3 helices of RAMP1 are shown in *pink*. Substitutions with pBpa at positions 94 and 121 allowing efficient cross-linking with CRLR are shown in *magenta* (Kusano et al. 2012a)

complex under physiologically relevant conditions (Kusano et al. 2012a, b; Hara-Yokoyama et al. 2012; Shimizu et al. 2016; Tagami et al. 2010) (Fig. 8.4).

We determined the crystal structure of the complex of extracellular domains of human membrane protein RAMP2 and CRLR (Kusano et al. 2012a). To probe the binding mode between the full-length RAMP2 and CRLR molecules by a photo-cross-linking approach, pBpa was separately incorporated at eight sites of the α2 and α3 helices of RAMP2, which interact with CRLR in the determined crystal structure. Each RAMP2 mutant was expressed as a full-length protein in HEK293 cells and cross-linked to co-expressed CRLR. The cross-linking patterns closely corresponded to the crystal structure of the extracellular domain complex. Two of the RAMP2 variants with pBpa at positions 94 and 121, each located adjacent to the binding interface of the α2 and α3 helices, respectively, were most efficiently cross-linked with CRLR. This result indicates that the helices of RAMP2 interact with CRLR on the cell membrane in the same manner as determined crystal structure. The same approach was applied to verify other protein complexes, such as interleukin-5 and its receptor complex or CD38 assembly on the mammalian cell surface (Kusano et al. 2012b; Hara-Yokoyama et al. 2012), multimeric voltage-gated sodium channel β subunits interacting in trans at cell–cell adhesion sites (Shimizu et al. 2016), and multi-subunit RNA–polymerase complexes in *E. coli* (Tagami et al. 2010). The validity of this strategy was supported by the crystal structure of a photo-cross-linked protein complex (Sato et al. 2011). We determined the crystal structure of the cross-linked complex of gankyrin and S6 protein at 2.05 Å resolution. The structure revealed that pBpa incorporated at position 85 of gankyrin formed a covalent bond exclusively with Glu356 in S6, showing the high selectivity of cross-link formation by pBpa. In addition, the cross-linked structure exhibited little structural distortion from the native complex structure, demonstrating that cross-linking with site-specifically incorporated pBpa preserves the native binding mode.

8.3.4 Precise Mapping of Binding Interface Between a Membrane Protein and Its Ligands

Recent dramatic development in membrane protein crystallography has enabled researchers to determine the full-length structure of multiple transmembrane proteins, including G protein-coupled receptors (GPCRs), and improved our understanding of ligands binding to receptors (Cherezov et al. 2010; Katritch et al. 2013). Nevertheless, it is still challenging to elucidate how a variety of ligands bind to and modulate GPCR structure and function in the native cellular membrane environment. To gain insight into ligands binding to GPCRs on the cell surface, Grunbeck et al. incorporated pBpa into eight different positions of the CXC chemokine receptor 4 (CXCR4) and analyzed cross-linking formation with the peptide antagonist of the receptor, T140 (Grunbeck et al. 2011). A cross-linked product was detected only when Phe189 of CXCR4 was replaced with pBpa, suggesting that T140 binds to the Phe189-containing extracellular loop or its vicinity. In subsequent studies, the authors elucidated the binding site of the small molecular ligand maraviroc on its target receptor CCR5 and antidepressant drugs on the human serotonin transporter (Grunbeck et al. 2012; Rannversson et al. 2016). Independently, Coin et al. attempted to determine the ligand-binding site on the corticotropin-releasing factor receptor type 1 (CRF-R1), which is a class B GPCR with a large extracellular N-terminal domain (NTD) (Coin et al. 2011). Since the NTD was previously identified as a major binding site for the C-terminal segments of the peptide ligand CRF (Wille et al. 1999), the authors first focused on this domain for the incorporation of pAzpa. The result of in vivo photo-cross-linking between eight pAzpa-containing CRF-R1 mutants and four different radiolabeled peptide ligands, astressin, urocortin 1 (Ucn1), CRF, and sauvagine, provided two important pieces of information about the binding site of the ligands on the receptor: the general cross-linking of the Tyr116pAzpa mutant indicated the hinge region between the N-terminal and transmembrane domains to be important for the binding of all of the ligands, and the exclusive cross-linking of the Tyr73pAzpa mutant to Ucn1 suggested a different binding mode of the ligands to the loop domain containing Tyr73, which was consistent with the previous observations that CRF binds the isolated NTD with low affinity and that sauvagine does not bind to the NTD at all (Perrin et al. 2001; Klose et al. 2005). In the following study, the authors systematically incorporated pAzpa into 140 different sites throughout the extracellular half of CRF-R1 to gain a wealth of information about the binding interface between the receptor and Ucn1 (Coin et al. 2013) because it is largely unknown how the N-terminus of peptide ligands interact with the seven transmembrane domains of class B GPCRs to trigger downstream signaling events (Parthier et al. 2009). They investigated cross-linking between each of the pAza-containing

CRF-R1 mutants and Ucn1 on the cell surface of HEK293T cells and found that 35 of the 140 mutant receptors produced cross-linking adducts, indicating that these positions of CRF-R1 constituted binding interfaces with Ucn1. A three-dimensional overview of the ligand-binding path, visualized by spotting the 35 cross-linking sites on the previously determined CRF-R1 structure, clearly demonstrated that the peptide ligand penetrated to a relatively deep and very broad pocket enclosed by the seven transmembrane helix bundles of the receptor.

8.4 Recent Advances in Technology

As described above, mass spectrometric analysis of cross-linked products has enabled researchers to identify direct binding partners of a target protein in living cells (Hino et al. 2011; Lin et al. 2011; Rahuel et al. 1998). Notably, the identification of the partner protein by database annotation requires at least one unique peptide derived from trypsin digestion of the cross-linked product; therefore, a peptide distant from the cross-linking site is usually used for the determination. Nevertheless, a cross-linked peptide, a heteropeptide covalently bonded via PAA, is still informative because it contains amino acid sequences derived from both of the interacting proteins, providing valuable information about the interface topology of protein complex (Forné et al. 2012; Hatanaka 2015; Trakselis et al. 2005; Sinz 2006). However, the attempts of direct mass analysis of cross-linked peptides have been hampered due to the complexity of the mass spectra.

Chen and coworkers have recently developed a novel type of PAA with an oxidatively cleavable selenium–carbon linker (Lin et al. 2014; Yang et al. 2016), achieving probe transfer from the cross-linker-incorporated target protein to the captured partner protein. Such a technique of photoaffinity labeling and probe transfer in combination can circumvent the complexity concomitant to direct mass analysis of a cross-linked peptide. With one of the developed PAAs, DiZHSeC, they established the chemical proteomic strategy termed IMAPP (In-situ cleavage and MS-label transfer After Protein Photo-cross-linking), which enables simultaneous identification of the captured proteins and the exact cross-linking sites (Yang et al. 2016). They incorporated DiZHSeC into the dimer interface of the *E. coli* acid chaperone protein HdeA and prepared cross-linked products with the wild-type HdeA. After the treatment with H_2O_2 to cleave the C–Se linker, the trypsin digest of the product was analyzed by mass spectrometry, searching a peptide having the transferred *N*-(4,4-bis-substituted-pentyl)acrylamide (NPAA) moiety with a mass corresponding to C_8H1_3NO. As a result, one peptide sequence was identified with the moiety exclusively bound to the unique glutamic acid residue, clearly demonstrating that the residue is proximally located to the DiZHSeC incorporated into the other protein at the dimer interface. Intriguingly, the NPAA group can additionally be tagged with a tetrazole-containing fluorogenic probe through an already established photo-click reaction (Yu and Lin 2014). The authors further extended the IMAPP technique to identify client proteins of HdeA in *E. coli* cells under acidic condition. Most of the total of 52 identified proteins are envelop proteins, consistent with the fact that HdeA is fundamental periplasm chaperone. This result highly contrasted the result obtained by a traditional coprecipitation approach, in which 402 of 767 candidate proteins binding with HdeA were cytoplasmic proteins, presumably products of false-positive detection.

Genetic code expansion of physiologically relevant or medically important mammalian cells has been another pressing issue. Usually, the delivery of the genes for aaRS and tRNA to encode PAA relies on plasmid transfection, although only a few cell lines are amenable to this method. Others and we have developed adenovirus, lentivirus, baculovirus, and adeno-associated virus vectors carrying the genes for aaRS and tRNA and achieved genetic encoding of unnatural amino acids, including PAAs, in human cancerous and primary endothelial cells, rat hippocampal neural stem cells, mouse fibroblast and embryonic stem cells, and mouse brain cells (Kita et al. 2016; Shen et al. 2011; Chatterjee et al. 2013; Ernst et al. 2016). Furthermore, we demonstrated that mTmdZLys was efficiently incorporated into the signaling protein GRB2 by using the adenovirus vector and that the cross-linkable GRB2 successfully captured its endogenous partner proteins, including EGFR and SHC, in human breast cancer cells and primary umbilical vein endothelial cells (Kita et al. 2016).

8.5 Conclusion

A genetic code expansion technology has enabled site specific incorporation of PAAs with a variety of sizes and photoreactivities into proteins in living cells. Thus generated photo-cross-linkable proteins can covalently capture their direct binding partners even if their interactions are weak and transient. This will simultaneously give detailed information about their binding interface because the introduced PAA exclusively reacts with closely located amino acid residues of the partner protein. These benefits overcome the drawbacks of conventional technologies, such as co-immunoprecipitation, used for identifying in vivo protein interactions or amino acid mutagenesis for surveying the binding interface of interacting proteins. Moreover, recent developments of the novel functional PAAs and the viral vectors for genetic encoding of PAAs will facilitate the reliable identification of biologically important protein–protein interactions of medical interest.

References

Ai H, Shen W, Sagi A et al (2011) Probing protein–protein interactions with a genetically encoded photo-crosslinking amino acid. ChembioChem 12:1854–1857

Ambrogelly A, Gundllapalli S, Herring S et al (2007) Pyrrolysine is not hardwired for cotranslational insertion at UAG codons. Proc Natl Acad Sci U S A 104:3141–3146

Bayley H, Knowles JR (1977) Photoaffinity labeling. Methods Enzymol 46:69–114

Blight SK, Larue RC, Mahapatra A et al (2004) Direct charging of tRNACUA with pyrrolysine in vitro and in vivo. Nature 431:333–335

Bruce J (2012) In vivo protein complex topologies: sights through a cross-linking lens. Proteomics 12:1565–1575
Brunner J (1993) New photolabeling and crosslinking methods. Annu Rev Biochem 62:483–514
Chatterjee A, Xiao H, Bollong M et al (2013) Efficient viral delivery system for unnatural amino acid mutagenesis in mammalian cells. Proc National Acad Sci USA 110:11803–11808
Cherezov V, Abola E, Stevens RC (2010) Recent progress in the structure determination of GPCRs, a membrane protein family with high potential as pharmaceutical targets. Methods Mol Biol 654:141–168
Chin JW, Schultz PG (2002) In vivo photocrosslinking with unnatural amino acid mutagenesis. ChembioChem 3:1135–1137
Chin JW, Martin AB, King DS et al (2002a) Addition of a photocrosslinking amino acid to the genetic code of Escherichia coli. Proc Natl Acad Sci U S A 99:11020–11024
Chin J, Santoro S, Martin A et al (2002b) Addition of *p*-azido-L-phenylalanine to the genetic code of *Escherichia coli*. J Am Chem Soc 124:9026–9027
Chin J, Cropp A, Anderson C et al (2003) An expanded eukaryotic genetic code. Science 301:964–967
Chou C, Uprety R, Davis L et al (2010) Genetically encoding an aliphatic diazirine for protein photocrosslinking. Chem Sci 2:480–483
Coin I, Perrin M, Vale W, Wang L (2011) Photo-cross-linkers incorporated into G-protein-coupled receptors in mammalian cells: a ligand comparison. Angew Chem Int Ed Eng 50:8077–8081
Coin I, Katritch V, Sun T et al (2013) Genetically encoded chemical probes in cells reveal the binding path of urocortin-I to CRF class B GPCR. Cell 155:1258–1269
Cunningham BC, Wells JA (1989) High-resolution epitope mapping of hGH-receptor interactions by alanine-scanning mutagenesis. Science 244:1081–1085
Davis L, Chin J (2012) Designer proteins: applications of genetic code expansion in cell biology. Nat Rev Mol Cell Biol 13:168–182
Dormán G, Prestwich G (1994) Benzophenone photophores in biochemistry. Biochemistry 33:5661–5673
Dunham WH, Mullin M, Gingras AC (2012) Affinity-purification coupled to mass spectrometry: basic principles and strategies. Proteomics 12:1576–1590
Ernst RJ, Krogager TP, Maywood ES et al (2016) Genetic code expansion in the mouse brain. Nat Chem Biol 12:776–778
Forné I, Ludwigsen J, Imhof A et al (2012) Probing the conformation of the ISWI ATPase domain with genetically encoded photoreactive crosslinkers and mass spectrometry. Mol Cell Proteomics 11:M111.012088
Freinkman E, Chng S-S, Kahne D (2011) The complex that inserts lipopolysaccharide into the bacterial outer membrane forms a two-protein plug-and-barrel. Proc Natl Acad Sci U S A 108:2486–2491
Grunbeck A, Huber T, Sachdev P, Sakmar TP (2011) Mapping the ligand-binding site on a G protein-coupled receptor (GPCR) using genetically encoded photocrosslinkers. Biochemistry 50:3411–3413
Grunbeck A, Huber T, Abrol R et al (2012) Genetically encoded photo-cross-linkers map the binding site of an allosteric drug on a G protein-coupled receptor. ACS Chem Biol 7:967–972
Guo L-T, Wang Y-S, Nakamura A et al (2014) Polyspecific pyrrolysyl-tRNA synthetases from directed evolution. Proc Natl Acad Sci U S A 111:16724–16729
Hancock SM, Uprety R, Deiters A, Chin JW (2010) Expanding the genetic code of yeast for incorporation of diverse unnatural amino acids via a pyrrolysyl-tRNA synthetase/tRNA pair. J Am Chem Soc 132:14819–14824
Hara-Yokoyama M, Kukimoto-Niino M, Terasawa K et al (2012) Tetrameric interaction of the ectoenzyme CD38 on the cell surface enables its catalytic and raft-association activities. Structure 20:1585–1595
Haslberger T, Weibezahn J, Zahn R et al (2007) M domains couple the ClpB threading motor with the DnaK chaperone activity. Mol Cell 25:247–260

Hatanaka Y (2015) Development and leading-edge application of innovative photoaffinity labeling. Chem Pharm Bull 63:1–12

Hino N, Okazaki Y, Kobayashi T et al (2005) Protein photo-cross-linking in mammalian cells by site-specific incorporation of a photoreactive amino acid. Nat Methods 2:201–206

Hino N, Oyama M, Sato A et al (2011) Genetic incorporation of a photo-crosslinkable amino acid reveals novel protein complexes with GRB2 in mammalian cells. J Mol Biol 406:343–353

Hong W, Jiao W, Hu J et al (2005) Periplasmic protein HdeA exhibits chaperone-like activity exclusively within stomach pH range by transforming into disordered conformation. J Biol Chem 280:22079–27034

Katritch V, Cherezov V, Stevens RC (2013) Structure-function of the G protein–coupled receptor superfamily. Annu Rev Pharmacol Toxicol 53:531–556

Kauer JC et al (1986) p-Benzoyl-L-phenylalanine, a new photoreactive amino acid. J Biol Chem 261:10695–10700

Kiga D, Sakamoto K, Kodama K et al (2002) An engineered *Escherichia coli* tyrosyl–tRNA synthetase for site-specific incorporation of an unnatural amino acid into proteins in eukaryotic translation and its application in a wheat germ cell-free system. Proc Natl Acad Sci USA 99:9715–9720

Kita A, Hino N, Higashi S et al (2016) Adenovirus vector-based incorporation of a photo-cross-linkable amino acid into proteins in human primary cells and cancerous cell lines. Sci Rep 6:36946

Klose J, Fechner K, Beyermann M et al (2005) Impact of N-terminal domains for corticotropin-releasing factor (CRF) receptor-ligand interactions. Biochemistry 44:1614–1623

Kusano S, Kukimoto-Niino M, Hino N et al (2012a) Structural basis for extracellular interactions between calcitonin receptor-like receptor and receptor activity-modifying protein 2 for adrenomedullin-specific binding. Protein Sci 21:199–210

Kusano S, Kukimoto-Niino M, Hino N et al (2012b) Structural basis of interleukin-5 dimer recognition by its α receptor. Protein Sci 21:850–864

Lacey V, Louie G, Noel J, Wang L (2013) Expanding the library and substrate diversity of the pyrrolysyl–tRNA synthetase to incorporate unnatural amino acids containing conjugated rings. ChembioChem 14:2100–2105

Lin S, Zhang Z, Xu H et al (2011) Site-specific incorporation of photo-cross-linker and bioorthogonal amino acids into enteric bacterial pathogens. J Am Chem Soc 133:20581–20587

Lin S, He D, Long T et al (2014) Genetically encoded cleavable protein photo-cross-linker. J Am Chem Soc 136:11860–11863

Liu W, Brock A, Chen S et al (2007) Genetic incorporation of unnatural amino acids into proteins in mammalian cells. Nat Methods 4:239–244

Markham K, Bai Y, Schmitt-Ulms G (2007) Co-immunoprecipitations revisited: an update on experimental concepts and their implementation for sensitive interactome investigations of endogenous proteins. Anal Bioanal Chem 389:461–473

Mogk A, Tomoyasu T, Goloubinoff P et al (1999) Identification of thermolabile *Escherichia coli* proteins: prevention and reversion of aggregation by DnaK and ClpB. EMBO J 18:6934–6949

Motohashi K, Watanabe Y, Yohda M, Yoshida M (1999) Heat-inactivated proteins are rescued by the DnaKJ-GrpE set and ClpB chaperones. Proc Natl Acad Sci U S A 96:7184–7189

Mukai T, Kobayashi T, Hino N et al (2008) Adding L-lysine derivatives to the genetic code of mammalian cells with engineered pyrrolysyl-tRNA synthetases. Biochem Biophys Res Commun 371:818–822

Okada H, Uezu A, Mason F et al (2011) SH3 domain-based phototrapping in living cells reveals rho family GAP signaling complexes. Sci Signal 4:rs13

Ong S-E, Blagoev B, Kratchmarova I et al (2002) Stable isotope labeling by amino acids in cell culture, SILAC, as a simple and accurate approach to expression proteomics. Mol Cell Proteomics 1:376–386

Parthier C, Reedtz-Runge S, Rudolph R (2009) Passing the baton in class B GPCRs: peptide hormone activation via helix induction? Trends Biochem Sci 34:303–310

Perrin MH, Fischer WH, Kunitake KS et al (2001) Expression, purification, and characterization of a soluble form of the first extracellular domain of the human type 1 corticotropin releasing factor receptor. J Biol Chem 276:31528–31534

Polycarpo C, Herring S, Bérubé A et al (2006) Pyrrolysine analogues as substrates for pyrrolysyl-tRNA synthetase. FEBS Lett 580:6695–6700

Rahuel J, García-Echeverría C, Furet P et al (1998) Structural basis for the high affinity of amino-aromatic SH2 phosphopeptide ligands. J Mol Biol 279:1013–1022

Rannversson H, Andersen J, Sørensen L et al (2016) Genetically encoded photocrosslinkers locate the high-affinity binding site of antidepressant drugs in the human serotonin transporter. Nat Commun 7:11261

Sakamoto K, Hayashi A, Sakamoto A et al (2002) Site-specific incorporation of an unnatural amino acid into proteins in mammalian cells. Nucleic Acids Res 30:4692–4699

Sato S, Mimasu S, Sato A et al (2011) Crystallographic study of a site-specifically cross-linked protein complex with a genetically incorporated photoreactive amino acid. Biochemistry 50:250–257

Schlieker C, Weibezahn J, Patzelt H et al (2004) Substrate recognition by the AAA+ chaperone ClpB. Nat Struct Mol Biol 11:607–615

Shen B, Xiang Z, Miller B et al (2011) Genetically encoding unnatural amino acids in neural stem cells and optically reporting voltage-sensitive domain changes in differentiated neurons. Stem Cells 29:1231–1240

Shimizu H, Miyazaki H, Ohsawa N et al (2016) Structure-based site-directed photo-crosslinking analyses of multimeric cell-adhesive interactions of voltage-gated sodium channel β subunits. Sci Rep 6:26618

Shiota T, Mabuchi H, Tanaka-Yamano S et al (2011) In vivo protein-interaction mapping of a mitochondrial translocator protein Tom22 at work. Proc Natl Acad Sci U S A 108:15179–15183

Shiota T, Imai K, Qiu J et al (2015) Molecular architecture of the active mitochondrial protein gate. Science 349:1544–1548

Sinz A (2006) Chemical cross-linking and mass spectrometry to map three-dimensional protein structures and protein-protein interactions. Mass Spectrom Rev 25:663–682

Sklar JG, Wu T, Kahne D, Silhavy TJ (2007) Defining the roles of the periplasmic chaperones SurA, Skp, and DegP in *Escherichia coli*. Genes Dev 21:2473–2484

Srinivasan G, James CM, Krzycki JA (2002) Pyrrolysine encoded by UAG in Archaea: charging of a UAG-decoding specialized tRNA. Science 296:1459–1462

Suchanek M, Radzikowska A, Thiele C (2005) Photo-leucine and photo-methionine allow identification of protein-protein interactions in living cells. Nat Methods 2:261–267

Tagami S, Sekine S-I, Kumarevel T et al (2010) Crystal structure of bacterial RNA polymerase bound with a transcription inhibitor protein. Nature 468:978–982

Takimoto JK, Adams KL, Xiang Z, Wang L (2009) Improving orthogonal tRNA-synthetase recognition for efficient unnatural amino acid incorporation and application in mammalian cells. Mol BioSyst 5:931–934

Tamura Y, Harada Y, Shiota T et al (2009) Tim23-Tim50 pair coordinates functions of translocators and motor proteins in mitochondrial protein import. J Cell Biol 184:129–141

Tanaka Y, Bond M, Kohler J (2008) Photocrosslinkers illuminate interactions in living cells. Mol BioSyst 4:473–480

Tapley TL, Körner JL, Barge MT et al (2009) Structural plasticity of an acid-activated chaperone allows promiscuous substrate binding. Proc Natl Acad Sci U S A 106:5557–5562

Tippmann E, Liu W, Summerer D et al (2007) A genetically encoded diazirine photocrosslinker in *Escherichia coli*. ChembioChem 8:2210–2214

Trakselis MA, Alley SC, Ishmael FT (2005) Identification and mapping of protein-protein interactions by a combination of cross-linking, cleavage, and proteomics. Bioconjug Chem 16:741–775

Vasilescu J, Guo X, Kast J (2004) Identification of protein-protein interactions using in vivo cross-linking and mass spectrometry. Proteomics 4:3845–3854

Wang L, Brock A, Herberich B, Schultz P (2001) Expanding the genetic code of *Escherichia coli*. Science 292:498–500

Wang Y-S, Fang X, Chen H-Y et al (2013) Genetic incorporation of twelve meta-substituted phenylalanine derivatives using a single pyrrolysyl-tRNA synthetase mutant. ACS Chem Biol 8:405–415

Weibezahn J, Schlieker C, Bukau B, Mogk A (2003) Characterization of a trap mutant of the AAA+ chaperone ClpB. J Biol Chem 278:32608–32617

Weibezahn J, Tessarz P, Schlieker C et al (2004) Thermotolerance requires refolding of aggregated proteins by substrate translocation through the central pore of ClpB. Cell 119:653–665

Weiss GA, Watanabe CK, Zhong A (2000) Rapid mapping of protein functional epitopes by combinatorial alanine scanning. Proc Natl Acad Sci U S A 97:8950–8954

Wille S, Sydow S, Palchaudhuri M et al (1999) Identification of amino acids in the N-terminal domain of corticotropin-releasing factor receptor 1 that are important determinants of high-affinity ligand binding. J Neurochem 72:388–395

Yamaguchi A, Matsuda T, Ohtake K et al (2016) Incorporation of a doubly functionalized synthetic amino acid into proteins for creating chemical and light-induced conjugates. Bioconjug Chem 27:198–206

Yanagisawa T, Ishii R, Fukunaga R et al (2008) Multistep engineering of pyrrolysyl-tRNA synthetase to genetically encode N(epsilon)-(o-azidobenzyloxycarbonyl) lysine for site-specific protein modification. Chem Biol 15:1187–1197

Yanagisawa T, Hino N, Iraha F et al (2012) Wide-range protein photo-crosslinking achieved by a genetically encoded N^{ε}-(benzyloxycarbonyl)lysine derivative with a diazirinyl moiety. Mol BioSyst 8:1131–1135

Yang T, Li X-M, Bao X et al (2015) Photo-lysine captures proteins that bind lysine post-translational modifications. Nat Chem Biol 12:70–72

Yang Y, Song H, He D et al (2016) Genetically encoded protein photocrosslinker with a transferable mass spectrometry-identifiable label. Nat Commun 7:12299

Yu Z, Lin Q (2014) Design of spiro[2.3]hex-1-ene, a genetically encodable double-strained alkene for superfast photoclick chemistry. J Am Chem Soc 136:4153–4156

Zhang M, Lin S, Song X et al (2011) A genetically incorporated crosslinker reveals chaperone cooperation in acid resistance. Nat Chem Biol 7:671–677

Chapter 9
Use of Hexose Photolabels to Reveal the Structure and Function of Glucose Transporters

Geoffrey D. Holman

Abstract Photoaffinity labelling probes which are hexose and bis-hexose derivatives have been expensively used to study the structure and physiological functions of the glucose transporters of the GLUT family. Substrate binding site locations have been probed using these derivatives. The glucose transport proteins are physiologically important in maintaining blood glucose levels, and they have a tissue-specific distribution that allows them to remove glucose from the bloodstream in a controlled manner. In heart, skeletal muscle and adipose tissue, the GLUT4 transporter is present both at the cell surface and in an intracellular membrane vesicle reservoir compartment. Insulin action leads to a translocation of this protein between these locations. Only the cell surface transporters are labelled by the impermeable photoaffinity probes, and these have therefore been used to determine the extent of response to insulin and to additional stimuli, including those that are due to changes in the cellular energy status that occurs during exercise. These responses can be monitored in human subjects with pathophysiological changes in glucose homeostasis as occurs in obesity and type 2 diabetes. It is also possible to carry out pulse-chase kinetic experiments in which the tagged GLUT4 movement between cellular compartments is followed over time. These studies have revealed that insulin action mainly leads to an acceleration of the exocytotic limb of the GLUT4 recycling pathway. For the use of photoprobes in kinetic experiments on intact cells, it was necessary to develop biotinylated photoaffinity probes that had very long spacer arms of approximately 20 Å.

Keywords Glucose transport • GLUT proteins • Insulin • Diazirine • GLUT4 translocation

---

G.D. Holman
Department of Biology and Biochemistry, University of Bath, Bath BA2 7AY, UK
e-mail: g.d.holman@bath.ac.uk

Y. Hatanaka, M. Hashimoto (eds.), *Photoaffinity Labeling for Structural Probing Within Protein*, DOI 10.1007/978-4-431-56569-7_9

9.1 Introduction

Photoreactive compounds are highly versatile reagents that can be specifically targeted to combine with almost any protein which has a reasonable low (sub-millimolar) affinity constant for the reagent. Here we review the use of a range of hexose reagents that have been designed to target and explore the structural and functional activity of hexose transport proteins, mainly the GLUT family of proteins.

9.2 Probing the Structure of the GLUT Family of Mammalian Hexose Transporters

The GLUT family of glucose transporters has 14 members that are coded for in the human genome (Fig. 9.1). The Class 1 family of the protein products of these genes have been extensively studied over many decades, but it is only recently that investigators have begun to investigate the Class 2 and Class 3 transporters, and there is

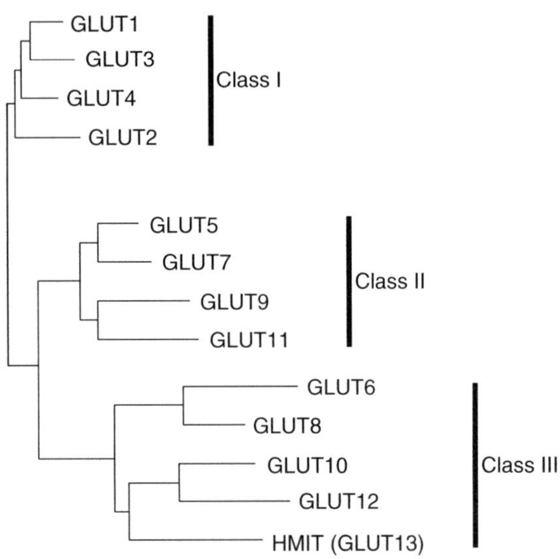

Fig. 9.1 Classification of GLUT genes. The GLUT genes appear to cluster into three classes. The extent to which they are similar to one another is indicated by the length of the linker arms in the tree diagram. Class 1 GLUT proteins have been most studied and are the most abundant. Class 2 and Class 3 transporters have not been extensively studied, but GLUT5 and GLUT7 transport both glucose and fructose. It is now thought that some of the Class 2 and Class 3 transporters may have a substrate preference for molecules other than glucose but which have a molecular similarity to glucose (such as myo-inositol and dehydroascorbic acid). The full characterisation of the substrate-specific and physiological functions of these transporters is incomplete

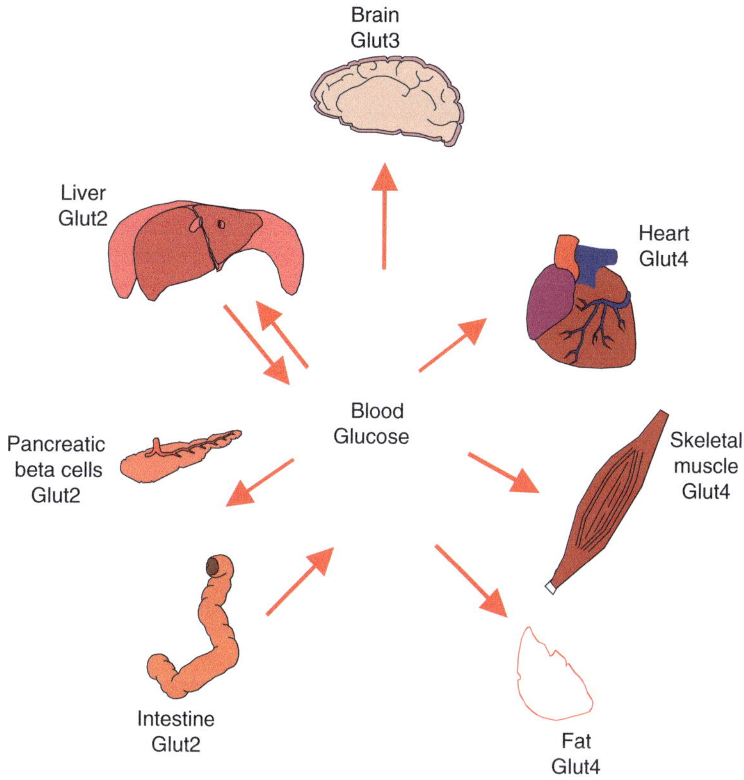

Fig. 9.2 Tissue-specific distributions of glucose transporter proteins. GLUT2 is present in the intestine, in pancreatic islets and in the liver. GLUT3 is abundant in the brain. GLUT4 is specifically expressed in insulin-responsive tissues of the heart and skeletal muscle and fat tissue. They are facilitative transporters and can potentially transport glucose both in and out of cells, but the normal direction of net movement is indicated by the *arrows*

still very little information on their specificity for hexose substrates or their separate functional role.

The transporters have a fairly specific tissue distribution (Fig. 9.2), but GLUT1 is present to some degree in all tissues. This isoform has been extensively studied in human erythrocytes which is a very rich source of this protein. The protein was purified from this source, and a portion of its primary protein sequence was determined that allowed cloning of the coding gene (Mueckler et al. 1985). The gene sequence then revealed the topography of the protein which has 12 transmembrane domains with both the N- and C-termini on the intracellular face of the membrane. There is a large cytoplasmic loop between transmembrane domains 6 and 7, and this divides the protein into two halves with a partial symmetry between the C-terminal and N-terminal halves suggesting an ancient gene duplication to produce the 12 transmembrane arrangements for optimum functional activity (Gould and Holman 1993). Expression studies have shown that neither the N- nor the C-terminal halves

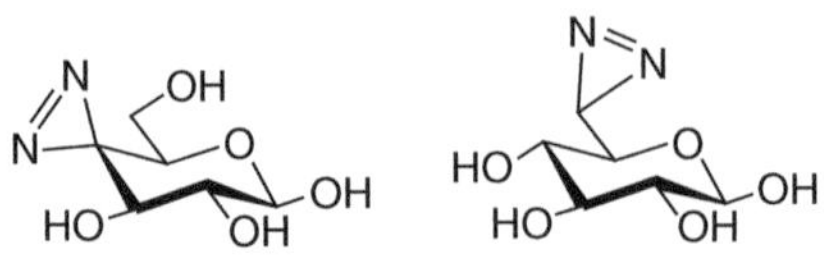

Fig. 9.3 Diazirine derivatives of glucose. Azi groups have been introduced at the C-4 and C-6 positions of D-glucose. The size of the group is similar to a hydroxyl, and these compounds are transported by the human erythrocyte GLUT1 transporter with affinity constants similar to that of glucose

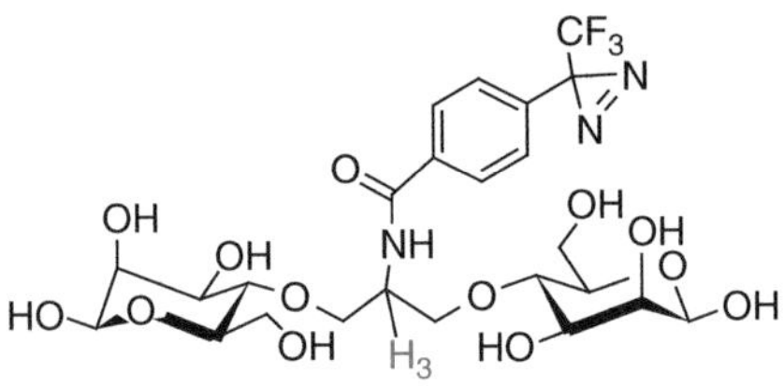

Fig. 9.4 Azi-trifluoroethyl-benzoyl-bis-D-mannose (ATB-BPMA). A tritiated version of this probe was prepared by reductive animation of the propanol spacer linking two mannose units. The bis-structure is designed so that the compounds are too large to be transported into cells but are still able to bind to the GLUT proteins from the exofacial cytoplasmic surface. Derivatives of this type have affinities for the GLUT proteins that are over tenfold higher than the parent sugars

alone can bind substrate, but expression of both halves in the same cell allows the two halves to assemble to form a functioning transporter (Cope et al. 1994). Several approaches have been used in attempts to identify regions of the protein that are important for function. Among the early studies of this type, photoaffinity labelling made important steps forwards. Our laboratory has used this approach extensively, and among our early labels was a series of hexose-substituted diazirines substituted at either the C-4 position or the C-6 position to replace the hydroxyl groups (Midgley et al. 1985) (Fig. 9.3). These derivatives were transported by GLUT1, but because of their relatively low affinity, they did not significantly label the protein.

Instead we designed a new range of analogues that are based on a bis-hexose structure (Fig. 9.4). The rationale for this design was to provide a substrate that combined only with the exofacial (cytoplasmic) side of the transporter. Studies had previously shown that to combine with this exofacial site, it was necessary to maintain H-bonding hydroxyls at C1, C3, C6 and the ring oxygen (Barnett et al. 1973). A C2 hydroxyl is not required, and the widely used substrate for this transporter 2-deoxy-D-glucose is readily transported with good affinity. Analogue studies had also shown a spatial requirement for binding. A substitution at the C1 end (such as a hydroxyl-methyl group) was not tolerated, while bulky groups could be inserted around the C4 region without loss of binding affinity (Barnett et al. 1975). These bulky substitutions did however prevent transport, but this was considered unnecessary when labelling the exofacial site. An additional rationale for the bis-hexose structure was that this would double the chances of successful binding and would effectively block any transport because of the bulkiness and the hydrophilicity of the molecule. The central amine group allowed a range of substituents with photoaffinity reactive groups to be attached. The photoaffinity reagents were developed that are based on the bis-hexose structure and containing a hydrophobic substituent have over 10-fold higher affinities for the GLUT proteins (around 200 μM) than the natural glucose substrate D-glucose (which has an affinity constant of about 5 mM and close to that concentration occurring in the blood steam). The central carbon of the bridge also allowed substation with tritium which could be used for detecting the tagged protein following labelling.

A tritiated 2-hydroxy-4 azidobenzoyl-substituted compound was found to be successful in labelling the GLUT1 transporter (Holman and Rees 1987). Similar studies were later carried out using a phenyldiazirine-substituted BMPA, ATB-BMPA (Fig. 9.4) (Clark and Holman 1990). The diazirine-based photolabels were found to give a cleaner signal with improved efficiency in the biological labelling experiments (Koumanov et al. 1998). Using these photolabelling bis-hexose derivatives, the labelling site within the tagged GLUT1 protein was determined by selectively cleaving the labelled protein with proteases and then identifying the labelled peptide fragments. This approach allowed identification of a fragment in the C-terminal half of the protein, around the top of transmembrane domain 8 and 9 of the sequence. The C-terminal half of the protein was also identified from mutagenesis studies as being critical for substrate recognition from the exofacial surface of the transporter. These early studies on the substrate specificity of GLUT transporters and the identification of regions important for substrate recognition have very recently been ratified by three-dimensional crystallographic data on the glucose transporters (Deng et al. 2015).

Additional members of the Class 1 GLUT proteins are differentially expressed with GLUT3 specific for brain, GLUT2 for liver and the developmentally related intestine and pancreas and GLUT4 for heart, skeletal muscle and adipose cells. All these proteins have been probed by photoaffinity labelling using bis-hexose photolabels which have revealed aspects of their cellular activities.

9.3 Studying the Insulin Stimulation of GLUT4

The GLUT4 is of particular importance in maintaining hormonal control of blood glucose levels. The GLUT4 tissues respond to insulin treatment that leads to increased GLUT4-mediated glucose transport activity. The response leads to increased storage of glucose carbons as fat in adipose tissue and as glycogen in muscle. As a consequence of this action, blood glucose is kept at the normally low concentration of about 7 mM, with fasting blood glucose of about 5 mM. It is now established that GLUT4 resides at the plasma membrane and in intracellular membrane vesicles (Cushman and Wardzala 1980). Insulin action results in a net translocation and net movement (see Sect. 9.3) of GLUT4 vesicles from the intracellular reservoir compartment to the plasma membrane. When the vesicles are incorporated

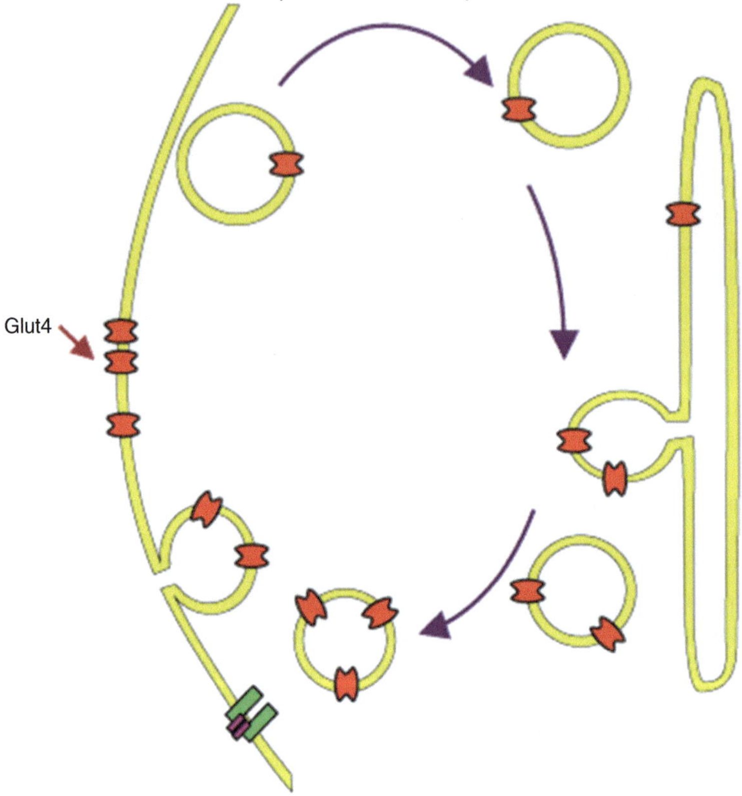

Fig. 9.5 Insulin-stimulated GLUT4 movement to the cell surface of target tissues. Insulin combines with its receptor at the cell surface, and this initiates a series of protein kinase signalling steps. These activated signalling intermediates activate the movement of membrane vesicles containing GLUT4 that moves to the plasma membrane. The higher level of transporters in the cell surface membrane facilitates the rapid movement of glucose into the cell

at the plasma membrane, more GLUT4 molecules are available to catalyse the transport of glucose into cells (Fig. 9.5). In some tissues the stimulation of glucose transport is very markedly increased, and stimulations of greater than tenfold have been observed in experimental systems (Holman et al. 1990).

The GLUT4 translocation process can become dysfunctional in obesity and type 2 diabetes, and therefore, it is important to have techniques for studying GLUT4 in the insulin target tissues. As the specificity for substrates is similar for all the Class 1 GLUT proteins (Gould and Holman 1993), it was possible to use the same photoaffinity labels to study GLUT4 to those that were developed for exploring the structure of GLUT1. Using the tritiated ATB-BMPA photolabel, it was possible to tag only those GLUT4 molecules that were at the surface of the insulin target cells while leaving the GLUT4 molecules that were in the intracellular compartment unlabelled. In this way drugs and hormone levels leading to increased cell surface could be explored. Using such approaches it was established that signalling intermediates downstream of the insulin receptor including components of the PI 3 kinase, the serine kinases Torc2 and Akt, and the Rab-GAPs TBC1D4/TBC1D1 sequential cascade of intermediates (Stockli et al. 2011; Holman and Kasuga 1997 for reviews) were essential for most of the stimulatory effect of insulin on GLUT4 movement to the cell surface of the target cells in the heart (Yang et al. 2002b) and skeletal muscle (Ryder et al. 2000) and adipose cells (Clarke et al. 1994). Other important stimuli of glucose transport have also been explored using the impermeable and cell surface-specific photoreactive labels. These include the stimulation of glucose transport that occurs in response to a change in the energy status (mainly the cellular AMP/ATP) ratio of target cells. Muscle activity during exercise and hypoxia both alter the AMP/ATP ratio and lead to increased glucose uptake, an effect which the cell uses to restore depletions in the energy balance. Using the photoaffinity labels, the combined effects of insulin and an exercise-type stimulus were shown to have an additive effect on the stimulation of the cell surface levels of GLUT4 (Yang and Holman 2005, 2006). This effect is consistent with an independent signalling pathway for the exercise and insulin stimulatory effects. It is now known that the exercise stimulatory effect occurs primarily through AMPK, the AMP-activated kinase. Since this pathway is distinct from the insulin signalling that occurs through the PI 3-kinase pathway, the stimulatory effect of a combined stimulus leads to additivity of the downstream response. This effect is tremendously important in combating defects in the insulin stimulatory pathway that occurs in obesity and type 2 diabetes. Reagents that stimulate the exercise pathway can alleviate the effects of an impaired insulin response. Among the reagents that are in clinical use to treat type 2 diabetes is metformin which is known to stimulate AMPK. This reagent has been studied using the photolabelling approach which has revealed that GLUT4 translocation to the surface is stimulated by metformin (Pryor et al. 2000; Yang and Holman 2006). Furthermore exercise and hypoxia can stimulate cell surface labelling in muscle samples from human type 2 diabetes subjects (Ryder et al. 2000). These cellular effects on GLUT4 have been mainly observed in experimental systems, but there is good evidence that exercise is beneficial for overcoming the effects of human type 2 diabetes. New drugs are continuing to be developed for subjects that have difficulty exercising. Among these drugs are those in the thiazolidinedione class of compounds that are primarily thought to stimulate the peroxisome proliferator-activated receptor gamma (PPAR-gamma) transcription factor that controls the production of selective target proteins. Using the photoaffinity labels, early studies on the thiazolidinedione called rosiglitazone in an obese mouse experimental system revealed effects on GLUT4 protein levels within the fat tissue became elevated in response to treatment (Young et al. 1995). Rosiglitazone was subsequently developed as a multimillion dollar treatment for type 2 diabetes, but in recent years this treatment has been withdrawn because of concerns over side effects. However other thiazolidine drugs continue to be developed.

9.4 Determination of the Rates of Subcellular Trafficking of GLUT4

In addition to determining the cell surface levels of GLUT4, it is important to understand the membrane trafficking pathway by which the levels are increased. This could occur by increased exocytosis from the reservoir compartment, reduced endocytosis from the plasma membrane or to both increased exocytosis and reduced endocytosis. In order to study this process and following GLUT4 movement between its subcellular compartments, we carried a series of experiments, all of which involve "pulse" tagging the GLUT4 with photolabels followed by a post-labelling "chase" in which the tagged transporter was allowed to move between compartments. Initially we combined cell surface photoaffinity labelling with a subsequent cell fractionation approach that was used to separate labelled samples into a plasma membrane and a subcellular vesicle fraction. These samples were taken at a series of time intervals after labelling so that the kinetics of movement between fractions was possible. These studies established that insulin mainly led to a large increase in the exocytosis of GLUT4 (approximately tenfold) with a much smaller effect on a reduction of endocytosis (less than twofold) (Satoh et al. 1993). This exocytosis effect was initially observed in rodent adipocyte preparations where good subcellular fractionation techniques had been developed that made such kinetic experiments possible. However, equivalent subcellular fractionation techniques were not available for experimentally intractable, but clinically important, tissues such as human muscle. A number of new developments were required for further studies.

9.5 Recent Developments in GLUT4 Photoaffinity Labelling

The need to improve the applicability and general ease of use of the GLUT4 photoaffinity reagents and the manifest utility of the biotin-streptavidin combination to isolate tagged protein suggested that the synthesis of a series of biotinylated diazirine photolabels based on the bis-hexose structure was warranted (Fig. 9.6). Hatanaka's group had developed a diazirine reagent that carried a biotin moiety (Park et al. 2005); we coupled this via an amide linkage to both BMPA (Koumanov et al. 1998) and later to the equivalent bis-glucose compound (Hashimoto et al. 2001a). The latter compound had slightly higher affinity than the bis-mannose compound, and this was expected as D-glucose has higher binding affinity for the GLUT proteins than does D-mannose. The latter compound has now been extensively used to study the GLUT proteins by many groups studying a range of tissues and reagents that can stimulate glucose transport. A cleavable version of the bis-glucose photolabel and this has further applicability in terms of ease of recovery of the tagged proteins from immobilised streptavidin (Hashimoto et al. 2001a).

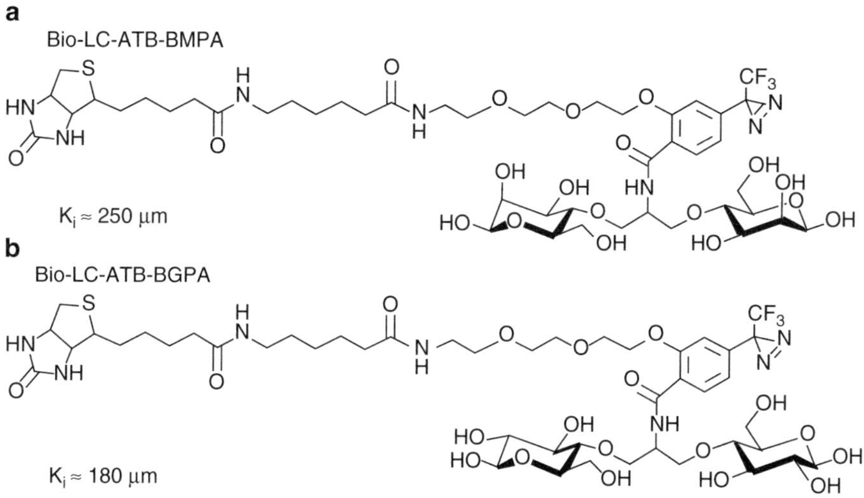

Fig. 9.6 Biotinylated derivatives of the bis-hexose photolabels. Bio-LC-ATB-BMPA is a bis-mannose derivative, and use of this compound has now been superseded by Bio-LC-ATB-BGPA. The latter has slightly higher affinity for the GLUT transporters as it is based on the hexose D-glucose which has roughly twofold higher affinity than D-mannose

We also wanted to use a biotinylated photolabel for studying the kinetics of GLUT4 subcellular traffic in insulin-responsive tissues, particularly in physiologically important tissues in which the technique of subcellular fractionation could not be applied. We reasoned that an approach to studying the translocation of GLUT4 would be to allow the tagged GLUT4 to traffic after labelling and then to artificially separate the two pools of biotin-tagged GLUT4 by quenching the biotin signal from the plasma membrane pool with soluble avidin (Fig. 9.7). Soluble avidin would be added to the cells after labelling and before cell disruption so that the biotin signal from the surface GLUT4 would not be precipitated by subsequent use of immobilised streptavidin. In effect only the GLUT4 that had moved from the surface to the internal compartment would be detected, and this signal would increase as more GLUT4 was internalised. The kinetic of movement could also be followed in reverse (from the internal reservoir GLUT4 to the plasma membrane) by first allowing GLUT4 to internalise and then following the kinetics of loss of signal from the internal compartment. However, we became aware of a difficulty in quenching in intact cells the surface signal with avidin when using the BGPA reagents that had the short amino-caproate chain link. Normally, and in the cell surface labelling experiments that have been referred to above, the immobilised streptavidin precipitation step is used in detergent-solubilised samples and under conditions in which the biotin moiety on the tagged GLUT4 is accessible. We reasoned that in the intact cell, the biotin-tagged GLUT4 is inaccessible as the biotin to avidin approach is hindered by the intact protein. This is consistent with recent crystal structures of the

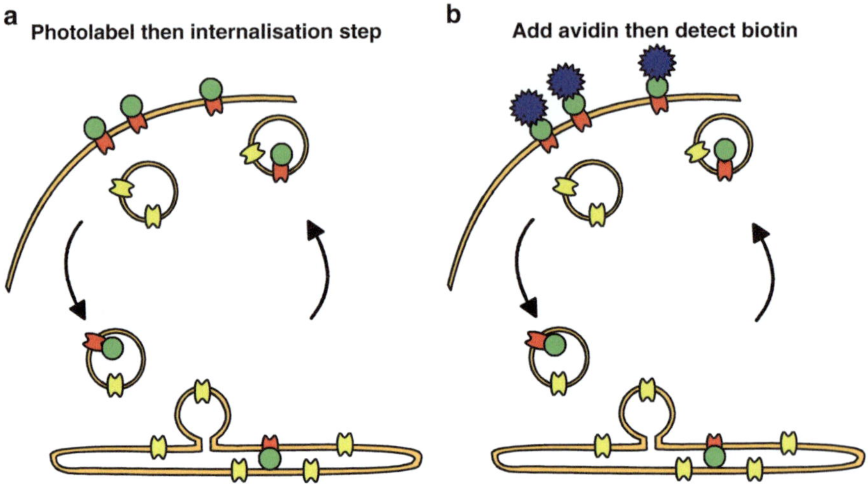

Fig. 9.7 Development of a method for studying the kinetics of GLUT4 membrane traffic. (**a**) GLUT4 is tagged with an impermeant photoaffinity probe and containing a biotin moiety only while at the cell surface. This pulse of labelling is followed by a chase period in which the tagged GLUT4 (*red* plus *green symbol*) is internalised and over a time course mixes with the bulk of the GLUT4 in the internal membrane vesicle compartment (*yellow symbols*). (**b**) At specific time intervals during the internalisation, any transporters remaining at the cell surface are quenched by adding avidin (*blue symbol*). In effect this allows selective recovery of only internal GLUT4 on immobilised streptavidin and allows the rates of internalisation to be determined. This experiment can also be carried out in reverse by allowing the bulk of the pulse labelled GLUT4 to return to the reservoir compartment and then stimulating the cells (e.g. with insulin) and following the kinetics of reduction of the internal signal as GLUT4 returns to the surface. Experiments of this type have been used to determine that the main effect of insulin on GLUT4 traffic is to increase the rate of exocytosis

GLUT proteins in which there is a deep substrate binding site, and if this deep site was labelled, the biotin would be inaccessible unless there was an extremely long chain between the hexose moiety held in the centre of the protein and the exofacial protein surface where streptavidin could approach without steric interference. We set up a simple assay for studying this problem using intact human erythrocytes (Fig. 9.8). A range of reagents was developed with extremely long spacer arms (Hashimoto et al. 2001b). The signal from biotin was detected by microscopy using soluble fluorescently tagged avidin. We found that only when the spacer was over 25 Å could a biotin signal be detected, and these data indicated that the compound GP15 (Fig. 9.8) would be suitable for the trafficking experiments that we planned to study in the insulin-responsive GLUT4 translocation system.

The GP15 reagent was then used successfully to study GLUT4 traffic in rodent adipose (Yang et al. 2002c) and heart cells (Yang and Holman 2005) and most recently in studying GLUT4 traffic in the previously experimentally intractable skeletal muscle tissue (Karlsson et al. 2009). In all cases the main feature of insulin action that was evident was a marked increase in the rate of exocytosis of the GLUT4

a

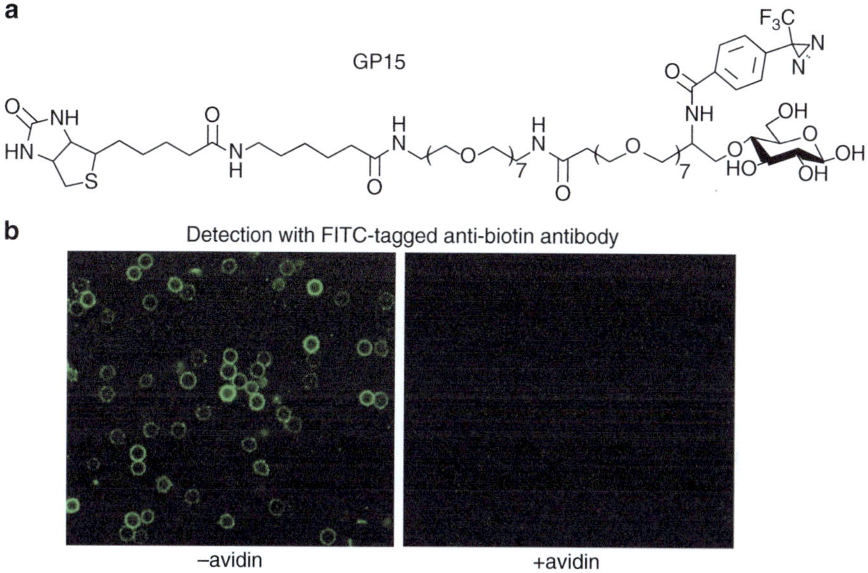

Fig. 9.8 Development of a reagent that interacts with avidin added directly to intact cells. We found that it was necessary that to label the GLUT transporters and form a complex with avidin in intact cells, an extremely long spacer (of approximately 20 Å) was required. (**a**) The structure of the reagent GP15. (**b**) Use of GP15 to label and then detect GLUT1 in human erythrocytes. Following labelling, the biotin signal from the tagged GLUT1 was detected by adding a fluorescently labelled avidin and then visualising the cells in a confocal microscope

from the reservoir compartment to the plasma membrane with virtually no detectable effect of insulin on the rate of endocytosis. These studies using GP15 were extended to study in a cell-free system the fusion of GLUT4 vesicles directly with the plasma membrane compartment (Koumanov et al. 2005). Together these studies have greatly narrowed down the site at which insulin-mediated acceleration of membrane traffic occurs to the exocytotic limb of the GLUT4 recycling circuit and more specifically to the fusion of GLUT4 vesicles with the plasma membrane.

9.6 Future Perspectives for Studies on GLUT Proteins Using Photoaffinity Labels

Currently developed photoprobes for the Class 1 GLUT transporters continue to be studied for physiological functional studies on the location and activity of these target transporters. Although experimental systems have now been developed for studying GLUT4 in which recombinant versions of the protein (with recombinant protein tag domains inserted) have been widely used, there remains a need to study the endogenous GLUT4, and here photolabelling is a useful resource. For example, many studies are now underway in which mouse gene knockout models for diseases such as obesity and type 2 diabetes have been developed. In these cases it is important to study changes in endogenous GLUT4.

Many of the Class 2 and Class 3 transporters have not been studied in detail, but it appears that some of these transporters have affinity for other substrates in addition to affinity for glucose. The Class 2 transporters GLUT5 and GLUT7 have affinity for D-fructose. Animals consume large quantities of D-fructose. Natural fructose is very sweet and consumed in fruit juices, while breakdown of sucrose provides equimolar amounts of glucose and fructose. It has become apparent in recent years that overconsumption of fructose may be detrimental to health, and further studies on the transporters that are responsible for the homeostasis of body fructose levels are warranted. Photoaffinity labels that are derivatives of D-fructose have been developed and characterised using an experimental system expressing GLUT5 (Yang et al. 2002a). However, further utility of these reagents and exploration of their reactivity with other members of the Class 2 and Class 3 transporters are yet to be explored.

Future development of the chemistry of the GLUT photoprobes also remains to be explored. Development of reagents that have particular utility in proteomics and mass spectroscopy would be useful. There have only been a small number of studies in which cleavable probes have been studied (Hashimoto et al. 2001a). The many benefits of such reagents would include the improvement of the ease of release of tagged proteins from the probe-protein adduct for their subsequent identification. Most studies described in this review have used antibodies to identify tagged proteins, but there is an increasing need to identify new photo-tagged proteins by mass spectroscopy. Reagents that incorporate mass tags within the structure of the photoprobes could be developed and explored. These could be useful for quantification of changes occurring in experimental situations and in physiological processes. Isotope-coded affinity tags (ICAT) moieties could be incorporated in the reagent structures so that changes of levels of recovered proteins under different experimental conditions could be compared directly from mass spectroscopy output of signature isotope ratios.

References

Barnett JEG, Holman GD, Munday KA (1973) Structural requirements for binding to the sugar transport system of the human erythrocyte. Biochem J 131:211–221

Barnett JEG, Holman GD, Chalkley RA, Munday KA (1975) Evidence for two asymmetric conformational states in the human erythrocyte sugar-transport system. Biochem J 145(3):417–429

Clark AE, Holman GD (1990) Exofacial photolabelling of the human erythrocyte glucose transporter with an azitrifluoroethylbenzoyl-substituted bismannose. Biochem J 269(3):615–622

Clarke JF, Young PW, Yonezawa K, Kasuga M, Holman GD (1994) Inhibition of the translocation of GLUT1 and GLUT4 in 3T3-L1 cells by the phosphatidylinositol 3-kinase inhibitor, wortmannin. Biochem J 300(Pt 3):631–635

Cope DL, Holman GD, Baldwin SA, Wolstenholme AJ (1994) Domain assembly of the glut1 glucose-transporter. Biochem J 300:291–294

Cushman SW, Wardzala LJ (1980) Potential mechanism of insulin action on glucose transport in the isolated rat adipose cell. Apparent translocation of intracellular transport systems to the plasma membrane. J Biol Chem 255:4758–4762

Deng D, Sun P, Yan C, Ke M, Jiang X, Xiong L, Ren W, Hirata K, Yamamoto M, Fan S, Yan N (2015) Molecular basis of ligand recognition and transport by glucose transporters. Nature 526(7573):391–396. doi:10.1038/nature14655

Gould GW, Holman GD (1993) The glucose-transporter family—structure, function and tissue-specific expression. Biochem J 295:329–341

Hashimoto M, Hatanaka Y, Yang J, Dhesi J, Holman GD (2001a) Synthesis of biotinylated bis(D-glucose) derivatives for glucose transporter photoaffinity labelling. Carbohydr Res 331(2):119–127

Hashimoto M, Yang J, Holman GD (2001b) Cell-surface recognition of biotinylated membrane proteins requires very long spacer arms: an example from glucose-transporter probes. ChembioChem 2(1):52–59. doi:10.1002/1439-7633(20010105)2:1<52::AID-CBIC52>3.0.CO;2-F

Holman GD, Kasuga M (1997) From receptor to transporter: insulin signalling to glucose transport. Diabetologia 40(9):991–1003. doi:10.1007/s001250050780

Holman GD, Rees WD (1987) Photolabelling of the hexose transporter at external and internal sites: fragmentation patterns and evidence for a conformational change. Biochim Biophys Acta 897(3):395–405

Holman GD, Kozka IJ, Clark AE, Flower CJ, Saltis J, Habberfield AD, Simpson IA, Cushman SW (1990) Cell surface labeling of glucose transporter isoform GLUT4 by bis-mannose photolabel. Correlation with stimulation of glucose transport in rat adipose cells by insulin and phorbol ester. J Biol Chem 265(30):18172–18179

Karlsson HK, Chibalin AV, Koistinen HA, Yang J, Koumanov F, Wallberg-Henriksson H, Zierath JR, Holman GD (2009) Kinetics of GLUT4 trafficking in rat and human skeletal muscle. Diabetes 58(4):847–854. doi:10.2337/db08-1539

Koumanov F, Yang J, Jones AE, Hatanaka Y, Holman GD (1998) Cell-surface biotinylation of GLUT4 using bis-mannose photolabels. Biochem J 330(Pt 3):1209–1215

Koumanov F, Jin B, Yang J, Holman GD (2005) Insulin signaling meets vesicle traffic of GLUT4 at a plasma-membrane-activated fusion step. Cell Metab 2(3):179–189. doi:10.1016/j.cmet.2005.08.007

Midgley PJW, Parkar BA, Holman GD, Thieme R, Lehmann J (1985) Transport-properties of photolabile sugar analogs. Biochim Biophys Acta 812:27–32

Mueckler M, Caruso C, Baldwin SA, Panico M, Blench I, Morris HR, Allard WJ, Lienhard GE, Lodish HF (1985) Sequence and structure of a human glucose transporter. Science 229(4717):941–945

Park JJ, Sadakane Y, Masuda K, Tomohiro T, Nakano T, Hatanaka Y (2005) Synthesis of diazirinyl photoprobe carrying a novel cleavable biotin. ChembioChem 6(5):814–818. doi:10.1002/cbic.200400342

Pryor PR, Liu SC, Clark AE, Yang J, Holman GD, Tosh D (2000) Chronic insulin effects on insulin signalling and GLUT4 endocytosis are reversed by metformin. Biochem J 348(Pt 1):83–91

Ryder JW, Yang J, Galuska D, Rincón J, Björnholm M, Krook A, Lund S, Pedersen O, Wallberg-Henriksson H, Zierath JR, Holman GD (2000) Use of a novel impermeable biotinylated photolabeling reagent to assess insulin and hypoxia-stimulated cell surface GLUT4 content in skeletal muscle from type 2 diabetic patients. Diabetes 49:647–654

Satoh S, Nishimura H, Clark AE, Kozka IJ, Vannucci SJ, Simpson IA, Quon MJ, Cushman SW, Holman GD (1993) Use of bis-mannose photolabel to elucidate insulin-regulated GLUT4 subcellular trafficking kinetics in rat adipose cells: evidence that exocytosis is a critical site of hormone action. J Biol Chem 268:17820–17829

Stockli J, Fazakerley DJ, James DE (2011) GLUT4 exocytosis. J Cell Sci 124(Pt 24):4147–4159. doi:10.1242/jcs.097063

Yang J, Holman GD (2005) Insulin and contraction stimulate exocytosis, but increased AMP-activated protein kinase activity resulting from oxidative metabolism stress slows endocytosis of GLUT4 in cardiomyocytes. J Biol Chem 280(6):4070–4078. doi:10.1074/jbc.M410213200

Yang J, Holman GD (2006) Long-term metformin treatment stimulates cardiomyocyte glucose transport through an AMP-activated protein kinase-dependent reduction in GLUT4 endocytosis. Endocrinology 147(6):2728–2736. doi:10.1210/en.2005-1433

Yang J, Dowden J, Tatibouet A, Hatanaka Y, Holman GD (2002a) Development of high-affinity ligands and photoaffinity labels for the D-fructose transporter GLUT5. Biochem J 367(Pt 2):533–539. doi:10.1042/BJ20020843

Yang J, Gillingham AK, Hodel A, Koumanov F, Woodward B, Holman GD (2002b) Insulin-stimulated cytosol alkalinization facilitates optimal activation of glucose transport in cardiomyocytes. Am J Physiol Endocrinol Metab 283(6):E1299–E1307. doi:10.1152/ajpendo.00341.2002

Yang J, Hodel A, Holman GD (2002c) Insulin and isoproterenol have opposing roles in the maintenance of cytosol pH and optimal fusion of GLUT4 vesicles with the plasma membrane. J Biol Chem 277(8):6559–6566. doi:10.1074/jbc.M108610200

Young PW, Cawthorne MA, Coyle PJ, Holder JC, Holman GD, Kozka IJ, Kirkham DM, Lister CA, Smith SA (1995) Repeat treatment of obese mice with BRL 49653, a new potent insulin sensitizer, enhances insulin action in white adipocytes. Association with increased insulin binding and cell-surface GLUT4 as measured by photoaffinity labeling. Diabetes 44(9):1087–1092

Chapter 10
Function of Peroxisome in Mammal and Analysis of the Fatty Acid Oxidation System by Photoaffinity Labeling

Tsuneo Imanaka, Kosuke Kawaguchi, and Masashi Morita

Abstract Peroxisomes play an essential role in a number of important metabolic pathways including oxidation of fatty acids, and synthesis of ether phospholipids and bile acids. Long, very long, and branched-chain fatty acid-CoA as well as intermediate metabolites for bile acid synthesis are transported into peroxisomes through ATP-binding cassette (ABC) transporters, ABCD1-3 on the membranes and oxidized by α- and β-oxidation enzymes. Mutation of these transporters and enzymes causes severe peroxisomal disorders. Characterization of molecular mechanism of the substrate transport and the enzyme reaction is an important issue to figure out the role of these proteins in lipid metabolism under physiological and pathological conditions. Recently precise structure of several enzymes involved in peroxisomal fatty acid oxidation has been revealed by the studies based on X-ray crystallography and NMR spectroscopy. However, the molecular mechanisms of these proteins, especially in terms of substrate binding, have not yet been elucidated in detail. Photoaffinity labeling has been a powerful tool to find specific region for the binding of the substrate using a ligand with a photoactivatable group. Here, we first review biogenesis and function of peroxisome, and then focus our attention to molecular recognition of substrate by peroxisomal proteins including ABC transporters by photoaffinity labeling.

Keywords ABC transporter • Fatty acid α- and β-oxidation enzymes • Photoaffinity labeling • Substrate recognition

---

T. Imanaka (✉)
Department of Biological Chemistry, Graduate School of Medicine and Pharmaceutical Sciences, University of Toyama, 2630 Sugitani, Toyama 930-0194, Japan

Faculty of Pharmaceutical Sciences, Hiroshima International University, 5-1-1, Hirokoshinkai, Kure, Hiroshima 737-0112, Japan
e-mail: imanaka@ps.hirokoku-u.ac.jp

K. Kawaguchi • M. Morita
Department of Biological Chemistry, Graduate School of Medicine and Pharmaceutical Sciences, University of Toyama, 2630 Sugitani, Toyama 930-0194, Japan

Y. Hatanaka, M. Hashimoto (eds.), *Photoaffinity Labeling for Structural Probing Within Protein*, DOI 10.1007/978-4-431-56569-7_10

Abbreviations

ABC	ATP-binding cassette
DHA	Docosahexaenoic acid
DHCA	Dihydroxycholestanoic acid
ER	Endoplasmic reticulum
LCFA	Long-chain fatty acid
PBD	Peroxisome biogenesis disorder
PMP	Peroxisomal membrane protein
PTS	Peroxisome targeting signal
RCDP	Rhizomelic chondrodysplasia punctata
THCA	Trihydroxycholestanoic acid
VLCFA	Very long-chain fatty acid
X-ALD	X-linked adrenoleukodystrophy

10.1 Introduction

Proteins are one of the major molecular components of cells and are involved in most biological processes taking place in living cells. In eukaryote cells, many biological processes proceed in subcellular compartments that are basically composed of organelle-specific proteins. Peroxisome is one of the organelles bounded by a single membrane and is present in almost all eukaryotic cells. The peroxisomes are involved in a variety of metabolic processes, including the oxidation of various fatty acids including very long-chain fatty acids (VLCFA), and the synthesis of ether phospholipids and bile acids in mammals (Waterham et al. 2016; Wanders et al. 2016; Vamecq et al. 2014). The substances metabolized in peroxisomes are first transported from cytosol through transporters on peroxisomal membranes and then metabolized by enzymes in the peroxisomes. Metabolized substances are transported outside of peroxisomes by other set of transporters and further transported to mitochondria or endoplasmic reticulum (ER) and subsequently metabolized as final products (Wanders et al. 2016). How such kind of metabolic flows are regulated is one of the important challenges.

To understand regulation of the metabolic process in the cells in detail, it is important to reveal three-dimensional structure of the proteins that is involved in the process. In addition, characterization of the interaction site of ligand on the proteins is also important. Recent development of several technologies such as X-ray crystallography, NMR spectroscopy, and mass spectrometric technique makes possible to understand the molecular detail of the functional mechanism of proteins. The binding analysis of small molecules with proteins has been also one of the important subjects for understanding the function of proteins. However, the molecular mechanisms of the proteins, especially in terms of substrate binding, have not yet been elucidated in detail. Photoaffinity labeling is an excellent method to identify the biomolecules that interact with proteins by applying a combination of covalent cross-linking and attachment of a detectable tag to the labeled protein (Masuda et al. 2015; Tomohiro et al. 2013; Hatanaka 2015).

Photoaffinity labeling has provided structural information of ligand binding at relatively high resolution based on the mapping of the labeled amino acid residues. It has been particularly useful in the analysis of molecular interactions in the transition state or weak binding state, which otherwise would be difficult to access using the X-ray crystallography and NMR techniques. In general, identification of the labeled sites has been achieved using tandem mass-based sequencing of purified and labeled peptides after digestion.

In this chapter, we first covered the role of peroxisomes in mammal and focused our attention on peroxisomal ABC transporters and fatty acid oxidation enzymes in the peroxisomes.

10.2 Role of Peroxisomes in Mammalian Cells

10.2.1 What Is Peroxisome?

Peroxisomes are subcellular organelles bounded by a single membrane. They are widely distributed in eukaryotic cells and are ubiquitous in mammalian cells. The size of peroxisomes is 0.1–1 μm although peroxisomes in different tissues vary in shape and size (Fig. 10.1). These organelles were first described in 1954 in the cytoplasm of proximal tubule cells of the mouse kidney using electron microscopy by Rhodin and referred to as "microbodies" (Rhodin 1954). However, most of the researchers had not become interested in the microbodies since the function had not yet been revealed. Then, biochemically, the microbodies were characterized for the first time in 1965 by de Duve and Baudhuin (1966). de Duve and Baudhuin isolated

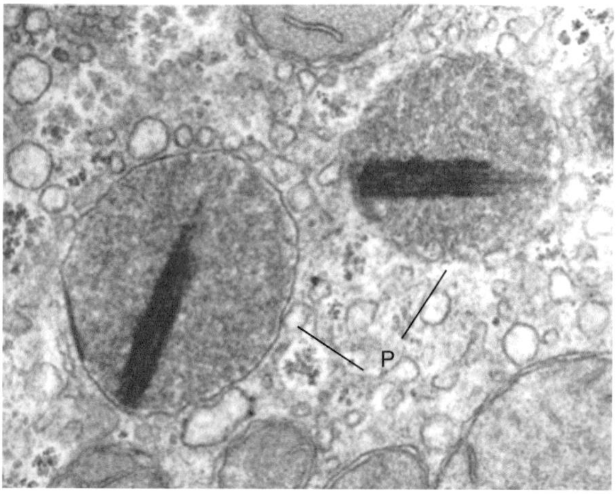

Fig. 10.1 Electron micrograph of rat liver peroxisomes. A core inside of peroxisome is complex of urate oxidase. P Peroxisome

microbodies from rat liver using differential and density gradient centrifugation techniques and defined microbodies as membrane-bound organelles containing several H_2O_2-producing oxidase and H_2O_2-degrading enzyme (catalase) and named peroxisomes (Bowers 1998). However, physiological function of peroxisomes in mammal including humans had still remained unknown.

In 1976 Lazarow and de Duve demonstrated that murine hepatic peroxisomes catalyzed fatty acid oxidation in a cyan insensitive manner and that oxidation was enhanced in animals by the peroxisome proliferator clofibrate (Lazarow and de Duve 1976). This was the first report on physiological function of peroxisomes in mammalian cells. Subsequently, it was revealed that treatment of rodents with hypolipidemic drugs like clofibrate and other structurally related compounds leads to a marked proliferation of peroxisomes and induction of certain peroxisomal enzymes like those involved in the β-oxidation of fatty acids. On the other hand, concerning peroxisomal disorders, it was first reported by Bowen et al. in 1964 that Zellweger's cerebro-hepato-renal syndrome could be considered to be the prototype of genetic disorders (Bowen et al. 1964). Later, Goldfischer et al. reported on the lack of peroxisomes in tissues from the patients (Goldfischer et al. 1973). Then, it is recognized that peroxisomes are essential for human's life.

Now, it is well known that peroxisomes are essential for lipid metabolisms and play a critical role in the homeostasis of multiple organs. Peroxisomes in central nervous system are smaller than that in liver and kidney, but are important for proper neurological development (Berger et al. 2016). The indispensable role of peroxisomes for human health and development is elucidated by the existence of a large number of inborn errors of peroxisomal metabolisms. At present, more than 120 peroxisome-related genes have been identified in human and more than 20 genetic diseases were classified as peroxisomal disorders (Waterham et al. 2016).

In addition, the peroxisomes are highly dynamic organelles and rapidly increase in number and size in response to the nutritional and environmental conditions. In human, it is known that physiological stimuli, such as starvation, diabetes mellitus, or high fat diets, induce peroxisomal proliferation (Vamecq et al. 2014).

10.2.2 Biogenesis of Peroxisome

It had been generally accepted that peroxisomes could form by growth and division from preexisting organelles. However, it is considered most likely at preset that peroxisomes are ER-derived organelles although diversity of biogenesis of peroxisome exists among yeast, mammals, and plants (Agrawal and Subramani 2016). The biosynthesis of peroxisomes in mammal involved three different processes, including the formation of the pre-peroxisome from ER, the import of peroxisomal membrane proteins (PMPs) and matrix proteins to pre-peroxisomes and peroxisomes, and the growth and division of peroxisome (Fig. 10.2a). It is known that many proteins so-called peroxin encoded by *PEX* genes are involved in peroxisome biogenesis. To date, 34 peroxins have been identified. The deficiency of one of the peroxins is associated with severe disease, termed peroxisome biogenesis disorder (PBD).

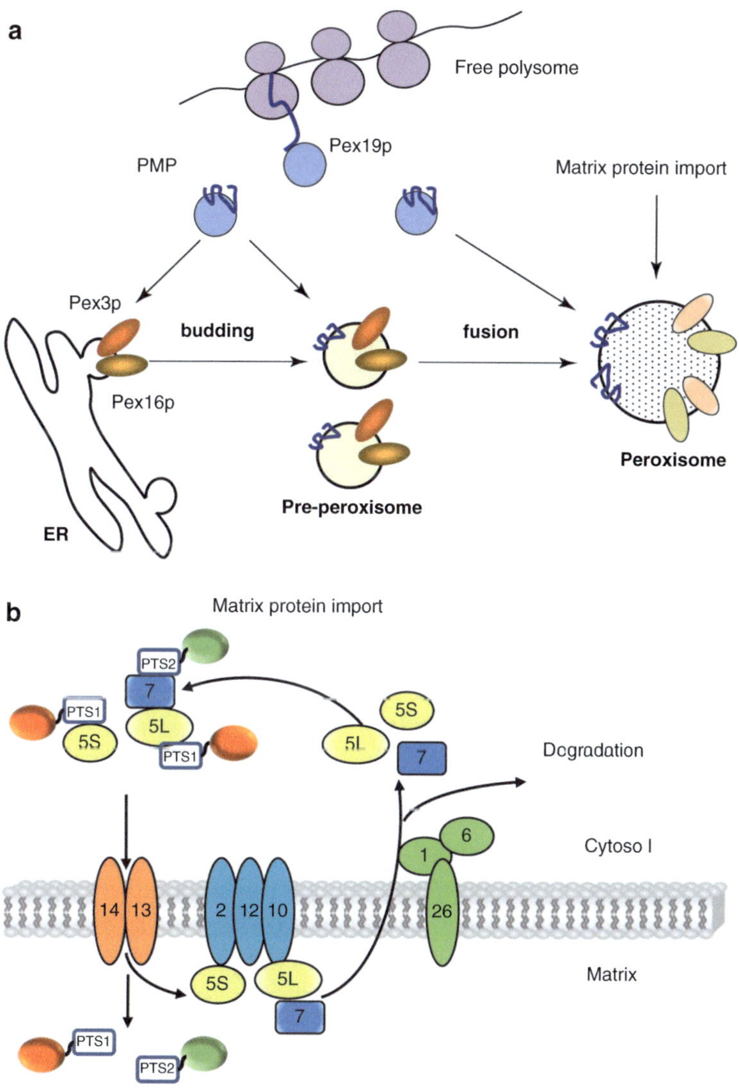

Fig. 10.2 Biogenesis of peroxisome. (**a**) A pre-peroxisome existing Pex3p and Pex16p is budded from ER. Newly synthesized PMPs and peroxisomal matrix proteins are posttranslationally imported to pre-peroxisome. After fusion of pre-peroxisomes mature peroxisome is developed. (**b**) Import of newly synthesized peroxisomal matrix proteins to peroxisome. Many peroxins are involved in the process. (**c**) Insertion of newly synthesized PMP into peroxisomal membranes

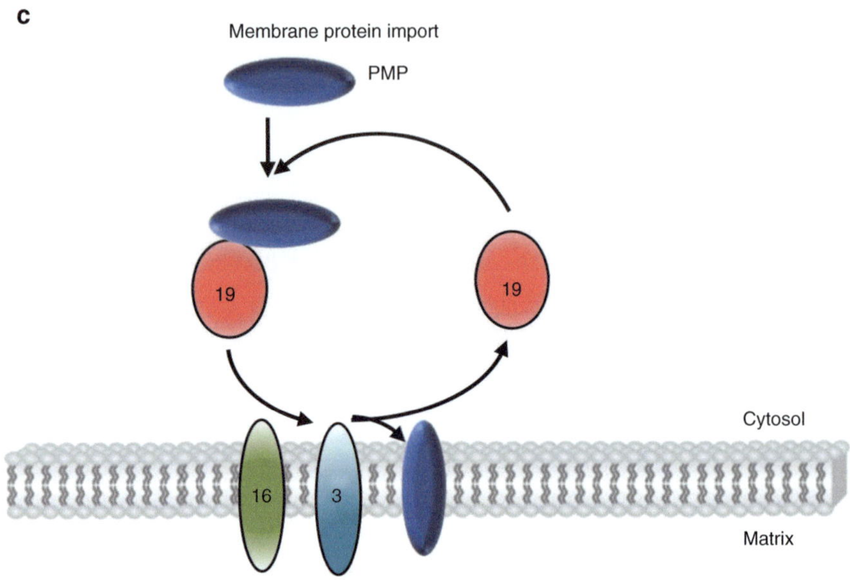

Fig. 10.2 (continued)

In contrast to mitochondria, peroxisomes do not contain DNA. Therefore, all peroxisomal proteins are encoded by genomic DNA. Peroxisomal proteins including PMPs are posttranslationally imported to peroxisomes. The transport of peroxisomal matrix proteins is mediated by specific targeting sequence known as peroxisome targeting signal (PTS). The majority of peroxisomal matrix proteins contain a PTS1, which is a COOH-terminal motif sequence consisting of the tripeptide: (S/A/C)-(K/R/H)-(L/M). On the other hand, PTS2 is a NH_2-terminus sequence consisting of the consensus sequence: (R/K)-(L/I/K)-X5-(Q/H)-(L/I/V). Newly synthesized matrix proteins with PTS1 motif signal are recognized by cytosolic protein Pex5p. Two types of Pex5p, Pex5pL and Pex5pS, exist in mammal and they are the result of differential splicing of the primary transcript of the *PEX5* gene. On the other hand, Pex7p recognizes the NH_2-terminal PTS2 motif. Pex7p needs to interact with Pex5pL to get target to the peroxisome (Fig. 10.2b). The matrix protein-loaded receptor proteins dock onto the peroxisomal docking complex constituted by Pex14p and Pex13p. After docking, the matrix proteins are imported into the peroxisomal lumen and the receptor proteins (Pex5p and Pex7p) are released from the peroxisomal membrane for another import cycle or directed to the proteasome for their degradation (Rucktaschel et al. 2011; Hasan et al. 2013). These latter two steps require the action of PMP such as Pex2p, Pex10p, and Pex12p, which are ubiquitin ligases that catalyze either the mono- or the poly-ubiquitination of Pex5p. Mono-ubiquitination promotes the recycling of Pex5p, while poly-ubiquitination makes Pex5p a substrate for proteasome-mediated degradation (Francisco et al. 2014). The cytosolic proteins named ATPases associated with diverse cellular activities (AAA), Pex1p and Pex6p form a hexamer on peroxisomal membrane by interaction with PMP Pex26p. The Pex1p-Pex6p-Pex26p complex is involved in the release of ubiquitinated Pex5p from the peroxisomal membrane (Fujiki et al. 2012).

Although much less is known about the biogenesis of the peroxisomal membrane, a different set of peroxins is reported to be required for this process in mammals. Pex19p, Pex3p, and Pex16p are involved in membrane assembly and insertion of PMPs (Fig. 10.2c). Pex19p is a mainly cytosolic protein that is capable of binding many different PMPs and has been postulated to serve as the receptor of newly synthesized PMPs (Sato et al. 2010; Morita and Imanaka 2012). The PMPs are bound to Pex19p via peroxisome targeting signal called mPTS in their sequences. Pex3p is located in the peroxisomal membrane and serves as docking site for PMP-loaded Pex19p. The Pex16p was shown to serve as the receptor for Pex19p loaded with Pex3p (Fujiki et al. 2012). Pex16p functions as a tethering factor for Pex3p, or as part of the putative membrane-insertion machinery. It has been reported that Pex3p and Pex16p were found in the ER, suggesting that peroxisomes may be derived from vesicles originating from the ER (Fig. 10.2a). Dysfunction of Pex3p, Pex16p, or Pex19p results in the absence of peroxisomes, which lead to the Zellweger syndrome.

10.2.3 Function of Peroxisome

Peroxisomes are indispensable organelle that catalyzes various metabolic pathways unique to this organelle. Human peroxisomes contain over 50 different enzymes, most of which are involved in unique peroxisomal metabolic process (Waterham et al. 2016). Among the metabolic process in peroxisome, the β-oxidation of fatty acids is virtually important catabolic pathway. Peroxisomes catalyze the β-oxidation of certain fatty acids including VLCFA, branched-chain fatty acids, bile acid intermediates such as dihydroxycholestanoic acid (DHCA) and trihydroxycholestanoic acid (THCA), long-chain dicarboxylic acids, 2-hydroxy fatty acids, and a number of prostanoids (see below for detail). These substances cannot be β-oxidized in mitochondria. On the other hand, mitochondria catalyze β-oxidation of long, medium, and short-chain fatty acids (C18 and shorter). Furthermore, peroxisome catalyzes the synthesis of ether-phospholipids known as plasmalogen that are important constituent in brain, and also catalyzes the α-oxidation of phytanic acid. In addition to lipid metabolism, peroxisomes also play a role in several non-lipid metabolic pathways including purine, polyamine, glyoxylate, and amino acid metabolism. Recently, new function for peroxisome has been demonstrated by Dixit et al. (2010). They have reported that peroxisomes as well as mitochondria act as intracellular signaling platforms in innate immunity and are important for early stage of antiviral signaling. Furthermore, Chu et al. has reported a novel peroxisomal function in cellular cholesterol transport (Chu et al. 2015). They demonstrated that peroxisomes have a role in the transport of free cholesterol from lysosomes to ER through lysosome-peroxisome membrane contacts. It is thus likely that peroxisomes are multifunctional organelles and interacted with other organelles such as mitochondria, lysosomes, ER, or lipid droplets to fulfill the diverse biological functions.

10.2.4 Peroxisomal Disorder

Peroxisomes are indispensable for the development and maintenance of human body (Waterham et al. 2016). The absence of peroxisomes results in the accumulation of non-degradation molecules such as VLCFA and phytanic acid, and in the reduction of physiologically essential molecules such as docosahexaenoic acid (DHA), bile acids, and plasmalogen. Mutation of genes encoding peroxisomal proteins and proteins involved in peroxisome biogenesis may cause a variety of human diseases. The disease can be grouped into two main classes of peroxisomal disorders, PBD and the single peroxisomal enzyme deficiency.

Peroxisome biogenesis includes assembly of peroxisomal membrane, transport of PMP and matrix proteins, and division and proliferation of peroxisomes (Fig. 10.2a). PBD is a genetically heterogeneous group of autosomal recessively inherited disorder with a specific defect in one of the peroxisome biogenesis processes. The majority of the peroxisome biogenesis disorders are caused by biallelic mutations. Mutations in any of the 13 different *PEX* genes have been identified that are encoding peroxins. Recently patients with mutations in genes involved in peroxisome fission were identified (Waterham et al. 2007; Huber et al. 2013; Ebberink et al. 2012). Clinically, PBD ranges from severe, lethal disorders to milder, late-onset progressive neurological disease. Severity of the diseases is dependent on importance of the proteins and how residual function of mutant proteins exists. Common to PBD are liver disease, variable neurodevelopmental delay, retinopathy, and perceptive deafness. Most severe cases are Zellweger syndrome, neonatal adrenoleukodystrophy, and infantile Refsum disease. The diseases are named based on the severity of clinical symptom. The severity depends on remaining function of mutant proteins and following mutations cause these diseases. Mutation in the *PEX3, PEX16,* and *PEX19* genes affect biogenesis of peroxisomal membranes and the *PEX1, PEX2, PEX5, PRX6, PEX10, PEX12, PEX13, PEX14*, and *PEX26* genes affect the import of matrix proteins into peroxisomes. In the case of *PEX7* mutation, it affects only the import of peroxisomal matrix proteins with PTS2. The mutation causes rhizomelic chondrodysplasia punctata (RCDP) type 1. These patients are often relatively mildly affected and show only a few of the symptoms that are typically associated with Zellweger syndrome. RCDP is a clinical phenotype and five genetically distinct subtypes have been reported at present. Type 1 is PBD and type 2, 3, and 4 are single peroxisomal enzyme deficiencies (see below). Recently, type 5 was found to be caused by the mutation of Pex5pL and the mutation of DLP/DRP1 and PEX11β that cause a novel class of peroxisomal fission disorders (Waterham et al. 2007; Ebberink et al. 2012).

Concerning the single peroxisomal enzyme deficiency, the diseases include defects of peroxisomal matrix enzymes as well as PMPs. A given single peroxisomal enzyme deficiency thus depends on the specific function of the protein in peroxisomal metabolism. Single enzyme deficiencies in peroxisomes also cause the biochemical abnormality, which frequently results in severe disease. Mutation of ABCD1 causes X-linked adrenoleukodystrophy (X-ALD). Deficiency of β-oxidation enzymes causes acyl-CoA oxidase deficiency, D-bifunctional protein (multifunctional protein 2, MFP2) deficiency, 2-methylacyl-CoA racemase deficiency, and sterol carrier protein X(SCPx) deficiency. Deficiency of α-oxidation enzyme, phytanoyl-CoA hydroxylase causes Refsum disease. Dysfunction of acyl-CoA: dihydroxyacetonephosphate acyltransferase (DHAPAT), alkyl-DHAP synthetase, and fatty acyl-CoA reductase that are essential for ether phospholipid synthesis causes RCDP type 2, 3, and 4, respectively. Dysfunction of glyoxylate detoxification and H_2O_2-metabolism causes hyperoxaluria type I and acatalasemia, respectively. Recently, ABCD3 deficiency has been reported to cause persisting hepatosplenomegaly and severe liver disease (Waterham et al. 2016).

10.3 ABC Transporters and Fatty Acid Transport and Oxidation in Peroxisomes

10.3.1 Transport of Substrate into Peroxisomes

Fatty acids exist mainly as CoA esters in cytoplasm. Therefore, fatty acid-CoA (acyl-CoA) is transported into peroxisome by transporters before oxidized by peroxisomal enzymes. In mammals, three ABC transporters belonging to subfamily D, ABCD1, ABCD2, and ABCD3 are known to exist on peroxisomal membranes predominately as a homodimer (Fig. 10.3) (Morita and Imanaka 2012). ABCD1 and ABCD2 have a high sequence similarity and both are involved in the transport of saturated and unsaturated VLCFA-CoA into peroxisomes (Fig. 10.3). ABCD1 is preferentially involved in the transport of saturated VLCFA. *ABCD1* mutation causes X-ALD that is biochemically characterized by the accumulation of pathognomonic amounts of saturated VLCFA in all tissues. Although ABCD2 has a functional redundancy with ABCD1, ABCD2 is known to have other specific roles in lipid metabolisms because of the differences in the expression patterns and phenotypes of *Abcd1*-deficient and *Abcd2*-deficient mice. The β-oxidation activity of poly-unsaturated VLCFA (C24:6), an immediate precursor of DHA, was reduced in *Abcd2*-deficient mice brain. In addition, ABCD2 expression is highly sensitive to dietary polyunsaturated fatty acids, suggesting that ABCD2, but not ABCD1, is involved in the transport of poly-unsaturated VLCFA such as C22:6-CoA and C24:6-CoA in relation to DHA metabolism (Fourcade et al. 2009; van Roermund et al. 2011; Genin et al. 2011).

In contrast, ABCD3 might have a function in the import of bile acid intermediates, pristanic acid and phytanic acid, because in *Abcd3*-deficient mice the bile acid precursors THCA and DHCA, as well as pristanic acid, were accumulated in plasma. The recent identification of ABCD3 deficiency in a patient with markedly elevated DHCA and THCA levels in plasma and the studies in the *Pmp70*(-/-) mouse has led to the conclusion that ABCD3 catalyzes the import of branched-chain acyl-CoA, THCA-CoA, and DHCA-CoA into peroxisomes (Fig. 10.3) (Ferdinandusse et al. 2015).

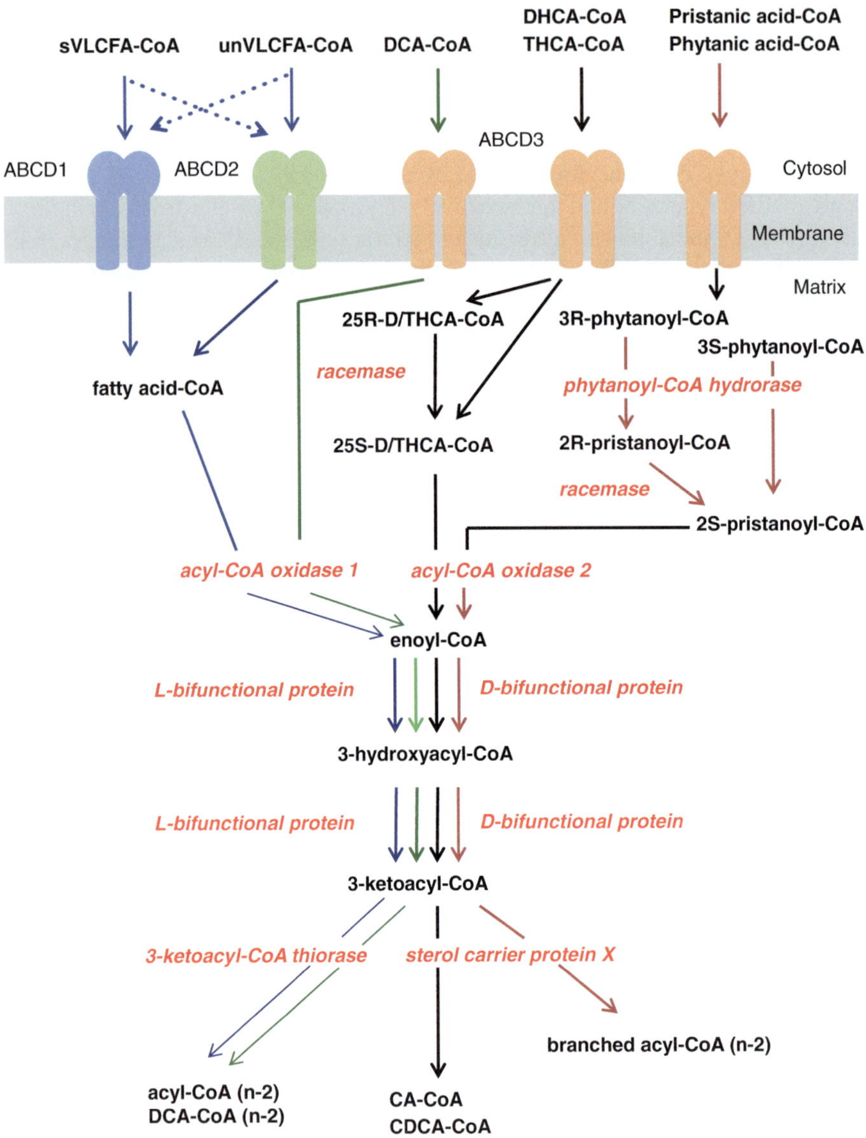

Fig. 10.3 Fatty acid oxidation pathway in peroxisome. A variety of fatty acid-CoA is transported into peroxisomes through three ABC transporters. 3R- and 3S-phytanic acid-CoA are metabolized to 2R- and 2S-pristanoyl-CoA and 2R-pristanoyl-CoA is converted to 2S-pristanoyl-CoA by 2-metyl-acyl-CoA rasemase. 25R-DHCA/THCA-CoA and 2R-pristanoyl-CoA are concerted to 25S-DHCA/THCA-CoA and 2S-prostanoyl-CoA by rasemase, respectively. These substrates are oxidized by fatty acid β-oxidation enzymes. sVLCFA; saturated VLCFA, unVLCFA; unsaturated VLCFA, CA; cholic acid, CDCA; chenodeoxycholic acid, DCA; dicarboxylic acid

10.3.2 Fatty Acid Oxidation in Peroxisomes

To date, about 50 peroxisomal enzymes have been identified, which participate in the fatty acid α- and β-oxidation, biosynthesis of ether phospholipids and bile acids, and metabolism of reactive oxygen species. Among them, the β-oxidation of fatty acid is the most prominent pathway. There are different types of fatty acids to be β-oxidized only in peroxisomes: VLCFA (such as C22:0, C24:0, and C26:0), polyunsaturated VLCFA (C24:6 n-3), long-chain dicarboxylic acids, 2-methyl branched-chain fatty acid (pristanic acid), phytanic acids (3,7,11,14-tetramethylhexadecanoic acid), and bile acid intermediates (DHCA and THCA). These fatty acids are imported into peroxisomes via ABCD1-3, degraded by one or more cycles of β-oxidation, and released to the cytosol for further metabolism.

The enzymes catalyzing the formation of DHCA and THCA are localized in different subcellular compartments including the cytosol, ER, and mitochondria. Cholesterol is converted to 25R-THCA and 25R-DHCA and they are activated to 25R-THCA-CoA and 25R-DHCA-CoA by the ER enzyme bile acid-CoA ligase and enter the peroxisome. Then 25R-THCA-CoA and 25R-DHCA-CoA are converted to 25S-THCA-CoA and 25S-DHCA-CoA by 2-methyl-acyl-CoA rasemase in peroxisomes, respectively.

Phytanic acid exists as 3R- and 3S-phytanic acids. Both phytanic acids are activated as CoA form and transported into peroxisomes by ABCD3. 3R- and 3S-phytanic acid-CoA are metabolized to 2R- and 2S-pristanoyl-CoA, respectively, and 2R-pristanoyl-CoA is converted to 2S-pristanoyl-CoA by 2-metyl-acyl-CoA rasemase. Phytanic acid-CoA is first α-oxidized by phytanoyl-CoA 2-hydroxylase, 2-hydroxyphytanoyl-CoA lyase, and pristanal dehydrogenase, and converted to pristanoyl-CoA and formyl-CoA before β-oxidation. 2R-Pristanoyl-CoA is oxidized by ACOX2 and enters β-oxidation cycle (Fig. 10.3) (Wanders et al. 2011).

Like mitochondria, the β-oxidation in peroxisomes is conducted by four steps consisting of dehydrogenation, hydration, dehydrogenation again, and thiolytic cleavage. In peroxisomes, the $FADH_2$ reacts with O_2 to produce H_2O_2, which is degraded by catalase into H_2O and O_2, while the reoxidation of $FADH_2$ is coupled to the electron transport chain to produce ATP in mitochondria (Wanders et al. 2016).

The first step for β-oxidation is catalyzed by acyl-CoA oxidases that interacted directly with molecular oxygen, generating H_2O_2. H_2O_2 is subsequently converted to molecular oxygen by catalase. There are at least three acyl-CoA oxidases; ACOX1 preferentially degrading saturated and unsaturated straight-chain fatty acids, ACOX2 with a high affinity for 2-methyl branched fatty acids such as pristanoyl-CoA and DHCA-CoA and THCA-CoA. ACOX3 is alternative splicing forms of ACOX1.

In the second and third steps, bifunctional proteins possessing two-enzymatic activities enoyl-CoA hydratase and 3-hydroxyacyl CoA dehydrogenase conduct the steps. They are L-bifunctional (L-PBE) and D-bifunctional enzyme (D-PBE), or alternatively termed multifunctional enzyme type 1 and type 2 (MEP1 and MEP2), respectively. It has been elucidated that D-PBE catalyzes the hydration and subsequent dehydrogenation of enoyl-CoA esters of VLCFAs, pristanic acid and DHCA and THCA. In contrasts, the physiological role of L-PBE is not fully understood but might be involved in the degradation of dicarboxylic enoyl-CoA esters with long to medium chain.

In the last step, two thiolases, 3-ketoacyl-CoA-thiolase 1 (pTH1) and SCPx (pTH2), encoded by *ACAA1* and *ACP2* genes, respectively, are involved in the oxidation of VLCFAs. pTH1 metabolizes only straight-chain fatty acids. The branched-chain fatty acids and bile acid precursors (pristanic acid, DHCA and THCA) are solely cleaved by pTH2. However, the straight-chain fatty acids are also cleaved by pTH2.

The chain-shortened acyl-CoAs can only be oxidized to CO_2 and H_2O in mitochondria. Therefore, once the fatty acid chains are shortened to medium-chain fatty acyl-CoA via peroxisomal β-oxidation, they are conjugated to carnitine and leave the peroxisomes for further β-oxidation in mitochondria. On the other hand, choloyl-CoA and deoxycholoyl-CoA produced in peroxisomes are converted to taurine- or glycine-conjugated cholic acid or deoxycholic acid by bile acids-CoA: amino acid N-acyltransferase and exported to cytosol.

10.4 Photoaffinity Labeling of Peroxisomal ABC Transporters

Peroxisomal ABC transporters have a predicted structure that is one half the size of the ABC transporter, with one transmembrane domain (TMD) and one nucleotide-binding domain (NBD) (Fig. 10.4). The hydropathy profile of the human peroxisomal

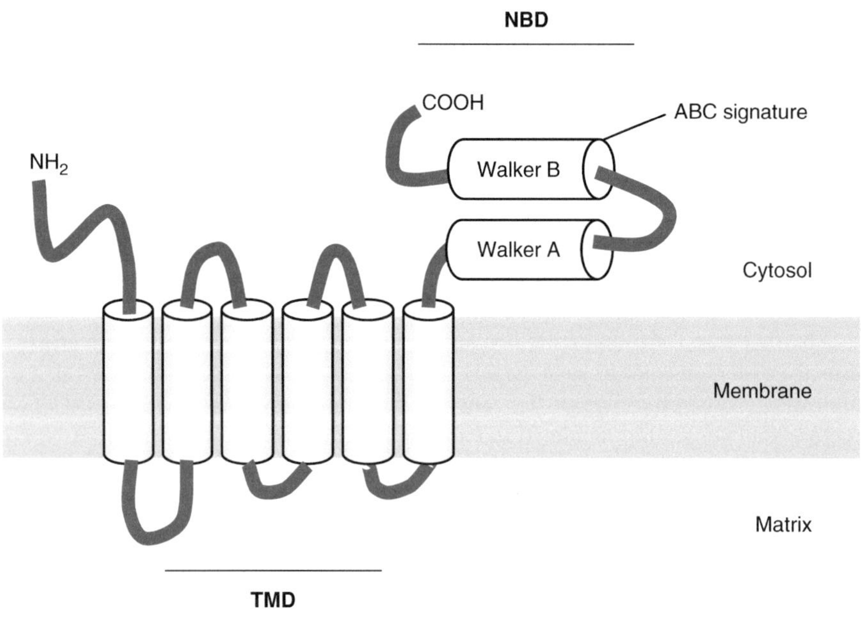

Fig. 10.4 Hypothesized structure of the peroxisomal ABC half-transporter. ABCD transporters are a half-size ABC transporter with one transmembrane domain (TMD) and one nucleotide-binding domain (NBD). Six transmembrane domains are located in the NH_2-terminal half of the transporter, and Walker A, B and ATP-binding cassette signature sequence (the ABC signature) is located in the COOH-terminal half of the transporter. In mammals, three ABC proteins belonging to subfamily D, ABCD1, ABCD2, and ABCD3 exist in peroxisomes, predominately as a homodimer

ABC transporters predicts that the NH_2 terminal half of the peroxisomal ABC transporters is hydrophobic, with six transmembrane segments, and the COOH-terminal half is hydrophilic, having NBD. TMD provides the passageway for the substrate across the membranes and NBD energizes the directional transport of these substrates by cycles of ATP binding and hydrolysis. NBD contains two conserved motifs: Walker A and Walker B, separated by 90–120 amino acids. A third, so-called ABC signature motif is situated upstream of the Walker B sequence.

10.4.1 Analysis of ATP Binding and Hydrolysis Cycle by Photoaffinity Ligands

Currently, one of the key outstanding questions is how peroxisomal ABC transporters recognize their substrates and import them into the lumen of peroxisomes during ATP binding and hydrolyzing cycle. Concerning ATP binding and hydrolysis, ATPase activity is quite low even in mammalian cells overexpressing ABCD1-3. Early studies, therefore, characterized the ATPase activity by photolabeling techniques. For this purpose, photoreactive ATP analog such as 8-azido-[α-^{32}P]ATP and 8-azido-[γ-^{32}P]ATP are available. Photolabeled proteins are identified by SDS-PAGE after immunoprecipitation and the amount of labeled compounds were quantified by image analyzer.

Tanaka et al. detected for the first time the ATP binding and hydrolyzing activities of native ABCD1 and ABCD3 in rat liver peroxisomes by photoaffinity labeling with 8-azido-[α-^{32}P]ATP and 8-azido-[γ-^{32}P]ATP (Tanaka et al. 2002). ABCD3 was photoaffinity labeled at similar efficiencies with 8-azido-[α-^{32}P]ATP and 8-azido-[γ-^{32}P]ATP when peroxisomes were incubated with these nucleotides at 37 °C in the absence of Mg^{2+} and exposed to UV light without removing unbound nucleotides. Addition of Mg^{2+} reduced the photoaffinity labeling of ABCD3 with 8-azido-[γ-^{32}P]ATP by 70%, whereas it reduced photoaffinity labeling with 8-azido-[α-^{32}P]ATP by only 20%. However, two-thirds of nucleotide (probably ADP) was dissociated during removal of unbound nucleotides. These results suggest that ATP binds to ABCD3 tightly in the absence of Mg^{2+}, the bound ATP is hydrolyzed to ADP in the presence of Mg^{2+}, and the produced ADP is dissociated from ABCD3, which allows ATP hydrolysis turnover.

Analysis of other ABC transporters also provides us useful information. ABCA1 on plasma membrane such as macrophages is involved in cholesterol efflux in the cells and important to maintain cholesterol homeostasis. It is known that ABCA1-mediated cholesterol efflux is highly regulated at the transcriptional level through the activity of the nuclear receptor liver X receptor (LXR). On the other hand, recently Hozoji-Inada et al. showed that in addition to its well-defined role in transcription, LXRβ directly binds to the C-terminal region of ABCA1 to mediate its posttranslational regulation (Hozoji-Inada et al. 2011). When ABCA1-LXRβ complex stably localizes to the plasma membranes in the macrophage-like cell line THP-1, apolipoprotein A-I (apoA-I) binding or cholesterol efflux does not occur through ABCA1. On the other hand, exogenously added LXR ligands promoted to dissociate ABCA1 to LXRβ, and subsequently apoA-I binding and cholesterol efflux from the cell occurred. Interestingly, ABCA1 was labeled with 8-azido-[α-^{32}P]ATP under the condition where the cells were only incubated with LXR ligand. Interaction of LXRβ to ABCA1 might inhibit the ATPase activity of ABCA1, and as a result, cholesterol efflux from the cells through ABCA1 does not occur. Therefore, photoreactive ATP analog is useful to investigate the state of ABC transporter whether it is active or inactive.

10.4.2 Substrate Transport by Peroxisomal ABC Transporter

Another important point is that how ABC transporter recognized its substrate and transport across membrane during ATP binding and hydrolysis cycle. Limited number of studies were reported biochemically about conformation change of peroxisomal ABC transporters. Actually, this type of conformational changes occurred in ABCD3 in the region close to the boundary between the transmembrane and NBD, and the helical domain between Walker A and B motifs (Kashiwayama et al. 2002). On the other hand, the TMD of ABCD1 is proposed to be involved in substrate recognition and to form a transport pathway across the peroxisomal membrane. Guimaraes et al. assessed the substrate-induced conformational alterations in ABCD1 with a protease-based assay, and found that LCFA- and VLCFA-CoA increased the sensitivity of the NH_2-terminal 44-kDa fragment of ABCD1 to Factor Xa, and this acyl-CoA-induced sensitivity was reversed by the presence of ATP-γS (Guimaraes et al. 2005). These findings suggest that the NH_2-terminal TMD of ABCD1 is involved in the recognition of these substrates, and undergoes a conformational change upon ATP binding to the COOH-terminal NBD of ABCD1.

To understand substrate recognition, precise structure of peroxisomal ABC transporters is required. Although the X-ray crystallographic structure of peroxisomal ABC transporters has not yet been obtained, recently the mouse Abcb1 (Mdr1, P-glycoprotein) was reported at 3.8 Å resolution (Aller et al. 2009). The structure of Abcb1 represents a nucleotide-free inward-facing conformation in the plane of the lipid bilayer. The inward-facing conformation results in a large internal cavity open to both the cytoplasm and the inner leaflet. The inward-facing conformation represents an initial stage of the transport cycle that is suitable for drug binding. The substrate enters the internal drug-binding pocket through an open portal and the residues in the drug-binding pocket interact with the substrate. ATP binds to the NBD, causing a large conformational change, thereby presenting the substrate and drug-binding site(s) to the outer leaflet/extracellular space. Based on these findings and the structure of bacterial multidrug ABC transporters Sav1866 (Dawson and Locher 2006) and MsbA (Ward et al. 2007), peroxisomal ABC transporters might capture acyl-CoA derivatives from cytosolic side and transport them to the lumen of peroxisome through ATP binding and hydrolysis (Fig. 10.5). The arrangement of TMDs that compose the cavity of acyl-CoA for the substrate, and the linkage between the TMD and NBD, might have a conformation that is different from other ABC transporters in order to transport acyl-CoA derivatives.

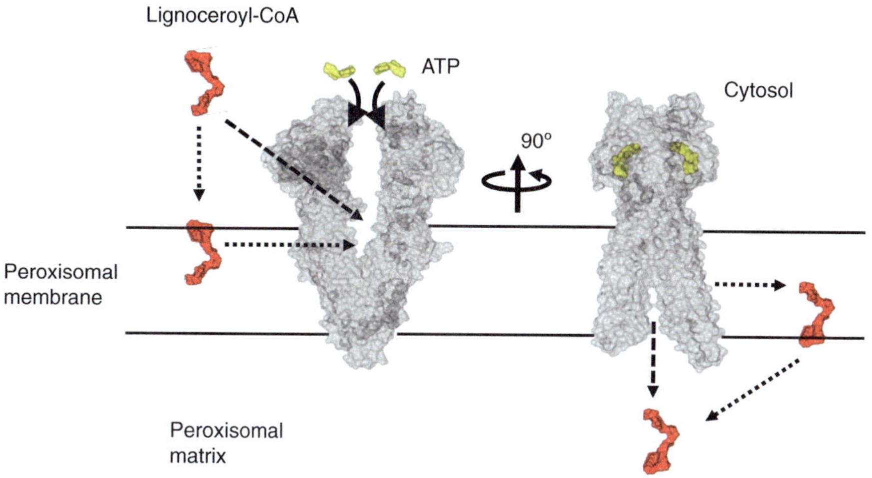

Fig. 10.5 A hypothetical model of substrate transport by the peroxisomal ABC transporter. Substrate (lignoceroyl-CoA) enters putative substrate-binding pocket of the ABC transporter from cytosol and/or leaflet of the lipid bilayer facing cytosol. ATP binds to the NBD, causing a large conformational change of the ABC transporter. The substrate is released into the lumen of peroxisome and/or the leaflet facing lumen of peroxisome. This model is drawn upon the base of structures of p-gp, Sav1866, and MsbA deposited in Protein Data Bank. A possible conformation of lignoceroyl-CoA in aqueous solution is also made up based on energy minimization calculation, and in lipid bilayer based on NMR study. The molecular-graphics software, CCP4mg was used to generate this model

10.4.3 Analysis of Substrate Binding by Photoaffinity Ligands

Photoaffinity techniques are useful for the characterization of substrate binding to the pocket in transporter. There is no report about substrate binding to peroxisomal ABC transporter at present. However, a couple of experiments have been done in other ABC transporters.

ABCB1 (human MDR, P-glycoprotein) transports a large number of structure unrelated compounds and plays an important role in normal physiology by protecting tissues from various toxic xenobiotics and endogenous metabolites. On the other hand, ABCB1 causes multidrug resistance in tumors by secretion of chemotherapeutic compounds from the cells. Pleban et al. used a set of photoaffinity ligands as propafenone type substrate and analyzed the substrate-binding domain(s) of ABCB1 with mass spectrometry (Pleban et al. 2005). They expressed ABCB1 in Sf9 insect cells and prepared plasma membrane vesicles from these cells and labeled with propafenone-type photoligands. After irradiation, labeled ABCB1 was separated by SDS-PAGE and the ABCB1 band was excised and subjected to in-gel digestion. The ensuing peptide fragments were eluted from the gel and identified by high-resolution MALDI-TOF mass spectrometry. The highest labeling was observed in transmembrane segments 3, 5, 8, and 11. A homology model for ABCB1 was generated on the basis of the dimeric crystal structure of

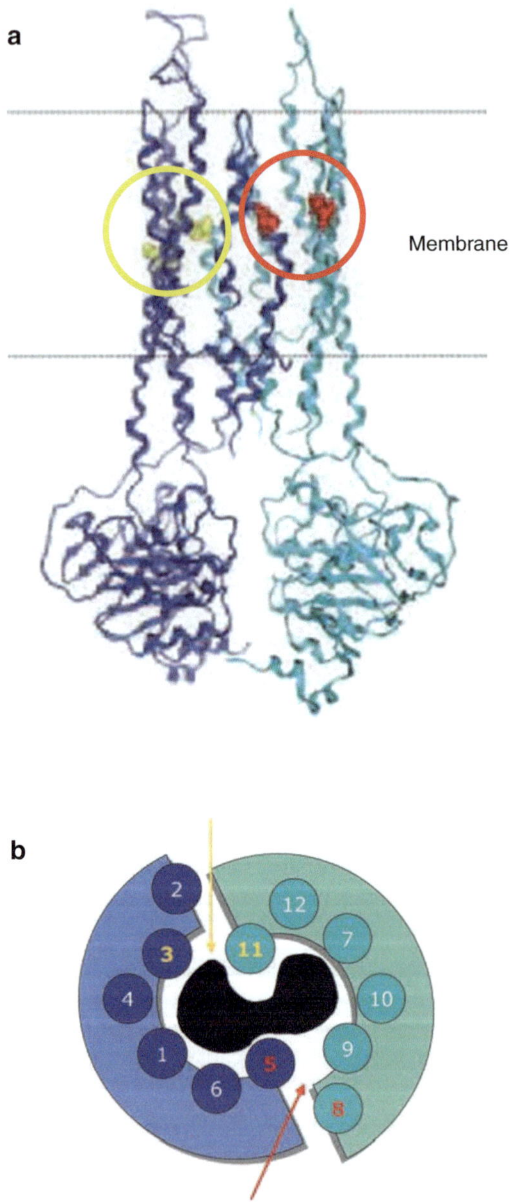

Fig. 10.6 Stereoimage of a side and top view of a refined ABCB1 model. (**a**) This model was generated to account for experimental evidence that the central pore is wider at the extra cytoplasmic face of the membrane than at the cytoplasmic face. The NH_2-terminal half of ABCB1 is depicted in *dark blue*, and the COOH-terminal half is in *cyan*. Residues with the highest labeling scores are depicted in *yellow* (residues Met197 in TM3 and Phe951 in TM11) and *red* (Ala311 in TM5 and Thr769 in TM8), respectively. These residues are located at the two TMD/TMD interfaces. The two residues depicted in yellow (Met197 and Phe951) have Cα-carbon distances of less than 9 Å, whereas the residues depicted in red (Ala311 and Thr769) are less than 14 Å apart. (**b**) The side view indicates that these four residues are located at similar depth within the membrane

MsbA (Ward et al. 2007). The labeling pattern was projected onto the three-dimensional atomic-detail model of ABCB1 to allow a visualization of the binding domain(s). Labeling is predicted to occur at the two TMDs and their interfaces formed between the NH_2- and COOH-terminal half of ABCB1 (Fig. 10.6). These interfaces are formed by TMD 3 and 11 on one hand and TMD 5 and 8 on the other hand. Considering available data on LmrA and AcrB, two bacterial multidrug efflux pumps, binding at domain interfaces may be a general feature of polyspecific drug efflux pumps. Recently large amount of peroxisomal ABC transporters became to be possible to express on the peroxisomes in yeast *P. pastoris* (unpublished observation). Therefore, substrate-binding site of peroxisomal ABC transporters seems to be characterized by photoaffinity techniques in the near future.

It is also important to know what kinds of substances including bioactive compounds are transported by ABC transporters. Ligands that interact specifically with the ABC transporters are useful for the characterization of the binding manner of these compounds. [^{125}I]iodoarylazidoprazosin is often used for showing binding of several drugs to ABC transporters by competitive manner. For example, sunitinib malate is a receptor tyrosine kinase inhibitor that inhibits cellular signaling of multiple targets such as the platelet-derived growth factor receptors. Shukla et al. investigated the effect of sunitinib on the ABC transporters ABCB1 and ABCG2, which are known to transport a wide variety of anticancer drugs (Shukla et al. 2009). They found that sunitinib inhibits ABCB1- and ABCG2-mediated efflux of fluorescent substrates in cells overexpressing these transporters. They further showed a direct interaction of sunitinib with the substrate-binding pocket of these transporters as it inhibited binding of the photoaffinity substrate [^{125}I]iodoarylazidoprazosin to ABCB1 with IC_{50} of 14.2 μM and ABCG2 with IC_{50} of 1.33 μM, respectively. In addition, sunitinib stimulated the ATP hydrolysis by both transporters in a concentration-dependent manner. Based on these findings they reported for the first time that sunitinib inhibits transport mediated by ABC drug transporters, which may affect the bioavailability of drugs co-administered with sunitinib. In other studies, [^{125}I]iodoarylazidoprazosin is also used for the showing binding of several drugs to ABCG2 by competitive manner (Brendel et al. 2007; Wu et al. 2007; Patel et al. 2013; Fujita et al. 2013; Kuang et al. 2012).

It is also interesting to use photoaffinity ligand to identify the target in the intact cells. Recently a representative of a new class of potent antimalarials with an unknown mode of action was described. To identify the molecular target of this class of antimalarials, Bruuner et al. employed a photoreactive affinity capture method to find parasite proteins specifically interacting with the capture compound in living parasitized cells (Brunner et al. 2013). The captured reagent retained the antimalarial properties of the parent molecule (ACT-213615) and accumulated within parasites. They identified several proteins interacting with the captured compound and established a functional interaction between ACT-213615 and *Plasmodium falciparum* (*Pf*) ABCB1 (MDR1). They supeculated that *Pf*MDR1 may play a role in the antimalarial activity of the piperazine-containing compound ACT-213615.

Recently, Kodan et al. revealed precise structure of human ABCB1 homolog, *Cm*ABCB1 from *Cyanidioschyzon merolae* (Fig. 10.7) (Kodan et al. 2014). They presented crystal structures of the ABCB1 homolog in two forms: unbound at 2.6-Å resolution and bound to a unique allosteric inhibitor at 2.4-Å resolution. The

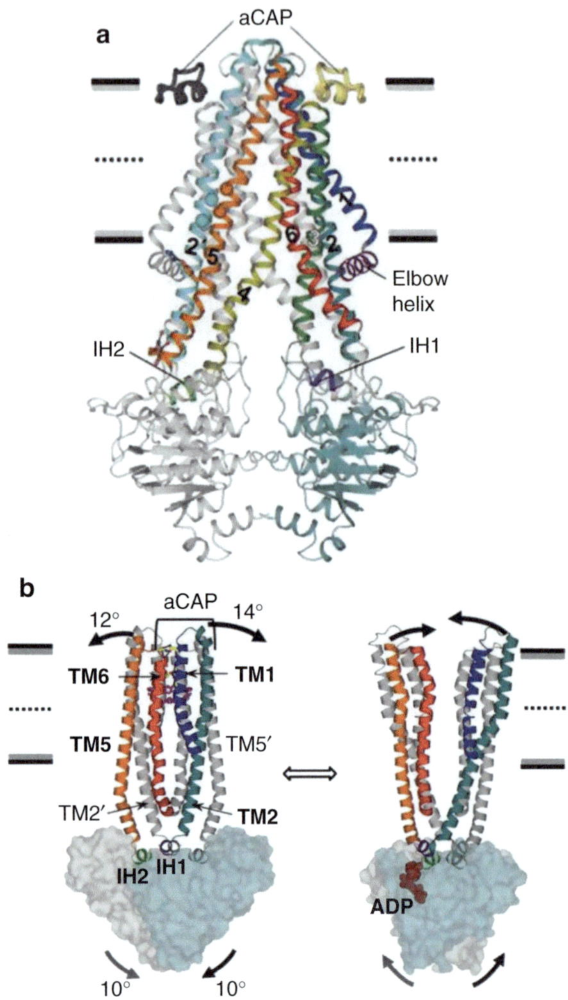

Fig. 10.7 Overall architecture of the *Cm*ABCB1-aCAP complex. (**a**) *Cm*ABCB1-aCAP complex structure viewed parallel to the plane of the membrane. *Cm*ABCB1 and bound aCAPs are depicted in cartoon representation. One subunit is displayed in multiple colors, and the other subunit is shown in gray except for TM2′ (*cyan*) in the lower panel and TM1′ (*light blue*) and TM6′ (*dark salmon*) in the upper panel. Horizontal *black bars* represent the expected positions of the hydrophilic surfaces of the lipid membrane; *gray bars* represent the expected positions of the hydrophobic surfaces. *Thick dashed lines* represent the middle of the membrane bilayer. (**b**) Conformational change of *Cm*ABCB1 during ATP binding and hydrolysis

inhibitor clamps the transmembrane helices from the outside, fixing the *Cm*ABCB1 structure in an inward-open conformation similar to the unbound structure, confirming that an outward-opening motion is required for ATP hydrolysis cycle. These structures, along with site-directed mutagenesis and transporter activity measurements, reveal the detailed architecture of the transporter, including a gate that opens to extracellular side and two gates that open to intramembranous region and the cytosolic side (Fig. 10.7). They propose that the motion of the nucleotide-binding domain drives those gating apparatuses via two short intracellular helices, IH1 and IH2, and two transmembrane helices, TM2 and TM5. It should be important to identify the region where the substrate binds with and without ATP by photoaffinity techniques.

10.5 Photoaffinity Labeling of Peroxisomal Fatty Acid β-Oxidation Enzymes

Peroxisomal fatty acid oxidation pathway and the enzyme involved in the process have become relatively clear. Photoaffinity studies are useful to identify the fatty acid-binding region of the proteins. In early studies Mangroo et al. found that peroxisomal acyl-CoA oxidase was specifically labeled by 11-diazirinophenoxy-undecanoyl-CoA in the lysate from *Candida tropicalis* (Mangroo et al. 1993). Rajasekharan et al. synthesized ^{32}P-labeled CoA of high specific activity and a photoaffinity dodecanoic acid analog, 12-((4-azidosalicyl)amino)dodecanoic acid and its CoA derivative (ASD-CoA). The synthesized photolabile reagents were tested as photoaffinity labels for acyl-CoA oxidase from Arthrobacter species (Rajasekharan et al. 1993). Acyl-CoA oxidase was incubated in the presence of acyl-CoA analogs and immediately photolyzed, which resulted in irreversible inhibition. Oleoyl-CoA and dodecanoyl-CoA protect the enzyme from photoactivated inhibition by 12-azidooleoyl-CoA and ASD-CoA, respectively. Analysis of photolyzed enzyme preparations by SDS-PAGE and autoradiography revealed that both analogs preferentially labeled a 54 kDa protein. These results demonstrate that the photoaffinity acyl-CoA analogs have potential application as probes to identify and characterize lipid biosynthetic enzymes and to identify the active site of these proteins. However, the precise molecular mechanism, especially in terms of substrate recognition, has not yet been understood. We will show our approach to characterize substrate binding to a β-oxidation enzyme in more detail below.

To characterize the properties of fatty acid oxidation enzymes in peroxisomes, we synthesized a novel photoreactive LCFA probe (Fig. 10.8) (Kashiwayama et al. 2010). The synthetic probe consists of LCFA, a diazirine-based photophore, and a

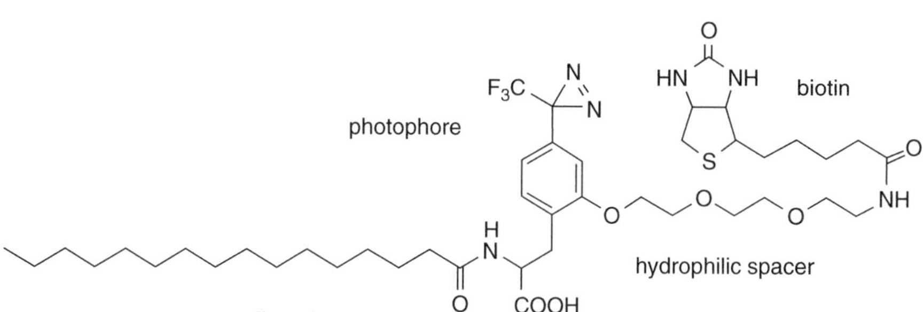

Fig. 10.8 Structure of the photoreactive LCFA probe

biotin moiety with a hydrophilic linker. (3-Trifluoromethyl)phenyldiazirine is one of the most promising photophores. By photo-irradiation, it produces a carbene species, a highly reactive intermediate, and immediately makes a covalent bond with spatially closed molecules. A biotin moiety was directly anchored to the photophore via a hydrophilic linker as a tag for detection and isolation of labeled proteins utilizing a biotin-avidin interaction, and is also useful to identify the labeled protein and its ligand-interacting site. Using the photoreactive LCFA analog as a probe, we performed photoaffinity-labeling experiment on purified rat liver peroxisomes.

The purified peroxisomes were incubated with the photoreactive fatty acid analog, and the labeled proteins were separated by SDS-PAGE, and detected by streptavidin-HRP. An 80-kDa protein was specifically labeled by the probe upon activation with UV irradiation (Fig. 10.9a, left side). Photo-incorporation of the probe within the 80-kDa protein was suppressed by the presence of palmitoyl-CoA in a concentration-dependent manner, but the labeling was not inhibited even in the presence of an excess amount of palmitic acid or CoA, suggesting that the 80-kDa protein has a potency to interact with palmitoyl-CoA, and the photoreactive LCFA probe could behave as palmitoyl-CoA rather than palmitic acid and could bind to the 80-kDa protein at the region responsible for its palmitoyl-CoA-binding site.

As we found that the 80-kDa protein labeled by the photoaffinity probe was a matrix protein of peroxisomes, we disrupted the peroxisomes with 0.1% Triton X-100 after photoaffinity labeling with the probe and isolated the labeled proteins by streptavidin agarose resin. As shown in Fig. 10.9a, right side, the 80-kDa protein was successfully purified. The probe-incorporated 80-kDa protein was digested with trypsin, and the resulting tryptic peptide mixture was analyzed by MALDI-TOF mass spectrometry. Several specific tryptic peptide fragments of the labeled protein were obtained, and MASCOT search engine revealed almost all of the major fragments were identified corresponding to MFE2-specific tryptic fragments. We assigned the fragment as a photoadduct fragment. MFE2, also called D-bifunctional protein, is a 79-kDa enzyme that possesses a typical peroxisomal matrix protein targeting signal (COOH-terminal Ala-Lys-Leu). MFE2 consists of a 3R-hydroxyacyl-CoA dehydrogenase (HD) domain, a 2E-enoyl-CoA hydratase 2 (H2) domain, and a sterol carrier protein 2-like (SCP2L) domain, and is known to catalyze the second and third steps of peroxisomal β-oxidation of fatty acids and their derivatives (Haapalainen et al. 2003).

The photoreactive LCFA probe could bind to MFE2 at its palmitoyl-CoA-binding site. Therefore, we tried to identify the individual amino acid residues labeled by the probe within MFE2 using recombinant MFE-His purified from *E. coli* expressing rat MFE2-His. MFE2-His was incubated with the probe and the labeled MFE2-His was purified by avidin resin (Fig. 10.9b). After digestion with trypsin, the resulting peptide mixture was analyzed by MALDI-TOF mass spectrometry. Various tryptic fragments with observed molecular masses that were compatible with the calculated molecular mass of MFE2 tryptic fragments were identified, both samples prepared from labeled and unlabeled MFE2. However, a peak at m/z 1345.39 was detected only in the sample prepared from labeled MFE2. The fragment was also detected in the cross-linked MFE2 purified from rat liver peroxisomes. The molecular mass of

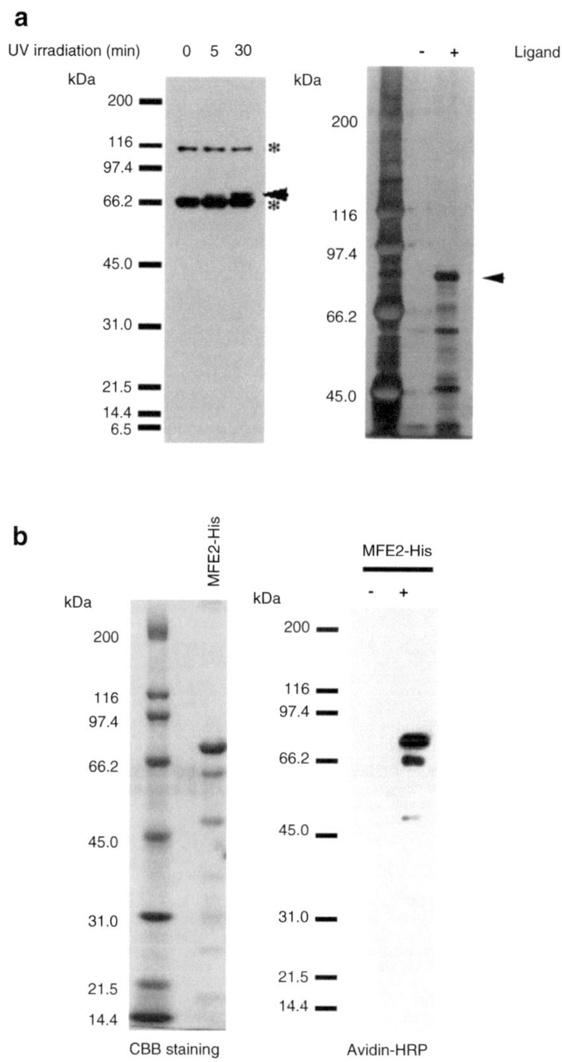

Fig. 10.9 (**a**, *left side*) Photoaffinity labeling of rat liver peroxisomes. Purified rat liver peroxisomes were incubated with the photoreactive LCFA probe for 2 h at 4 °C. After UV irradiation at 360 nm for 0–30 min at 0 °C, the labeled proteins were separated by SDS-PAGE and detected by streptavidin-HRP. The arrowhead indicates the 80-kDa protein labeled by the photoaffinity probe. Asterisks indicate nonspecific bands. (**a**, *right side*) Purified rat liver peroxisomes (1 mg protein) were labeled by the photoreactive LCFA probe. After solubilization with 0.1% Triton X-100, labeled proteins were purified by SoftLink SoftRelease Avidin Resin. The purified proteins were separated on a 5–10% SDS-polyacrylamide gradient gel, and stained with silver staining. The arrowhead and the asterisk indicate the 80-kDa protein and the 65-kDa protein labeled by the photoaffinity probe, respectively. (**b**) Photoaffinity labeling of purified MFE2-His. Purified MFE2-His were incubated with the photoreactive LCFA probe for 2 h at 4 °C After UV irradiation at 360 nm for 30 min at 0 °C, labeled proteins were separated by SDA-PAGE, and detected by streptavidin-HRP. The asterisk indicates nonspecific bands

the fragment did not agree with any of the theoretical MFE2 fragments digested with trypsin, instead, it matched exactly with the mass of tryptic peptide of MFE2 (residues Trp249-Glu250-Arg251; *m*/*z* 489.23) modified with the probe (*m*/*z* 1345.69 = 489.23 + 856.46). On the other hand, the tryptic fragment corresponding to the MFE2 sequence Trp249-Glu250-Arg251 (*m*/*z* 489.24) was absent from the labeled MFE2-His, but observed in the control experiment using the tryptic digestion of the unlabeled MFE2-His.

To confirm that the amino acid region (Trp249-Glu250-Arg251) within MFE2 is involved in the ligand binding, we purified the MFE2 mutants replacing these amino acids with Ala, and analyzed the ligand-binding efficiency. As shown in Fig. 10.10, wild-type MFE2 was labeled by the photoaffinity probe. However, mutant MFE2(W249A) and MFE2(R251A) showed the decreased labeling efficiency. The labeling efficiencies of mutant MFE2(W249A) and MFE2(R251A) with the probe

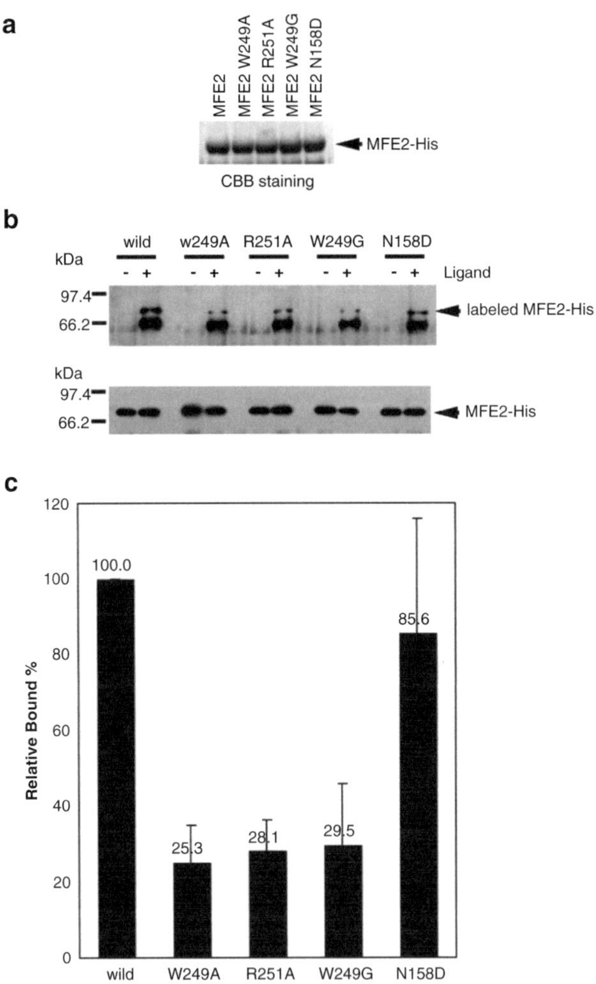

Fig. 10.10 Photoaffinity labeling of mutant MFE2-His. (**a**) Comparison of the purities of mutant MFE2-His. (**b**) Wild-type and mutant MFE2-His were labeled by the photoreactive LCFA probe. The labeled proteins were separated by SDS-PAGE, and detected by streptavidin-HRP. (**c**) The amount of probe-incorporated MFE2s was quantified with a LAS4000 luminoanalyzer and the relative labeling percentage was expressed as a ratio of the labeling percentage of each mutant MFE2 with that of wild-type MFE2

decreased to ~30% compared to that of wild-type MFE2. These three amino acid residues lie within the HD domain of MFE2, and a disease-causing mutation has been identified in this region (Huyghe et al. 2006). The mutant MFE2(W249G) also showed the decreased labeling efficiency while MFE2(N158D), another disease-causing mutant existing out of the labeled region (Huyghe et al. 2006), showed the slightly affected labeling efficiency for the photoreactive LCFA probe in the same conditions. Purified MFE2 HD domain was also labeled by the photoaffinity ligand, and MFE2 HD(W249A) showed the decreased labeling efficiency. These data indicate that the amino acid region within MFE2 (Trp249-Glu250-Arg251) is important for the ligand binding.

The photophore of the photoreactive LCFA probe cross-linked with MFE2 at the three amino acid region of MFE2 HD domain (Trp249-Glu250-Arg251). The crystal structure of HD domain of rat MFE2 has been solved (Haapalainen et al. 2003). Based on the three-dimensional structure, these residues lie on the β strand located on the top of dimerization interface of the subunits (Fig. 10.11a). The location is apart from the catalytic triad of MFE2 HD domain (Ser151, Tyr164, Lys168). However, these resi-

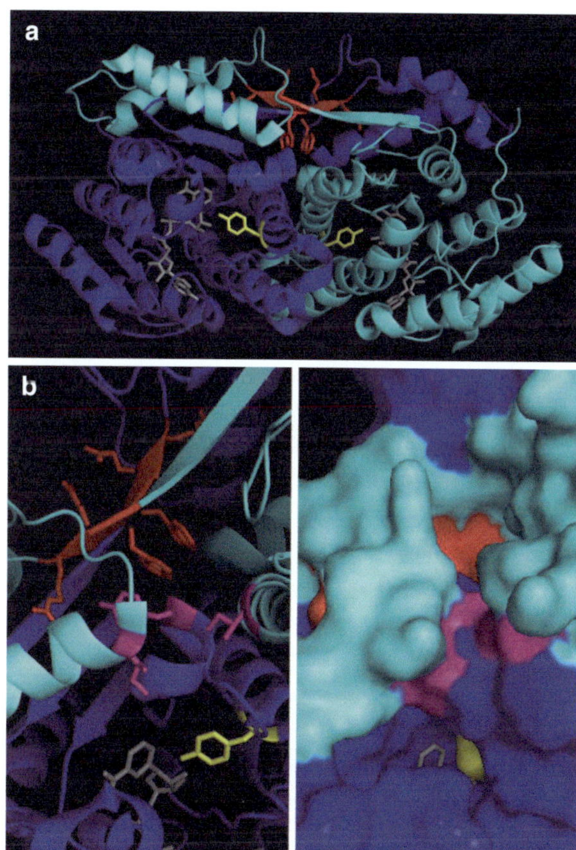

Fig. 10.11 Three-dimensional structure of MFE2 HD domain. (**a**) Binary structure of rat MFE2HD (accession code 1GZ6; Protein Data Bank). The structure consists of two identical monomers colored *blue* and *cyan*, respectively. The catalytic residue (Y164) is colored *yellow*, and the probe-incorporated fragment (W249-R251) is colored *red*. The bound NAD^+ are shown as a stick model and colored *gray*. (**b**) The zoom-in view of the molecular surface of the putative substrate-binding cavity. The catalytic residue (Y164) and the probe-incorporated fragment (W249-R251) are colored as described in (**a**). Aromatic residue (Y156) and hydrophobic residues (I160, I288) forming the hydrophobic cavity are colored magenta. Images were generated using PyMOL

dues also locate in the tip of the hydrophobic cavity leading to the active site (Fig. 10.11b). Therefore, we speculated that MFE2 could anchor the photophore of the LCFA derivative ligand on the top of the hydrophobic cavity (Trp249-Glu250-Arg251) and bury the hydrophobic probe along with the hydrophobic cavity with heading the fatty acyl tail of the ligand toward the active site. To challenge this hypothetical substrate-binding mode of MFE2 HD, we disrupted the hydrophobic residues (Ile180, Ile288) and the aromatic residue (Tyr156) forming the hydrophobic cavity of MFE2 HD domain and examined the photoaffinity labeling of these mutant proteins. These data indicate that the hydrophobic cavity leading to the active site of MFE2 HD domain could be assigned as one of the substrate-binding sites of MFE2 (Fig. 10.11b).

The three amino acid region (Trp249-Glu250-Arg251), which we identified as a photoprobe anchoring site, locates on the dimer interface of the MFE2 HD domains. Among them, Trp249 of both monomers is positioned in close proximity and the side chains of the two tryptophan residues lie parallel with each other. Therefore, Trp249 is suggested to be important for the homodimerization of the domains. Ferdinandusse et al. identified a W249G mutation in a patient with MFE2 deficiency, and they considered the mutation to be responsible for dimerization based on structural analysis (Ferdinandusse et al. 2006). In this study, MFE2(W249A) and MFE2(W249G) indeed affected the ligand-binding efficiency. However, these mutants as well as wild-type MFE2 exhibited a molecular mass corresponding to a dimer on sucrose density gradient centrifugation. These data indicate that the tryptophan residue could be responsible for substrate binding and not solely for dimerization.

Recently crystal structure of a full-length MFE2 of *Drosophila melanogaster* was revealed (Haataja et al. 2011). This is a first report of crystallization of MFE2. The structure explains the necessity of dimerization and the lack of substrate channel for the sequential enzyme reactions. They demonstrated that the turnover number and the catalytic efficiency for separate hydratase and dehydrogenase enzymes are of the same order of magnitude as for the native full-length proteins. The date does not support moving of substrate through a channel. How substrate is transferred from the active site of hydratase to that of dehaydrogenase is interesting. In addition, deficiency of MFE2 is associated with a spectrum of neurological disorders encompassing development and degenerative pathologies (Verheijden et al. 2014). Impaired recognition of substrate must be elucidated.

10.6 Concluding Remarks

Further investigation is needed to understand how peroxisomal ABC proteins and fatty acid oxidation enzymes recognize their substrates. However, considerable progress has been made in the research reported over the last few years. The expression of human peroxisomal ABC transporters in yeast has clearly demonstrated the substrate specificity of the peroxisomal ABC transporters. To understand the detailed transport of substrate by the peroxisomal ABC transporters, X-ray crystallographic studies together with photoaffinity studies will be required. Dysfunction of ABCD1 and peroxisomal fatty acid oxidation enzymes such as MFE2 results in severe peroxisomal diseases. VLCFA seems to be closely associated with the disease. Photoaffinity approaches may be required to identify the regulatory mechanisms in association with the dysfunction of peroxisomal fatty oxidation system.

Acknowledgments This research was supported in part by a Grant-in-Aid for Intractable Diseases from the Ministry of Health, Labour and Welfare of Japan, and for Scientific Research from the Ministry of Education, Culture, Sports, Science and Technology of Japan (20590054, 23590072, 26460063). Pacific Edit reviewed the manuscript prior to submission.

References

Agrawal G, Subramani S (2016) De novo peroxisome biogenesis: evolving concepts and conundrums. Biochim Biophys Acta 1863:892–901

Aller SG, Yu J, Ward A, Weng Y, Chittaboina S, Zhuo R, Harrell PM, Trinh YT, Zhang Q, Urbatsch IL, Chang G (2009) Structure of P-glycoprotein reveals a molecular basis for poly-specific drug binding. Science 323:1718–1722

Berger J, Dorninger F, Forss-Petter S, Kunze M (2016) Peroxisomes in brain development and function. Biochim Biophys Acta 1863:934–955

Bowen P, Lee CS, Zellweger H, Lindenberg R (1964) A familial syndrome of multiple congenital defects. Bull Johns Hopkins Hosp 114:402–414

Bowers WE (1998) Christian de Duve and the discovery of lysosomes and peroxisomes. Trends Cell Biol 8:330–333

Brendel C, Scharenberg C, Dohse M, Robey RW, Bates SE, Shukla S, Ambudkar SV, Wang Y, Wennemuth G, Burchert A, Boudriot U, Neubauer A (2007) Imatinib mesylate and nilotinib (AMN107) exhibit high-affinity interaction with ABCG2 on primitive hematopoietic stem cells. Leukemia 21:1267–1275

Brunner R, Ng CL, Aissaoui H, Akabas MH, Boss C, Brun R, Callaghan PS, Corminboeuf O, Fidock DA, Frame IJ, Heidmann B, Le Bihan A, Jeno P, Mattheis C, Moes S, Muller IB, Paguio M, Roepe PD, Siegrist R, Voss T, Welford RW, Wittlin S, Binkert C (2013) UV-triggered affinity capture identifies interactions between the plasmodium falciparum multidrug resistance protein 1 (PfMDR1) and antimalarial agents in live parasitized cells. J Biol Chem 288:22576–22583

Chu BB, Liao YC, Qi W, Xie C, Du X, Wang J, Yang H, Miao HH, Li BL, Song BL (2015) Cholesterol transport through lysosome-peroxisome membrane contacts. Cell 161:291–306

Dawson RJ, Locher KP (2006) Structure of a bacterial multidrug ABC transporter. Nature 443:180–185

De Duve C, Baudhuin P (1966) Peroxisomes (microbodies and related particles). Physiol Rev 46:323–357

Dixit E, Boulant S, Zhang Y, Lee AS, Odendall C, Shum B, Hacohen N, Chen ZJ, Whelan SP, Fransen M, Nibert ML, Superti-Furga G, Kagan JC (2010) Peroxisomes are signaling platforms for antiviral innate immunity. Cell 141:668–681

Ebberink MS, Koster J, Visser G, Spronsen F, Stolte-Dijkstra I, Smit GP, Fock JM, Kemp S, Wanders RJ, Waterham HR (2012) A novel defect of peroxisome division due to a homozygous non-sense mutation in the PEX11beta gene. J Med Genet 49:307–313

Ferdinandusse S, Jimenez-Sanchez G, Koster J, Denis S, Van Roermund CW, Silva-Zolezzi I, Moser AB, Visser WF, Gulluoglu M, Durmaz O, Demirkol M, Waterham HR, Gokcay G, Wanders RJ, Valle D (2015) A novel bile acid biosynthesis defect due to a deficiency of peroxisomal ABCD3. Hum Mol Genet 24:361–370

Ferdinandusse S, Ylianttila MS, Gloerich J, Koski MK, Oostheim W, Waterham HR, Hiltunen JK, Wanders RJ, Glumoff T (2006) Mutational spectrum of D-bifunctional protein deficiency and structure-based genotype-phenotype analysis. Am J Hum Genet 78:112–24

Fourcade S, Ruiz M, Camps C, Schluter A, Houten SM, Mooyer PA, Pampols T, Dacremont G, Wanders RJ, Giros M, Pujol A (2009) A key role for the peroxisomal ABCD2 transporter in fatty acid homeostasis. Am J Physiol Endocrinol Metab 296:E211–E221

Francisco T, Rodrigues TA, Pinto MP, Carvalho AF, Azevedo JE, Grou CP (2014) Ubiquitin in the peroxisomal protein import pathway. Biochimie 98:29–35

Fujiki Y, Nashiro C, Miyata N, Tamura S, Okumoto K (2012) New insights into dynamic and functional assembly of the AAA peroxins, Pex1p and Pex6p, and their membrane receptor Pex26p in shuttling of PTS1-receptor Pex5p during peroxisome biogenesis. Biochim Biophys Acta 1823:145–149

Fujita Y, Noguchi K, Suzuki T, Katayama K, Sugimoto Y (2013) Biochemical interaction of anti-HCV telaprevir with the ABC transporters P-glycoprotein and breast cancer resistance protein. BMC Res Notes 6:445

Genin EC, Geillon F, Gondcaille C, Athias A, Gambert P, Trompier D, Savary S (2011) Substrate specificity overlap and interaction between adrenoleukodystrophy protein (ALDP/ABCD1) and adrenoleukodystrophy-related protein (ALDRP/ABCD2). J Biol Chem 286:8075–8084

Goldfischer S, Moore CL, Johnson AB, Spiro AJ, Valsamis MP, Wisniewski HK, Ritch RH, Norton WT, Rapin I, Gartner LM (1973) Peroxisomal and mitochondrial defects in the cerebro-hepato-renal syndrome. Science 182:62–64

Guimaraes CP, Sa-Miranda C, Azevedo JE (2005) Probing substrate-induced conformational alterations in adrenoleukodystrophy protein by proteolysis. J Hum Genet 50:99–105

Haapalainen AM, Koski MK, Qin YM, Hiltunen JK, Glumoff T (2003) Binary structure of the two-domain (3R)-hydroxyacyl-CoA dehydrogenase from rat peroxisomal multifunctional enzyme type 2 at 2.38 Å resolution. Structure 11:87–97

Haataja TJ, Koski MK, Hiltunen JK, Glumoff T (2011) Peroxisomal multifunctional enzyme type 2 from the fruitfly: dehydrogenase and hydratase act as separate entities, as revealed by structure and kinetics. Biochem J 435:771–781

Hasan S, Platta HW, Erdmann R (2013) Import of proteins into the peroxisomal matrix. Front Physiol 4:261

Hatanaka Y (2015) Development and leading-edge application of innovative photoaffinity labeling. Chem Pharm Bull (Tokyo) 63:1–12

Hozoji-Inada M, Munehira Y, Nagao K, Kioka N, Ueda K (2011) Liver X receptor beta (LXRbeta) interacts directly with ATP-binding cassette A1 (ABCA1) to promote high density lipoprotein formation during acute cholesterol accumulation. J Biol Chem 286:20117–20124

Huber N, Guimaraes S, Schrader M, Suter U, Niemann A (2013) Charcot-Marie-tooth disease-associated mutants of GDAP1 dissociate its roles in peroxisomal and mitochondrial fission. EMBO Rep 14:545–552

Huyghe S, Mannaerts GP, Baes M, Van Veldhoven PP (2006) Peroxisomal multifunctional protein-2: the enzyme, the patients and the knockout mouse model. Biochim Biophys Acta 1761:973–994

Kashiwayama Y, Morita M, Kamijo K, Imanaka T (2002) Nucleotide-induced conformational changes of PMP70, an ATP binding cassette transporter on rat liver peroxisomal membranes. Biochem Biophys Res Commun 291:1245–1251

Kashiwayama Y, Tomohiro T, Narita K, Suzumura M, Glumoff T, Hiltunen JK, Van Veldhoven PP, Hatanaka Y, Imanaka T (2010) Identification of a substrate-binding site in a peroxisomal beta-oxidation enzyme by photoaffinity labeling with a novel palmitoyl derivative. J Biol Chem 285:26315–26325

Kodan A, Yamaguchi T, Nakatsu T, Sakiyama K, Hipolito CJ, Fujioka A, Hirokane R, Ikeguchi K, Watanabe B, Hiratake J, Kimura Y, Suga H, Ueda K, Kato H (2014) Structural basis for gating mechanisms of a eukaryotic P-glycoprotein homolog. Proc Natl Acad Sci U S A 111:4049–4054

Kuang YH, Patel JP, Sodani K, Wu CP, Liao LQ, Patel A, Tiwari AK, Dai CL, Chen X, Fu LW, Ambudkar SV, Korlipara VL, Chen ZS (2012) OSI-930 analogues as novel reversal agents for ABCG2-mediated multidrug resistance. Biochem Pharmacol 84:766–774

Lazarow PB, De Duve C (1976) A fatty acyl-CoA oxidizing system in rat liver peroxisomes; enhancement by clofibrate, a hypolipidemic drug. Proc Natl Acad Sci U S A 73(6):2043

Mangroo D, Steele L, Rachubinski RA, Gerber GE (1993) Specific labeling of Candida tropicalis peroxisomal proteins with photoreactive fatty-acid derivatives. Biochim Biophys Acta 1168:280–284

Masuda S, Tomohiro T, Yamaguchi S, Morimoto S, Hatanaka Y (2015) Structure-assisted ligand-binding analysis using fluorogenic photoaffinity labeling. Bioorg Med Chem Lett 25(8):1675

Morita M, Imanaka T (2012) Peroxisomal ABC transporters: structure, function and role in disease. Biochim Biophys Acta 1822:1387–1396

Patel A, Tiwari AK, Chufan EE, Sodani K, Anreddy N, Singh S, Ambudkar SV, Stephani R, Chen ZS (2013) PD173074, a selective FGFR inhibitor, reverses ABCB1-mediated drug resistance in cancer cells. Cancer Chemother Pharmacol 72:189–199

Pleban K, Kopp S, Csaszar E, Peer M, Hrebicek T, Rizzi A, Ecker GF, Chiba P (2005) P-glycoprotein substrate binding domains are located at the transmembrane domain/transmembrane domain interfaces: a combined photoaffinity labeling-protein homology modeling approach. Mol Pharmacol 67:365–374

Rajasekharan R, Marians RC, Shockey JM, Kemp JD (1993) Photoaffinity labeling of acyl-CoA oxidase with 12-azidooleoyl-CoA and 12-[(4-azidosalicyl)amino]dodecanoyl-CoA. Biochemistry 32:12386–12391

Rhodin JAG (1954) Correlation of ultrastructural organization: and function in normal and experimentally changed proximal convoluted tubule cells of the mouse kidney: an electron microscopic study. Karolinska Institutet

van Roermund CW, Visser WF, Ijlst L, Waterham HR, Wanders RJ (2011) Differential substrate specificities of human ABCD1 and ABCD2 in peroxisomal fatty acid beta-oxidation. Biochim Biophys Acta 1811:148–152

Rucktaschel R, Girzalsky W, Erdmann R (2011) Protein import machineries of peroxisomes. Biochim Biophys Acta 1808:892–900

Sato Y, Shibata H, Nakatsu T, Nakano H, Kashiwayama Y, Imanaka T, Kato H (2010) Structural basis for docking of peroxisomal membrane protein carrier Pex19p onto its receptor Pex3p. EMBO J 29:4083–4093

Shukla S, Robey RW, Bates SE, Ambudkar SV (2009) Sunitinib (Sutent, SU11248), a small-molecule receptor tyrosine kinase inhibitor, blocks function of the ATP-binding cassette (ABC) transporters P-glycoprotein (ABCB1) and ABCG2. Drug Metab Dispos 37:359–365

Tanaka AR, Tanabe K, Morita M, Kurisu M, Kasiwayama Y, Matsuo M, Kioka N, Amachi T, Imanaka T, Ueda K (2002) ATP binding/hydrolysis by and phosphorylation of peroxisomal ATP-binding cassette proteins PMP70 (ABCD3) and adrenoleukodystrophy protein (ABCD1). J Biol Chem 277:40142–40147

Tomohiro T, Inoguchi H, Masuda S, Hatanaka Y (2013) Affinity-based fluorogenic labeling of ATP-binding proteins with sequential photoactivatable cross-linkers. Bioorg Med Chem Lett 23:5605–5608

Vamecq J, Cherkaoui-Malki M, Andreoletti P, Latruffe N (2014) The human peroxisome in health and disease: the story of an oddity becoming a vital organelle. Biochimie 98:4–15

Verheijden S, Beckers L, De Munter S, Van Veldhoven PP, Baes M (2014) Central nervous system pathology in MFP2 deficiency: insights from general and conditional knockout mouse models. Biochimie 98:119–26

Wanders RJ, Komen J, Ferdinandusse S (2011) Phytanic acid metabolism in health and disease. Biochim Biophys Acta 1811:498–507

Wanders RJ, Waterham HR, Ferdinandusse S (2016) Metabolic interplay between peroxisomes and other subcellular organelles including mitochondria and the endoplasmic reticulum. Front Cell Dev Biol 3:83

Ward A, Reyes CL, Yu J, Roth CB, Chang G (2007) Flexibility in the ABC transporter MsbA: alternating access with a twist. Proc Natl Acad Sci U S A 104:19005–19010

Waterham HR, Koster J, van Roermund CW, Mooyer PA, Wanders RJ, Leonard JV (2007) A lethal defect of mitochondrial and peroxisomal fission. N Engl J Med 356:1736–1741

Waterham HR, Ferdinandusse S, Wanders RJ (2016) Human disorders of peroxisome metabolism and biogenesis. Biochim Biophys Acta 1863:922–933

Wu CP, Shukla S, Calcagno AM, Hall MD, Gottesman MM, Ambudkar SV (2007) Evidence for dual mode of action of a thiosemicarbazone, NSC73306: a potent substrate of the multidrug resistance linked ABCG2 transporter. Mol Cancer Ther 6:3287–3296

Chapter 11
Photoaffinity Labeling Methods to Explore Internalization Mechanisms of Arginine-Rich Cell-Penetrating Peptides

Gen Tanaka, Yoshimasa Kawaguchi, Keiko Kuwata, Toshihide Takeuchi, Ikuhiko Nakase, and Shiroh Futaki

Abstract Peptides with membrane permeability have been used extensively as carriers for intracellular delivery of bioactive molecules, and endocytosis plays an important role in the cellular uptake of these peptides. Elucidating the mechanisms of uptake will contribute to the development of more sophisticated delivery systems. In this chapter, we introduce a novel approach for identifying receptors involved in the cellular uptake of oligoarginines, a representative class of cell-penetrating peptides (CPPs). Photoaffinity probes with oligoarginines together with trifluoromethyl diazirines as photoreactive moieties and biotin as an isolation tag were designed. We identified C-X-C chemokine receptor type 4 (CXCR4) as a receptor for dodecaarginine (R12) uptake via macropinocytosis, a typical uptake pathway for arginine-rich CPPs. Interestingly, CXCR4 does not play a significant role in the uptake of octaarginine (R8) and HIV-1 TAT peptides, suggesting that more than one form of macropinocytosis is involved in arginine-rich CPP uptake. The cytoplasmic protein LanCL1 was also identified because of its ability to promote R8 cellular uptake. A novel probe with a diazobenzene linker, which allowed chemoselective cleavage of the photocrosslinked proteins from affinity beads, facilitated isolation of the target proteins. Using this probe, syndecan-4 was identified as a receptor involved in R8 uptake via clathrin-mediated endocytosis.

Keywords Trifluoromethyl diazirine • Arginine-rich cell-penetrating peptide • Cellular uptake receptor • CXCR4 • Syndecan-4

G. Tanaka • Y. Kawaguchi • T. Takeuchi • I. Nakase • S. Futaki (✉)
Institute for Chemical Research, Kyoto University, Uji, Kyoto 611-0011, Japan
e-mail: futaki@scl.kyoto-u.ac.jp

K. Kuwata
Institute of Transformative Bio-Molecules (WPI-ITbM), Nagoya University, Furo-cho, Chikusa-ku, Nagoya 464-8602, Japan

Y. Hatanaka, M. Hashimoto (eds.), *Photoaffinity Labeling for Structural Probing Within Protein*, DOI 10.1007/978-4-431-56569-7_11

11.1 Introduction

It is now possible to achieve the intracellular delivery of bioactive molecules using membrane-permeable peptides as a delivery tool (Futaki 2006; Wender et al. 2008; Bechara and Sagan 2013; Stanzl et al. 2013). These peptides have collectively been called cell-penetrating peptides (CPPs) or protein transduction domains. The intracellular delivery of membrane-impermeable high-molecular-weight compounds or particles has also been achieved via chemical conjugation or stable complex formation with these peptides (Nakase et al. 2012; Boisguérin et al. 2015; Kurrikoff et al. 2016). As a result of the technical convenience and ease of obtaining the expected bioactivities, the number of applications of these peptides has been increasing steadily (Stanzl et al. 2013). As described later in this chapter, there are two methods of internalization of CPPs and their conjugates: endocytosis and direct penetration through the plasma membrane (Futaki et al. 2013; Brock 2014; Cleal et al. 2013). The former mechanism may be explained from biological and physiological perspectives and the latter from physicochemical ones. Depending on the physicochemical properties of CPPs and their conjugates and the cellular conditions, intracellular delivery is based on a balance of these two pathways. When the size of the delivered molecules increases, the endocytic contribution increases (Tünnemann et al. 2006). CPPs can be divided into several classes based on their physicochemical properties (Futaki 2006). The HIV-1 TAT peptide (Vivès et al. 1997) and oligoarginines (Rothbard et al. 2000; Futaki et al. 2001), as well as other CPPs rich in arginine, are among the most frequently employed classes of CPPs, and it has been asserted that one of the reasons for the efficient internalization of arginine-rich CPPs is the involvement of actin-driven fluid-phase endocytosis, or macropinocytosis (Wadia et al. 2004; Nakase et al. 2004). This form of endocytosis is generally not active, except in specific cell types such as macrophages, but can be activated by specific external stimuli (Marechal et al. 2001; Meier et al. 2002; Falcone et al. 2006; Mercer and Helenius 2009). Macropinocytosis is thought to be activated by the interaction of arginine-rich CPPs with the cell, leading to enhanced endocytic uptake of the peptides and their conjugated cargo. Elucidation of the mechanism behind this activation should allow the establishment of more efficient and sophisticated intracellular delivery systems. Macropinocytosis is activated by interaction of arginine-rich CPPs on cell surfaces (Nakase et al. 2004; Kaplan et al. 2005), which suggests that there are receptors that stimulate the macropinocytic uptake of arginine-rich CPPs and their conjugates. Identification of these macropinocytosis-inducing receptors will not only deepen our understanding of the methods of cellular uptake of arginine-rich CPPs and their conjugates, but could also lead to the development of alternative ligands for receptor activation as intracellular delivery vectors. This may also lead to the development of a targeting system for the efficient delivery of drugs and other bioactive molecules to specific cells in the body. Identification of the cellular uptake receptors for arginine-rich CPPs is a challenge in this field, and photocrosslinking can be a powerful approach. In this chapter, we introduce our efforts to identify endocytic receptors, which include identification of the C-X-C chemokine receptor type 4 (CXCR4) as a macropinocytic receptor for dodecaarginine (R12) (Tanaka et al. 2012). Lanthionine synthetase component C-like protein 1 (LanCL1) was also identified as a potential intracellular target of octaarginine (R8) (Kawaguchi et al. 2013). Finally, syndecan-4 was identified as a receptor for the clathrin-mediated endocytosis of R8 (Kawaguchi et al. 2016). Biotin-tagged R12 or R8 peptides bearing trifluoromethyl diazirine phenylalanine (Nakashima et al. 2006) (biotin-TmdPhe-R12 and biotin-TmdPhe-R8, respectively) were used in the identification of CXCR4 and LanCL1. The biotin tag allowed the isolation of proteins crosslinked with R12 and R8 by affinity purification with streptavidin beads (Hofmann and Kiso 1976). During the course of these studies, nonspecific binding of cellular proteins to the streptavidin beads made the identification of crosslinked products considerably more difficult. Therefore, two additional modifications were employed in the identification of syndecan-4 and its role in clathrin-mediated endocytosis of R8. To reduce cytoplasmic and nuclear protein contamination, membrane fractions were separated from cell lysates and subjected to affinity purification using streptavidin beads. In addition, diazobenzene was employed as a cleavable moiety in the photoaffinity probe (Verhelst et al. 2007; Yang et al. 2010), which allowed the selective elution of photocrosslinked products from the streptavidin beads. These modifications were of considerable help in the identification of crosslinked products.

11.2 Internalization Methods of Arginine-Rich CPPs and Their Conjugates with Cargo Molecules

The two major internalization methods of arginine-rich CPPs and their conjugates are direct plasma membrane penetration and endocytic uptake. To achieve direct penetration of arginine-rich CPPs through the plasma membrane, it is thought that the accumulation of arginine-rich CPPs on the plasma membrane is important (Nakase et al. 2004; Fretz et al. 2007; Kosuge et al. 2008). Here, the interaction of arginine-rich CPPs with membrane-associated proteoglycans, membrane proteins whose extracellular domains have been modified with sulfated polysaccharides, is important (Nakase et al. 2007; Kosuge et al. 2008). Arginine residues can form complexes with sulfate moieties via electrostatic interaction and hydrogen bonding. Therefore, membrane-associated proteoglycans are good candidates for the recruitment of arginine-rich CPPs to the cell surface. Arginine-rich CPPs can also interact with serum proteins (Kosuge et al. 2008), although the presence of serum reduces the ability of arginine-rich CPPs to cross the plasma membrane. In other words, the efficacy of membrane penetration is higher in the absence of serum. Alternatively, hydrophobic counter anions such as pyrenebutyrate facilitate the translocation of arginine-rich CPPs by complex formation with the carboxylate in pyrenebutyrate and the guanidino function of arginine, which neutralizes the charge (Perret et al. 2005; Takeuchi et al. 2006). Direct penetration of arginine-rich CPPs is not simply induced by passive diffusion, and the accumulation of arginine-rich CPPs is important to some extent. Once the cell-surface concentration of arginine-rich CPPs exceeds a threshold concentration, direct penetration of arginine-rich CPPs is induced (Fretz et al. 2007; Kosuge et al. 2008). Influx of arginine-rich CPPs through the plasma membrane does not occur uniformly across the cell surface, but usually occurs at a number of specific points on the plasma membrane (Takeuchi et al. 2006; Kosuge et al. 2008). The cell interior generally has a lower voltage than the cell surface, and this negative membrane potential is considered to function as a driving force for peptide influx (Rothbard et al. 2004). The influx of arginine-rich CPPs through the plasma membrane may be accompanied by pulse-like temporal perturbation of the membrane, allowing the penetration of arginine-rich CPPs accompanied by relatively small cargo. However, efficacy decreases as the size of the cargo increases, making endocytosis the predominant pathway for the cellular internalization of arginine-rich CPPs.

More than one pathway is involved in the endocytic uptake of arginine-rich CPPs and their conjugates (Jones 2007). In addition to clathrin-mediated and caveola-mediated endocytosis, macropinocytosis plays a significant role in the cellular uptake of arginine-rich CPPs and their conjugates (Nakase et al. 2004, 2008; Wadia et al. 2004). As described above, macropinocytosis is actin-driven fluid-phase endocytosis. We have reported that the interaction of arginine-rich CPPs with cell-surface molecules induces actin reorganization and eventual macropinocytosis (Nakase et al. 2004; Kosuge et al. 2008). This is accompanied by intracellular signaling and Rac protein activation. The role of macropinocytosis in the cellular uptake of arginine-rich CPPs becomes more significant at peptide concentrations in the micromolar range. Membrane-associated proteoglycans are important for the macropinocytic uptake of arginine-rich CPPs and the conjugates (Nakase et al. 2007), and the interaction of arginine-rich CPPs with membrane-associated proteoglycans may lead to their aggregation, triggering cellular signaling and macropinocytosis. Alternatively, macropinocytosis may be activated by the interaction of arginine-rich CPPs with specific receptors. It is also possible that arginine-rich CPPs may need to be coupled with membrane-associated proteoglycans in order to bind receptors, as is the case with fibroblast growth factor 2 (FGF2) receptor activation (Tkachenko et al. 2004).

Although some membrane proteins have been reported as receptors for proteins and peptides related to arginine-rich CPPs, no receptor has been implicated in the cellular uptake of arginine-rich CPPs. For example, human immunodeficiency virus type 1 (HIV-1) TAT peptide is a representative arginine-rich CPP that consists of the basic domain (positions 48–60) of the full-length Tat protein (86 residues), a transcriptional regulator protein of HIV-1. Vascular endothelial growth factor (VEGF) receptor is activated by the full-length Tat protein, but not by the TAT peptide (Mitola et al. 1997). The full-length Tat protein binds to CXCR4 and acts as an antagonist against HIV infection via positions 11–50, which include its cysteine-rich domain (Xiao et al. 2000). However, the basic domain corresponding to the TAT CPP sequence is not responsible for the interaction with CXCR4. It has been reported that interaction of arginine-rich peptides with membrane-associated proteoglycans, including syndecans, induces actin organization and macropinocytosis via activation of the protein kinase Cα and Rac1 (Nakase et al. 2004, 2014). Identification of the receptors that activate macropinocytosis or alternative forms of endocytosis by interacting with arginine-rich CPPs should provide a better understanding of the methods of internalization of these peptides. It should also allow the development of more sophisticated delivery systems, including photocrosslinking methods, which are described here.

11.3 A Photocrosslinking Approach for the Identification of Macropinocytosis-Inducing Receptors

11.3.1 CXCR4 Is a Macropinocytic Uptake Receptor for Dodecaarginine (R12)

Crosslinking reactions using a photoreactive probe are a powerful tool for the identification of cell-surface receptors. Diazirines and benzophenones are among the most frequently employed photoreactive groups (Tomohiro et al. 2005). Diazirines generate carbene intermediates upon relatively short exposure to ultraviolet (UV) irradiation (typically less than a few minutes) compared with the time required for benzophenone activation (typically in the order of tens of minutes) (Hatanaka 2015). Since UV radiation is toxic to cells, this shorter dose of UV irradiation is an advantage for the identification of oligoarginine cell-surface receptors. Photoactivation of diazirines creates reactive carbene intermediates that covalently crosslink with nearby molecules. Surrounding water molecules readily quench the carbene intermediates. The use of diazirine may allow the capture of transient cell-surface interactions. Oligoarginines interact with cell-surface molecules electrostatically and through hydrogen bonding, although these interactions can be weak and transient. Although diazirine may yield crosslinked products that represent non-specific interactions, we believe that the shorter reaction time of diazirine will be suitable for the identification of oligoarginine cell surface receptors.

To identify receptors, we first focused on dodecaarginine (R12), which is internalized more efficiently than the R8 and TAT peptides (Kosuge et al. 2008). A biotin-tagged R12 peptide bearing trifluoromethyl diazirine phenylalanine (biotin-TmdPhe-R12) was designed (Tanaka et al. 2012). Aryl-3-phenyl-3-trifluoromethyl diazirine derivatives are effective photocrosslinking agents (Brunner et al. 1980)

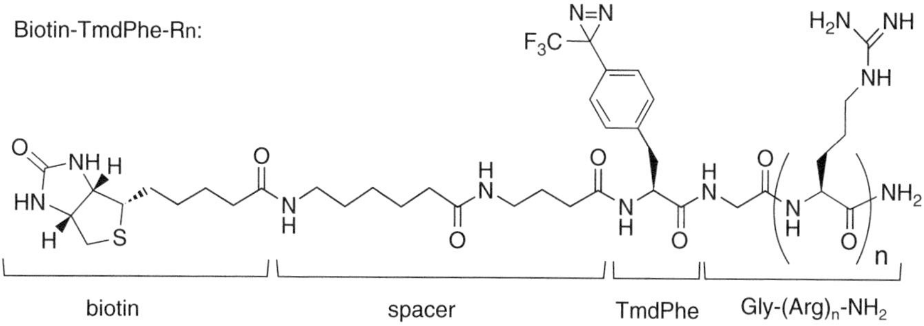

Fig. 11.1 Structures of Biotin-TmdPhe-Rn (n = 12, 8, 4)

(Fig. 11.1). Nakashima et al. (2006) developed the trifluoromethyldiazirinephenylalanine (TmdPhe) derivative as a diazirine-containing building block compatible with solid-phase peptide synthesis. TmdPhe was attached to the N-terminus of R12 through a glycine residue spacer. The peptide also contained an amide structure to avoid a negative charge at the C-terminus and to reduce possible proteolytic degradation. For the isolation of crosslinked products, biotin was incorporated at the N-terminus of the peptide with γ-aminobutyric acid (GABA) as a spacer. The biotin-TmdPhe-R12 was synthesized using standard 9-fluorenylmethyloxycarbonyl (Fmoc) solid-phase peptide synthesis on Rink amide resin, followed by on-resin biotinylation with biotinamidohexanoic acid N-hydroxysuccinimide ester. The removal of protecting groups and peptide detachment from the solid-phase resin was performed prior to high-performance liquid chromatography (HPLC) purification, which yielded pure biotin-TmdPhe-R12. The structure was confirmed by matrix-assisted laser desorption ionization time-of-flight mass spectrometry (MALDI-TOFMS). The biotin-TmdPhe-R12 was thus prepared without difficulty.

To perform the crosslinking, HeLa cells were pretreated with 0.5 μM biotin-TmdPhe-R12 at 37 °C for 30 s to allow the peptide to interact with cell-surface molecules. Cells were then irradiated with UV light (365 nm) at 4 °C for 3 min. Since many receptors internalize into cells via endocytosis upon ligand binding, treatment was performed at 4 °C to prevent this internalization. Treatment at 4 °C was also important to avoid UV damage to the cells. When the peptide concentration exceeds a threshold level, R12 may directly translocate through the plasma membrane (Kosuge et al. 2008). A peptide concentration of 0.5 μM and treatment time of 3 min were thus set, with the application of live cell imaging using FITC-labeled TmdPhe-R12, so that no significant signal of internalized FITC-labeled TmdPhe-R12 was observed. The biotin-TmdPhe-R4 peptide was used as a negative control since tetraarginine (R4) has little CPP activity (Kosuge et al. 2008). Following UV irradiation, the cells were washed with phosphate-buffered saline, lysed, and dialyzed. The crosslinked products were then collected using streptavidin magnetic spheres and subjected to sodium dodecyl sulfate polyacrylamide gel electrophoresis (SDS-PAGE) analysis and silver staining.

Many protein bands were obtained that were also observed in the biotin-TmdPhe-R4-treated cells. There were no obvious bands corresponding to crosslinked products because of nonspecific interactions with the streptavidin magnetic spheres. However, proteins greater than 240 kDa were specifically observed for biotin-TmdPhe-R12-treated but not biotin-TmdPhe-R4-treated cells. Trypsinization of these bands followed by MALDI-TOFMS analysis using peptide mass fingerprinting (PMF) revealed that these four bands were likely derived from the cytoskeletal protein myosin-9 (non-muscle myosin heavy chain IIA). The involvement of myosin-9 in CXCR4 endocytosis together with cell-surface colocalization has been suggested (Rey et al. 2007; Arii et al. 2010). CXCR4 is a C-X-C chemokine receptor (C-X-C chemokine receptor type 4) and is also a co-receptor for HIV-1 infection (Schramm et al. 2000).

The involvement of CXCR4 in the cellular uptake of R12 was confirmed by western blot analysis of the proteins pulled-down with magnetic spheres from biotin-TmdPhe-R12-treated cells. We observed positive staining with an anti-CXCR4 antibody, suggesting that CXCR4 forms a complex with the proteins cross-linked with biotin-TmdPhe-R12 (e.g., myosin-9). No anti-CXCR4-positive bands were observed in the samples treated with biotin-TmdPhe-R4, suggesting that CXCR4 may interact with biotin-TmdPhe-R12 on the cell surface.

siRNA knockdown of CXCR4 reduced the uptake of R12, confirming the involvement of CXCR4 in the cellular uptake of the R12 peptide. Treatment of HeLa cells with the CXCR4-specific antagonist FC131 [cyclo(-D-Tyr-Arg-Arg-Nal-Gly-), (Nal = 3-(2-naphtyl)alanine)] (Fujii et al. 2003) also led to a decrease in R12 uptake. Interestingly, no significant reduction in the cellular uptake of R8 or TAT was observed in CXCR4 knockdown cells and FC131-treated cells, indicating that CXCR4 was not responsible for the uptake of these peptides. The R8 and TAT peptides are typical arginine-rich CPPs with fewer arginine residues in their sequences than R12, suggesting the complexity of the mechanisms of cellular uptake, even among arginine-rich peptides. Cell-surface colocalization of CXCR4 with R12 was also shown by confocal laser scanning microscopy (CLSM) analysis.

Stromal cell-derived factor 1α (SDF-1α) is a natural ligand of CXCR4, which, upon binding, induces actin reorganization via the Gi protein signaling pathway (Möhle et al. 2001). R12 has also been shown to induce actin reorganization, although this was not observed in CXCR4-knocked-down cells. Involvement of the Gi protein signaling pathway in R12-mediated actin reorganization was also suggested by R12 treatment in the presence of the pertussis toxin, an inhibitor of the Gi protein signaling pathway (Phillips and Ager 2002), which resulted in reduced cellular uptake of R12 and actin reorganization. Therefore, CXCR4 serves as an R12 receptor that promotes actin organization and cellular uptake of R12 via the Gi protein signaling pathway.

CLSM analysis of R12-treated cells showed intracellular colocalization of R12 and CXCR4, suggesting that the binding of R12 with CXCR4 led to the internalization of CXCR4. We also observed colocalization of the CXCR4 signal with dextran (70 kDa), a macropinosome marker, in the presence of the R12 peptide, suggesting that the R12 peptide induced macropinocytosis-dependent CXCR4 internalization. As described above, macropinocytosis is fluid-phase endocytosis. R12 treatment resulted in increased cellular uptake of dextran, which functions as a fluid-phase solute. Stimulation of CXCR4 with its natural ligand, SDF-1α or HIV-1 envelope glycoprotein 120 (gp120), also triggered macropinocytosis and receptor internalization. Although the receptor that we identified was not an endocytic receptor responsible for the uptake of the more commonly employed R8 and TAT peptides, our photocrosslinking approach identified a cell-surface receptor responsible for R12 uptake, which is the first example of an arginine-rich CPP cellular uptake receptor. These results not only provide insight into the cellular uptake machinery of arginine-rich CPPs, but also have implications for the involvement of CXCR4 in HIV infection.

11.3.2 LanCL1 as a Potential Intracellular Target to Stimulate Cellular Uptake of R8

As described in the previous section, cellular uptake of the R8 and TAT peptides was not increased through interaction with the CXCR4 receptor. Macropinocytosis is known to be involved in the uptake of R8 and TAT (Nakase et al. 2004; Wadia et al. 2004), suggesting that more than one mechanism is involved in the macropinocytic uptake of arginine-rich CPPs. Identification of these receptors and cellular uptake pathways will benefit the development of novel CPP-mediated delivery systems. Macropinocytosis is thought to be involved in the internalization of various therapeutic macromolecules and in virus infection, so an improved understanding of these pathways will also impact these areas of research.

We performed a photocrosslinking study using an R8 peptide fused to trifluoromethyl diazirine phenylalanine (TmdPhe) as the photocrosslinking agent. The R8 peptide derivative used for photocrosslinking (biotin-TmdPhe-R8) in our study was designed and prepared using Fmoc solid-phase peptide synthesis in a manner similar to that in the case of biotin-TmdPhe-R12 (Kawaguchi et al. 2013) (Fig. 11.1). Human embryonic kidney 293T (HEK293T) cells were treated with 2.5 μM biotin-TmdPhe-R8 for 30 s at 37 °C prior to UV irradiation (365 nm) at 4 °C for 3 min. Considering that membrane potential is a driving force for direct penetration of the plasma membrane by R8 peptides (Rothbard et al. 2004), a potassium-rich HEPES buffer (pH 7.4), known to reduce the membrane potential (Hirose et al. 2012), was employed. We also expected that the use of this buffer would allow the peptides to interact with the plasma membrane for a longer period of time without direct penetration. Cells were then washed with a potassium-rich HEPES buffer containing heparin prior to cell lysis, and the photocrosslinked proteins were collected using streptavidin-coated magnetic particles. The bound proteins were analyzed by SDS-PAGE and silver staining, as well as by western blotting, using an anti-biotin primary antibody and a horseradish peroxidase (HRP)-conjugated secondary antibody, and detected with chemiluminescence.

A band on the western blot specific to biotin-TmdPhe-R8-treated cells was detected at approximately 37 kDa. The corresponding band in the silver-stained gel was excised and subjected to trypsin digestion. MALDI-TOFMS/PMF analysis indicated it was the cytosolic, peripheral membrane protein lanthionine synthetase component C-like protein 1 (LanCL1) (Mayer et al. 2001). Although the biological function of LanCL1 is unclear, it has been reported to bind specifically to the SRC homology 3 (SH3) domain of the epidermal growth factor (EGF) receptor kinase substrate 8 (EPS8) signaling protein (Zhang et al. 2009; Zhong et al. 2012); involvement of EPS8 in nerve growth factor (NGF)-induced neurite outgrowth has been suggested.

To analyze the effect of LanCL1 on the cellular uptake of R8, the corresponding gene was amplified from the HeLa cell cDNA library using polymerase chain reaction and subcloned into an expression vector. Overexpression of LanCL1 in HeLa cells resulted in a 25% increase in the cellular uptake of Alexa Fluor 488-labeled R8 (R8-Alexa), suggesting that LanCL1 promotes the uptake of R8.

The initial focus of this study was to isolate membrane proteins that function as receptors of R8 and promote its cellular uptake. Here, we have isolated the cytosolic, peripheral membrane protein LanCL1, which promotes R8 uptake. It is possible that some of the R8 peptides directly penetrate the plasma membrane to interact with LanCL1, which may then affect the ability of LanCL1 to promote R8 cellular uptake. Further studies will be necessary to address in detail the contribution of LanCL1 to uptake of the R8 peptide.

11.3.3 Syndecan-4 as an Uptake Receptor for Clathrin-Mediated Endocytosis of R8

We have already described the possible involvement of LanCL1 in uptake of the R8 peptide. However, LanCL1 is a cytosolic protein and not a cell-surface receptor. One possible reason for the difficulty in identifying cell-surface receptors is the nonspecific binding of cellular proteins to the streptavidin-coated magnetic particles, as seen with the identification of CXCR4 and LanCL1. Since the amount of target protein that was crosslinked to the R8 peptide was considerably smaller than the amount of nonspecifically bound proteins, these nonspecific proteins may mask target proteins, making their isolation difficult. We decided to address this issue by selectively separating the target proteins from the nonspecifically bound proteins on the affinity beads. In addition, fractionation of the membrane proteins from the cytoplasmic proteins may further improve our results. Therefore, a photoaffinity probe of the R8 peptide with a diazobenzene moiety, in addition to trifluoromethyl benzoyl diazirine and biotin, was designed (Kawaguchi et al. 2016) (Fig. 11.2). A diazobenzene moiety, which is cleavable with the addition of reducing agents such as $Na_2S_2O_4$, was introduced into the linker for selective detachment of the crosslinked target proteins from the affinity beads (Verhelst et al. 2007; Yang et al. 2010).

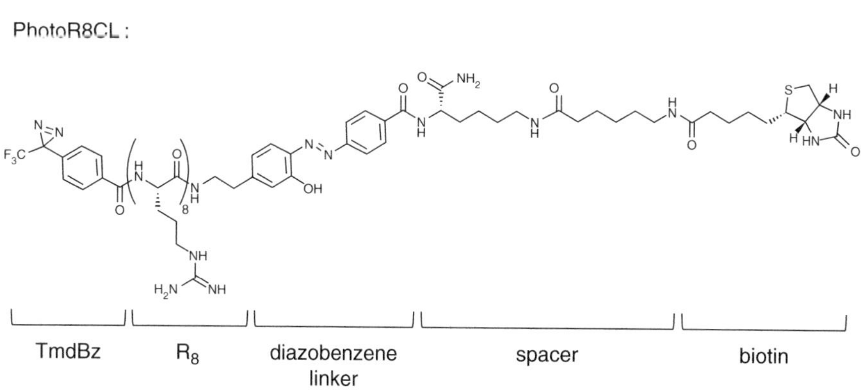

Fig. 11.2 Structure of PhotoR8CL

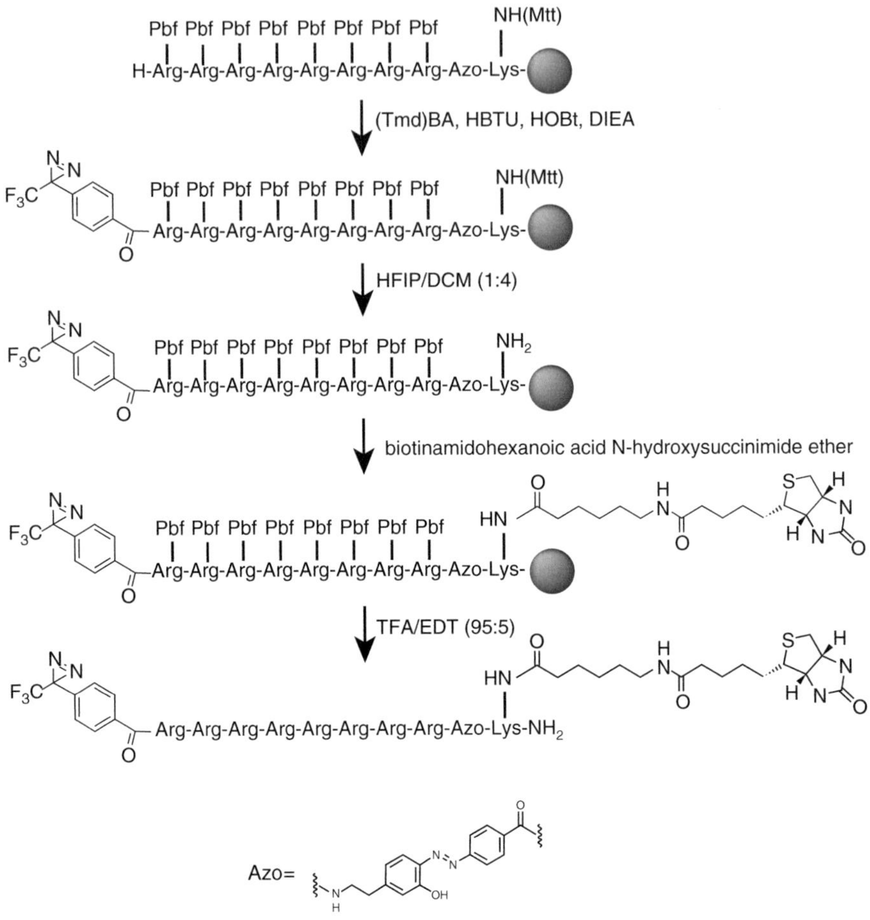

Fig. 11.3 Scheme of PhotoR8CL synthesis. [Pbf = N-ω-(2,2,4,6,7-pentamethyldihydrobenzofuran-5-sulfonyl)]

The photoaffinity probe (PhotoR8CL) was constructed by Fmoc solid-phase peptide synthesis (Fig. 11.3). To introduce the diazobenzene derivative into the peptide chain, 4-[5-(*N*-fluorenylmethyloxycarbonyl-2-aminoethyl)-2-hydroxyphenylazo]-benzoic acid (Fmoc-Azo-OH) was employed (Verhelst et al. 2007). A diazirine derivative 4-[3-(trifluoromethyl)-3H-diazirin-3-yl]benzoic acid (TmdBz) (Nassal 1983; Hatanaka et al. 1993) was also incorporated at the N-terminus of the peptide chain. After construction of the protected peptide segment, the methyltrityl (Mtt) side-chain protecting group of the C-terminal lysine was selectively removed by treatment with 1,1,1,3,3-hexafluoroisopropanol (HFIP)/dichloromethane (DCM) (1:4) (Azuma et al. 2014). A biotin tag was then added to the resin by treatment with biotinamidohexanoic acid N-hydroxysuccinimide ester. Cleavage of the peptide from the resin and deprotection of the side-chain protecting groups were conducted by treatment

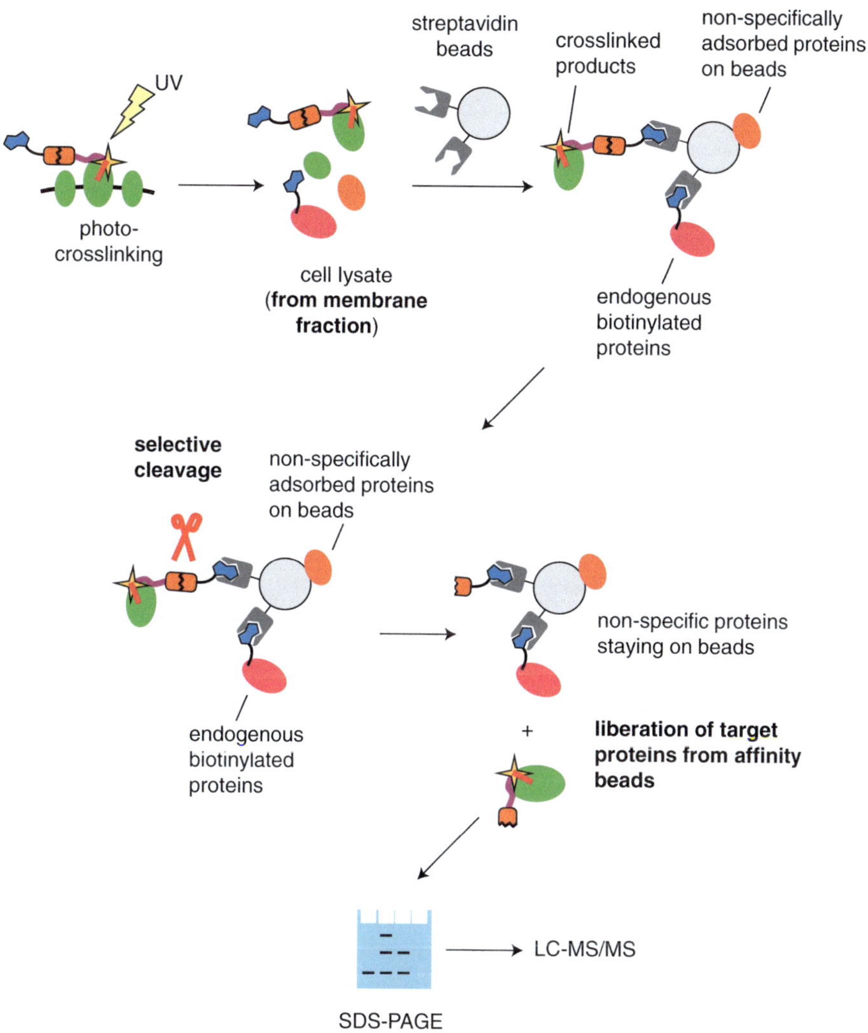

Fig. 11.4 PhotoR8CL facilitates identification of photocrosslinked products

with trifluoroacetic acid (TFA)/ethanedithiol (EDT) (95:5). HPLC purification yielded PhotoR8CL. Model experiments showed that the diazobenzene linker was cleavable even following UV irradiation, indicating the orthogonality of the photo-crosslinking and cleavage reactions.

The internalization of arginine-rich CPPs is highly dependent on the concentration at which they are administered. Endocytic pathways play a crucial role in the cellular uptake of arginine-rich CPPs at a relatively low concentration (1 μM or less) (Kosuge et al. 2008). To identify endocytic receptors responsible for the cellular uptake of R8, crosslinking experiments were conducted using a relatively low con-

Table 11.1 Candidate proteins for cellular uptake receptors of R8

Band No.	Protein names	Description
1	Glypican-1 (GPC1)	Proteoglycan
	Syndecan-2 (SDC2)	Proteoglycan
	Syndecan-4 (SDC4)	Proteoglycan
	Chondroitin sulfate proteoglycan 4 (CSPG4)	Proteoglycan
	Transforming growth factor beta receptor type 3 (TGFR3)	Proteoglycan
	Collagen alpha-1(V) chain (COL5A1)	Collagen
2	Glypican-5 (GPC5)	Proteoglycan
	Myoferlin (MYOF)	Ferlin
	Laminin subunit gamma-1 (LAMC1)	Laminin
3	Zinc transporter ZIP10 (ZIP10)	Transporter
4	Transferrin receptor protein 1 (CD71)	Receptor
	Semaphorin-3C (SEMA3C)	Semaphorin
5	4F2 cell-surface antigen heavy chain (CD98)	Solute carrier
6	Choline transporter-like protein 2 (CTL2)	Transporter
7	Interferon-induced transmembrane protein 1 (IFITM1)	Antiviral protein
	Interferon-induced transmembrane protein 3 (IFITM3)	Antiviral protein
	Vesicle-associated membrane protein 1 (VAMP1)	v-SNARE

centration of PhotoR8CL (Fig. 11.4). HeLa cells were treated with 1 μM PhotoR8CL for 5 min at 4 °C using potassium-rich buffer and the crosslinking was performed by UV irradiation at 365 nm for 10 min at 4 °C. Following cell lysis, the membrane fraction was collected by ultracentrifugation in order to separate the membrane proteins from cytoplasmic and nuclear proteins. The crosslinked products were then collected from the membrane fractions using streptavidin beads and eluted following $Na_2S_2O_4$-dependent cleavage of the diazobenzene linker.

The eluted proteins were separated by SDS-PAGE, which showed seven bands specific to the PhotoR8CL-treated cells compared with the untreated control cells. Therefore, we achieved efficient purification of photocrosslinked proteins by fractionation of the membrane proteins with $Na_2S_2O_4$-mediated cleavage of the diazobenzene linker, and this was accompanied by a significant reduction in nonspecifically bound proteins. Consistent with this, large amounts of nonspecific proteins were observed when the $Na_2S_2O_4$-treated affinity beads were boiled and subjected to SDS-PAGE analysis, regardless of PhotoR8CL treatment. Liquid chromatography-tandem mass spectrometry (LC-MS/MS) analysis suggested 17 distinct membrane proteins that interact with the R8 peptide, including proteoglycans, transporters, and membrane-traffic-related proteins (Table 11.1).

siRNA knockdown experiments were conducted to reduce the expression of the above proteins. A 30% decrease in uptake of the R8 peptide was observed when syndecan-4 was knocked down. When fluorescently labeled R8 was endocytosed by the cells and analyzed by confocal microscopy, the endosome-trapped peptide yielded punctate signals in the cytoplasm. Significant colocalization of syndecan-4 and R8 in the endosomes was also observed. In syndecan-4-knockdown cells, there was a significant decrease in punctate cytoplasmic R8 signals. Overexpression of syndecan-4 also resulted in a significant increase in uptake of the R8 peptide, suggesting that syndecan-4 is involved in the endocytic uptake of R8. Importantly, the role of syndecan-4 in R8 uptake is a function of its extracellular concentration. Knockdown of syndecan-4 had no apparent effect on the uptake of R8 when HeLa cells were treated with 10 μM R8, suggesting that syndecan-4 plays a significant role in R8 uptake only when the concentration of the R8 peptide is relatively low.

Studies using pharmacological inhibitors have suggested that syndecan-4 is primarily involved in R8 internalization via clathrin-mediated endocytosis, although it has been reported that clathrin/caveola-mediated endocytosis and macropinocytosis also contribute to the cellular uptake of arginine-rich CPPs under physiological conditions. The role of syndecan-4 in the internalization of bioactive proteins fused with R8 has also been confirmed with a Cre-loxP recombination assay (Ghosh and Van Duyne 2002). Treatment of cells with Cre fused to R8 (CreR8) yielded a significant increase in the number of cells that underwent gene recombination. Moreover, there was a considerable decrease in the number of cells that underwent gene recombination when syndecan-4 was knocked down. The effects of pharmacological inhibitors on the endocytic uptake of CreR8 were also analyzed. We conclude that syndecan-4 is involved not only in cellular uptake of the R8 peptide, but also in R8-mediated intracellular delivery of Cre recombinase, via clathrin-mediated endocytosis, since CreR8 must move from the endosome into the cytoplasm in order to exert its function.

11.4 Summary and Perspectives

The photocrosslinking studies for receptor identification involved (a) formation of the initial covalent bond with the target proteins, (b) isolation of the photocrosslinked products by affinity purification, and (c) protein identification by mass spectrometry. In our initial studies using the trifluoromethyl diazirine derivative to identify CXCR4 and LanCL1, large amounts of nonspecific proteins that bound to the affinity beads hampered the identification of target proteins. Considering the need for the efficient identification of crosslinked proteins, we developed an approach using a photoaffinity probe equipped with a diazobenzene moiety. Diazobenzene has been used as a cleavable linker unit for the purification of proteins using affinity beads functionalized with small molecule ligands, allowing effective isolation via selective cleavage of the target proteins from the affinity beads while nonspecific proteins remain bound. We have demonstrated that introduction of this cleavable moiety in the photoaffinity probe significantly facilitated the isolation of photocrosslinked products via chemoselective cleavage from the affinity beads. We have therefore demonstrated the potential usefulness of these cleavable linkers in the design of photocrosslinking probes.

Acknowledgments This work was supported in part by a Grant-in-Aid of The FUGAKU TRUST FOR MEDICINAL RESEARCH to S.F. This work was also supported by the Collaborative Research Program of Institute for Chemical Research, Kyoto University. Y.K. is grateful for Japan Society for the Promotion of Science (JSPS) Fellowship for Young Scientists. ITbM is supported by the World Premier International Research Center Initiative (WPI), Japan.

References

Arii J, Goto H, Suenaga T et al (2010) Non-muscle myosin IIA is a functional entry receptor for herpes simplex virus-1. Nature 467:859–862

Azuma Y, Kükenshöner T, Ma G et al (2014) Controlling leucine-zipper partner recognition in cells through modification of a-g interactions. Chem Commun 50:6364–6367

Bechara C, Sagan S (2013) Cell-penetrating peptides: 20 years later, where do we stand? FEBS Lett 587:1693–1702

Boisguérin P, Deshayes S, Gait MJ et al (2015) Delivery of therapeutic oligonucleotides with cell penetrating peptides. Adv Drug Deliv Rev 87:52–67

Brock R (2014) The uptake of arginine-rich cell-penetrating peptides: putting the puzzle together. Bioconjug Chem 25:863–868

Brunner J, Senn H, Richards FM (1980) 3-Trifluoromethyl-3-phenyldiazirine. A new carbene generating group for photolabeling reagents. J Biol Chem 255:3313–3318

Cleal K, He L, Watson PD, Jones AT (2013) Endocytosis, intracellular traffic and fate of cell penetrating peptide based conjugates and nanoparticles. Curr Pharm Des 19:2878–2894

Falcone S, Cocucci E, Podini P et al (2006) Macropinocytosis: regulated coordination of endocytic and exocytic membrane traffic events. J Cell Sci 119:4758–4769

Fretz MM, Penning NA, Al-Taei S et al (2007) Temperature-, concentration- and cholesterol-dependent translocation of L- and D-octa-arginine across the plasma and nuclear membrane of $CD34^+$ leukaemia cells. Biochem J 403:335–342

Fujii N, Oishi S, Hiramatsu K et al (2003) Molecular-size reduction of a potent CXCR4-chemokine antagonist using orthogonal combination of conformation- and sequence-based libraries. Angew Chem Int Ed Eng 42:3251–3253

Futaki S (2006) Oligoarginine vectors for intracellular delivery: design and cellular-uptake mechanisms. Biopolymers 84:241–249

Futaki S, Suzuki T, Ohashi W et al (2001) Arginine-rich peptides. An abundant source of membrane-permeable peptides having potential as carriers for intracellular protein delivery. J Biol Chem 276:5836–5840

Futaki S, Hirose H, Nakase I (2013) Arginine-rich peptides: methods of translocation through biological membranes. Curr Pharm Des 19:2863–2868

Ghosh K, Van Duyne GD (2002) Cre-loxP biochemistry. Methods 28:374–383

Hatanaka Y (2015) Development and leading-edge application of innovative photoaffinity labeling. Chem Pharm Bull 63:1–12

Hatanaka Y, Nakayama H, Kanaoka Y (1993) An improved synthesis of 4-[3-(Trifluoromethyl)-3H-diazirin-3-yl] benzoic acid for photoaffinity-labeling. Heterocycles 35:997–1004

Hirose H, Takeuchi T, Osakada H et al (2012) Transient focal membrane deformation induced by arginine-rich peptides leads to their direct penetration into cells. Mol Ther 20:984–993

Hofmann K, Kiso Y (1976) An approach to the targeted attachment of peptides and proteins to solid supports. Proc Natl Acad Sci U S A 73:3516–3518

Jones AT (2007) Macropinocytosis: searching for an endocytic identity and role in the uptake of cell penetrating peptides. J Cell Mol Med 11:670–684

Kaplan IM, Wadia JS, Dowdy SF (2005) Cationic TAT peptide transduction domain enters cells by macropinocytosis. J Control Release 102:247–253

Kawaguchi Y, Tanaka G, Nakase I et al (2013) Identification of cellular proteins interacting with octaarginine (R8) cell-penetrating peptide by photo-crosslinking. Bioorg Med Chem Lett 23:3738–3740

Kawaguchi Y, Takeuchi T, Kuwata K et al (2016) Syndecan-4 is a receptor for clathrin-mediated endocytosis of arginine-rich cell-penetrating peptides. Bioconjug Chem 27:1119–1130

Kosuge M, Takeuchi T, Nakase I et al (2008) Cellular internalization and distribution of arginine-rich peptides as a function of extracellular peptide concentration, serum, and plasma membrane associated proteoglycans. Bioconjug Chem 19:656–664

Kurrikoff K, Gestin M, Langel Ü (2016) Recent in vivo advances in cell-penetrating peptide-assisted drug delivery. Expert Opin Drug Deliv 13:373–387

Marechal V, Prevost M-C, Petit C et al (2001) Human immunodeficiency virus type 1 entry into macrophages mediated by macropinocytosis. J Virol 75:11166–11177

Mayer H, Bauer H, Breuss J et al (2001) Characterization of rat LANCL1, a novel member of the lanthionine synthetase C-like protein family, highly expressed in testis and brain. Gene 269:73–80

Meier O, Boucke K, Hammer SV et al (2002) Adenovirus triggers macropinocytosis and endosomal leakage together with its clathrin-mediated uptake. J Cell Biol 158:1119–1131

Mercer J, Helenius A (2009) Virus entry by macropinocytosis. Nat Cell Biol 11:510–520

Mitola S, Sozzani S, Luini W et al (1997) Tat-human immunodeficiency virus-1 induces human monocyte chemotaxis by activation of vascular endothelial growth factor receptor-1. Blood 90:1365–1372

Möhle R, Bautz F, Denzlinger C et al (2001) Transendothelial migration of hematopoietic progenitor cells. Role of chemotactic factors. Ann N Y Acad Sci 938:26–34

Nakase I, Niwa M, Takeuchi T et al (2004) Cellular uptake of arginine-rich peptides: roles for macropinocytosis and actin rearrangement. Mol Ther 10:1011–1022

Nakase I, Tadokoro A, Kawabata N et al (2007) Interaction of arginine-rich peptides with membrane-associated proteoglycans is crucial for induction of actin organization and macropinocytosis. Biochemistry 46:492–501

Nakase I, Takeuchi T, Tanaka G et al (2008) Methodological and cellular aspects that govern the internalization mechanisms of arginine-rich cell-penetrating peptides. Adv Drug Deliv Rev 60:598–607

Nakase I, Akita H, Kogure K et al (2012) Efficient intracellular delivery of nucleic acid pharmaceuticals using cell-penetrating peptides. Acc Chem Res 45:1132–1139

Nakase I, Osaki K, Tanaka G et al (2014) Molecular interplays involved in the cellular uptake of octaarginine on cell surfaces and the importance of syndecan-4 cytoplasmic V domain for the activation of protein kinase Cα. Biochem Biophys Res Commun 446:857–862

Nakashima H, Hashimoto M, Sadakane Y et al (2006) Simple and versatile method for tagging phenyldiazirine photophores. J Am Chem Soc 128:15092–15093

Nassal M (1983) 4-(1-Azi-2,2,2-trifluoroethyl)benzoic acid, a highly photolabile carbene generating label readily fixable to biochemical agents. Liebigs Ann Chem:1510–1523

Perret F, Nishihara M, Takeuchi T et al (2005) Anionic fullerenes, calixarenes, coronenes, and pyrenes as activators of oligo/polyarginines in model membranes and live cells. J Am Chem Soc 127:1114–1115

Phillips R, Ager A (2002) Activation of pertussis toxin-sensitive CXCL12 (SDF-1) receptors mediates transendothelial migration of T lymphocytes across lymph node high endothelial cells. Eur J Immunol 32:837–847

Rey M, Valenzuela-Fernández A, Urzainqui A et al (2007) Myosin IIA is involved in the endocytosis of CXCR4 induced by SDF-1α. J Cell Sci 120:1126–1133

Rothbard JB, Garlington S, Lin Q et al (2000) Conjugation of arginine oligomers to cyclosporin a facilitates topical delivery and inhibition of inflammation. Nat Med 6:1253–1257

Rothbard JB, Jessop TC, Lewis RS et al (2004) Role of membrane potential and hydrogen bonding in the mechanism of translocation of guanidinium-rich peptides into cells. J Am Chem Soc 126:9506–9507

Schramm B, Penn ML, Speck RF et al (2000) Viral entry through CXCR4 is a pathogenic factor and therapeutic target in human immunodeficiency virus type 1 disease. J Virol 74:184–192

Stanzl EG, Trantow BM, Vargas JR et al (2013) Fifteen years of cell-penetrating, guanidinium-rich molecular transporters: basic science, research tools, and clinical applications. Acc Chem Res 46:2944–2954

Takeuchi T, Kosuge M, Tadokoro A et al (2006) Direct and rapid cytosolic delivery using cell-penetrating peptides mediated by pyrenebutyrate. ACS Chem Biol 1:299–303

Tanaka G, Nakase I, Fukuda Y et al (2012) CXCR4 stimulates macropinocytosis: implications for cellular uptake of arginine-rich cell-penetrating peptides and HIV. Chem Biol 19:1437–1446

Tkachenko E, Lutgens E, Stan R-V et al (2004) Fibroblast growth factor 2 endocytosis in endothelial cells proceed via syndecan-4-dependent activation of Rac1 and a Cdc42-dependent macropinocytic pathway. J Cell Sci 117:3189–3199

Tomohiro T, Hashimoto M, Hatanaka Y (2005) Cross-linking chemistry and biology: development of multifunctional photoaffinity probes. Chem Rec 5:385–395

Tünnemann G, Martin RM, Haupt S, Patsch C, Edenhofer F, Cardoso MC (2006) Cargo-dependent mode of uptake and bioavailability of TAT-containing proteins and peptides in living cells. FASEB J 20:1775–1784

Verhelst SH, Fonović M, Bogyo M (2007) A mild chemically cleavable linker system for functional proteomic applications. Angew Chem Int E 46:1284–1286

Vivès E, Brodin P, Lebleu B (1997) A truncated HIV-1 tat protein basic domain rapidly translocates through the plasma membrane and accumulates in the cell nucleus. J Biol Chem 272:16010–16017

Wadia JS, Stan RV, Dowdy SF (2004) Transducible TAT-HA fusogenic peptide enhances escape of TAT-fusion proteins after lipid raft macropinocytosis. Nat Med 10:310–315

Wender PA, Galliher WC, Goun EA et al (2008) The design of guanidinium-rich transporters and their internalization mechanisms. Adv Drug Deliv Rev 60:452–472

Xiao H, Neuveut C, Tiffany HL et al (2000) Selective CXCR4 antagonism by Tat: implications for in vivo expansion of coreceptor use by HIV-1. Proc Natl Acad Sci U S A 97:11466–11471

Yang YY, Grammel M, Raghavan AS et al (2010) Comparative analysis of cleavable azobenzene-based affinity tags for bioorthogonal chemical proteomics. Chem Biol 17:1212–1222

Zhang W, Wang L, Liu Y et al (2009) Structure of human lanthionine synthetase C-like protein 1 and its interaction with Eps8 and glutathione. Genes Dev 23:1387–1392

Zhong WX, Wang YB, Peng L et al (2012) Lanthionine synthetase C-like protein 1 interacts with and inhibits cystathionine β-synthase: a target for neuronal antioxidant defense. J Biol Chem 287:34189–34201

Chapter 12
Photoaffinity Labeling in Drug Discovery Research

Hideyuki Muranaka, Takaki Momose, Chiaki Handa, and Tomonaga Ozawa

Abstract Photoaffinity labeling (PAL) is a method that forms an irreversible covalent bond between photoreactive ligands and neighboring amino acids under the irradiation of light. PAL is utilized in the pharmacological and biochemical identification of ligand target molecules and ligand binding sites. Recent technological advances in mass spectrometry have enabled measurement of the mass of intact proteins and peptides with extremely high accuracy. Mass spectrometry has also been adopted in PAL to analyze labeled proteins and identify crosslink amino acid, although this application has been mostly for soluble proteins and reports on the successful identification of crosslink amino acids in GPCRs by mass spectrometry are scarce. In this chapter, we describe in detail our PAL technique that determines crosslink amino acid using the human adenosine A_{2A} receptor as a representative class A GPCR.

Keyword Photoaffinity labeling • Diazirine • Crosslink • Mass spectrometry • GPCR • A_{2A} adenosine receptor

12.1 Adenosine Receptors as Drug Discovery Targets

The G protein-coupled receptor (GPCR) superfamily comprises one of the largest groups of eukaryotic cell-surface membrane receptors. It has been estimated that GPCRs make up more than 30% of marketed drug targets (Rask-Andersen et al. 2011). GPCRs share a common seven-transmembrane (TM) topology and are involved in cellular responses to a variety of extracellular signals, ranging from light, ions, odor, taste, neurotransmitters, lipids, peptides, and hormones

H. Muranaka (✉) • T. Momose • C. Handa • T. Ozawa
Central Research Laboratories, Kissei Pharmaceutical Co., Ltd.,
4365-1 Kashiwabara, Hotaka, Azumino, Nagano 399-8304, Japan
e-mail: hideyuki_muranaka@pharm.kissei.co.jp

Y. Hatanaka, M. Hashimoto (eds.), *Photoaffinity Labeling for Structural Probing Within Protein*, DOI 10.1007/978-4-431-56569-7_12

(Gether 2000). The total number of functional GPCRs in the human genome including those for taste and olfaction is estimated at more than 800. GPCRs are classified into five main classes based on amino acid sequence: rhodopsin (class A), secretin (class B), glutamate (class C), adhesion, and frizzled receptors (Fredriksson et al. 2003). Class A contains the largest number of receptors (approximately 200 non-sensory receptors with known ligands) and is subdivided into four main groups: α, β, γ, and δ. Adenosine receptors (ARs) belong to the α-group of class A and consist of four subtypes, A_1, A_{2A}, A_{2B}, and A_3, that mediate signaling of the natural ligand adenosine (**1**) via different G proteins (A_1 and A_3 through G_i and A_2 through G_s) (Olah and Stiles 2000). ARs are widely expressed in humans and have been implicated in numerous physiological and pathological effects. Accordingly, a number of AR agonists and antagonists have been or are being evaluated in clinical studies (Chen et al. 2013). Representative examples of such compounds that are approved or in clinical development are depicted in Fig. 12.1. Trabodenoson (**2**), a selective A_1 agonist, is currently undergoing clinical trials for primary open-angle glaucoma (Myers et al. 2016). Regadenoson (**3**), a selective A_{2A}AR agonist, was approved by the Food and Drug Administration for vasodilator stress myocardial perfusion imaging (Al Jaroudi and Iskandrian 2009). Piclidenoson (**4**) and namodenoson (**5**), which are selective A_3 agonists, are now being evaluated for the treatment of psoriasis and rheumatoid arthritis (Fishman and Cohen 2016) or hepatocellular carcinoma (Stemmer et al. 2013), respectively. Among ARs, the A_{2A}AR represents an attractive therapeutic target for the treatment of neurodegenerative and psychiatric disorders, such as Parkinson's disease (PD) and Alzheimer's disease (AD). A_{2A}ARs are highly expressed in the striatum of the brain, immune cells of the spleen, thymus, leukocytes, and blood platelets, and to a lesser extent the heart, lung, and blood vessels (Fredholm et al. 2001, 2011). A_{2A}AR antagonists have emerged as the leading non-dopaminergic drugs for the treatment of PD. Notably, istradefylline (**6**) and preladenant (**7**) are extensively studied selective A_{2A}AR antagonists in the clinical field; while istradefylline was approved in Japan as the first A_{2A}AR antagonist in 2013 for adjunctive treatment of PD, the development of preladenant (**7**) was discontinued due to a lack of efficacy in three phase III clinical trials despite excellent in vivo and positive phase II results. Other clinical trials of A_{2A}AR antagonists, including tozadenant (**8**) and V-81444 (structure not disclosed), are in progress (Pinna 2014). Caffeine (**9**) is a naturally occurring AR antagonist, particularly of A_{2A} receptors, and a link between caffeine consumption and a reduced risk of PD and AD has been reported (Chen et al. 2013). ARs are indeed promising targets in drug discovery.

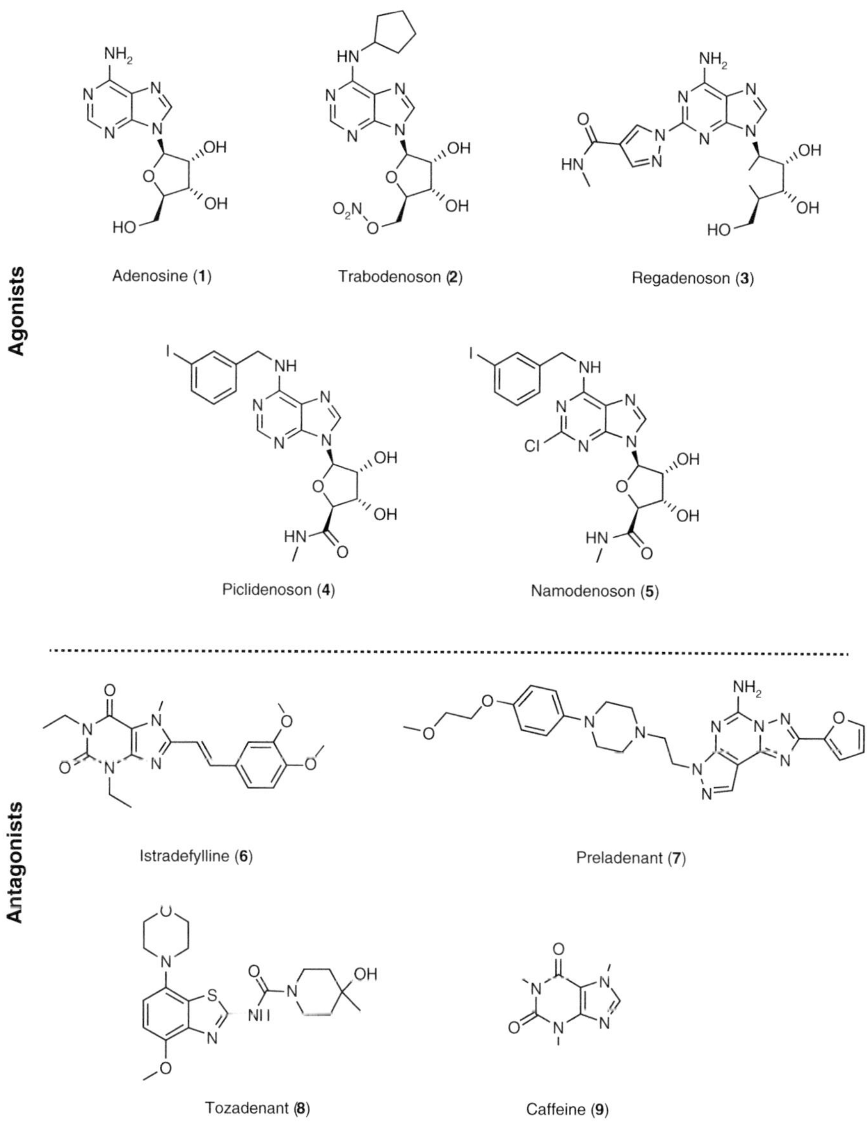

Fig. 12.1 Representative AR agonists and antagonists that are approved or in clinical development

12.1.1 Crystallographic Structures of A_{2A}ARs

The first X-ray structure of an A_{2A}AR obtained with the high-affinity antagonist ZM241385 (**10**) was published by Jaakola et al. in 2008 to reveal the inactive states of the receptor (Jaakola et al. 2008). Since then, over ten crystal structures of A_{2A}ARs derived not only with antagonists such as XAC (**11**) and caffeine (**9**), but also with agonists including adenosine (**1**) and NECA (**12**) have been published by different groups (Fig. 12.2) (Yuan et al. 2015).

Structure-based docking using these structures has been surprisingly successful (Shoichet and Kobilka 2012); the results of two docking screens showed high hit rates (35% and 41%, respectively, of the molecules tested), and affinities were in the sub-micro molar range or higher (32 nM and 200 nM, respectively). Similar results were obtained in docking studies of β_2 adrenergic, D3 dopamine, and H1 histamine receptors. In this decade, nearly 30 distinct GPCR structures have been characterized among the major class A, B, C, and frizzled receptors, which have significantly contributed to the understanding of drug action and novel compound design in drug discovery (Shonberg et al. 2015; Lee et al. 2015; Jazayeri et al. 2015). Although the

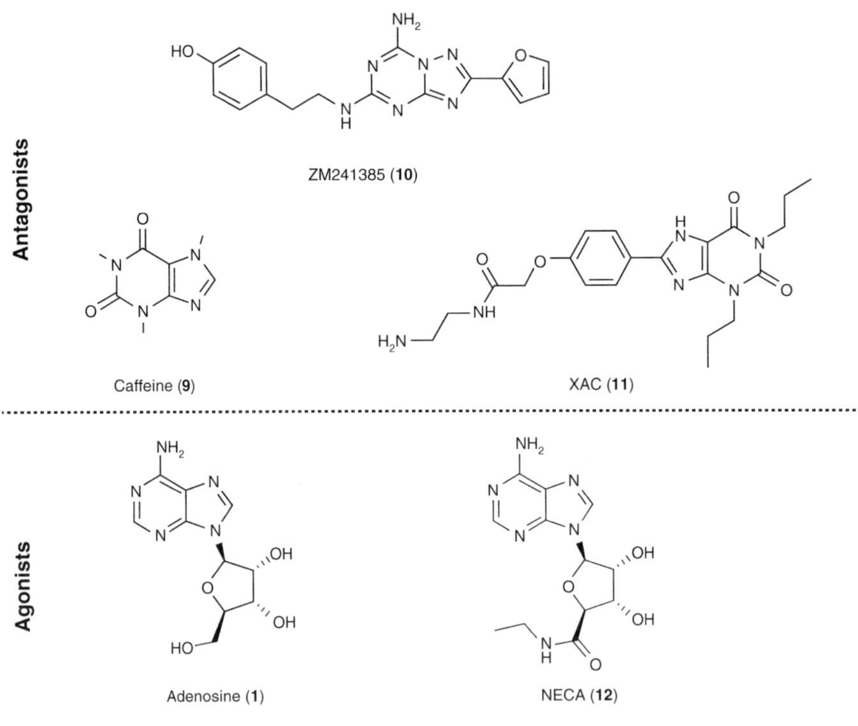

Fig. 12.2 X-ray crystal structures of representative A_{2A}AR antagonists and agonists

number of identified GPCR structures has been growing rapidly, they represent a small proportion of the pharmacologically relevant GPCRs (approximately 300 non-olfactory GPCRs in the human genome) (Fredriksson et al. 2003). In the drug discovery process, detailed analysis of receptor-ligand interactions is crucial prior to medicinal chemistry efforts. Especially when novel chemotypes are obtained by high throughput screening (HTS) and/or fragment-based drug discovery (FBDD), experimental evidence of these interactions becomes paramount. For the structural study, a high level of recombinant GPCR production was achieved in a range of expression systems that included bacteria, yeast, insect, and mammalian cells (Chiu et al. 2008). However, few laboratories can successfully determine their high resolution crystal structures. An alternative approach to structural characterization is photoaffinity labeling (PAL). PAL provides an experimental evidence of receptor–ligand interactions. Moreover, the only equipment required for this technique is a UV-lamp and conventional high-resolution mass spectrometer. If PAL can be developed to elucidate of GPCR-ligand interactions at the amino acid resolution, it will become possible to quickly "screen" receptor–ligand interactions at a pace needed for highly competitive drug discovery efforts.

12.2 Photoaffinity Labeling

PAL was first reported by Westheimer et al. (Singh et al. 1962) over 50 years ago, after which three major photophores, arylazide (**13**), diazirine (**14**), and benzophenone (**15**) have emerged for PAL assays (Fig. 12.3). The advantages and disadvantages of each photophore are described in detail in earlier chapters of this book and in some excellent reviews (Hatanaka 2015; Hatanaka and Sadakane 2002; Hashimoto and Hatanaka 2008). PAL has proven to be a powerful tool for the identification of ligand target proteins as well as for the investigation of protein-protein interactions and ligand binding sites. It has also been widely employed for the characterization of GPCRs for over 40 years; numerous studies have successfully utilized PAL to confirm the localization of GPCRs (Grunbeck and Sakmar 2013).

Arylazide (**13**) Diazirine (**14**) Benzophenone (**15**)

Fig. 12.3 Three major photophors

12.2.1 Synthetic Photoaffinity Probes for GPCR Labeling

As a general PAL strategy for GPCRs, photoaffinity probes with radioactive isotopes are used for biochemical characterization by autoradiography, and then Edman degradation is employed to identify crosslink positions in some case (Fig. 12.4). As described earlier in this chapter, class A GPCRs is the most populated subgroup, with endogenous ligands that include a wide variety of neurotransmitters, peptides, hormones, and other small molecules. In addition, many synthetic photoaffinity probes have been prepared and extensively studied for biochemical and pharmacological characterization. In this section, we describe several PAL applications on class A GPCRs.

Adrenergic receptors have been one of the most extensively studied class A GPCR by PAL. For example, [^{125}I] aryl azidoprazosin (**16**), an α_1 adrenergic receptor antagonist, selectively labeled a rat heart α_1 receptor (Terman and Insel 1986). [^{125}I]Iodoaminoflisopolol, [^{125}I]IAmF (**17**), a β_2 adrenergic receptor antagonist having a unique fluorenone scaffold that also acts as a photoreactive center, labeled the adrenergic β_2 receptor expressed in a baculovirus/Sf9 system. The labeling position of [^{125}I]IAmF (**17**) was successfully determined by radiochemical release sequencing (Wu and Ruoho 2000; Wu et al. 2001). Salometerol is a long-acting β_2 adrenergic receptor agonist used in the treatment of asthma. A salmeterol-based photoaffinity probe, [^{125}I]IAS (**18**), labeled a specific region (TM6 and TM7) of the β_2 adrenergic receptor (Rong et al. 1999). A variety of photoaffinity ligands, including agonists and antagonists, have been reported for ARs as well. Patel et al. revealed that ^{125}I-azido- BW-A844U (**19**), a selective adenosine A_1 receptor antagonist, specifically photoincorporated into the 34 kDa polypeptide of the adenosine A_1 receptor in bovine brain membranes (Patel et al. 1988), and the photoreactive adenosine A_1 receptor agonist [^{125}I]AHPIA (**20**) successfully labeled the A_1 receptor in both rat and bovine brain-coated vesicles (González-Calero et al. 1992). Piersen et al. performed PAL of canine A_{2A}AR overexpressed in COS M6 cells using ^{125}I-azido-PAPA-APEC (**21**) and tracked the crosslinked TM5 region. The authors suggested that further sequencing analysis and mass spectroscopic studies would be necessary

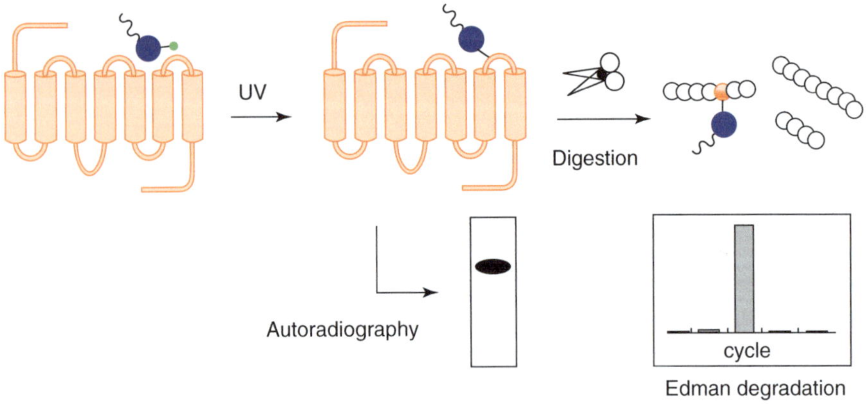

Fig. 12.4 Schematic representation of GPCR characterization by PAL

to determine the individual amino acid(s) where ^{125}I-azido-PAPA-APEC (**21**) incorporated (Piersen et al. 1994). With the above representing only a fraction of the studies targeting class A GPCRs, it can be said that PAL is an important tool for GPCR characterization (Fig. 12.5).

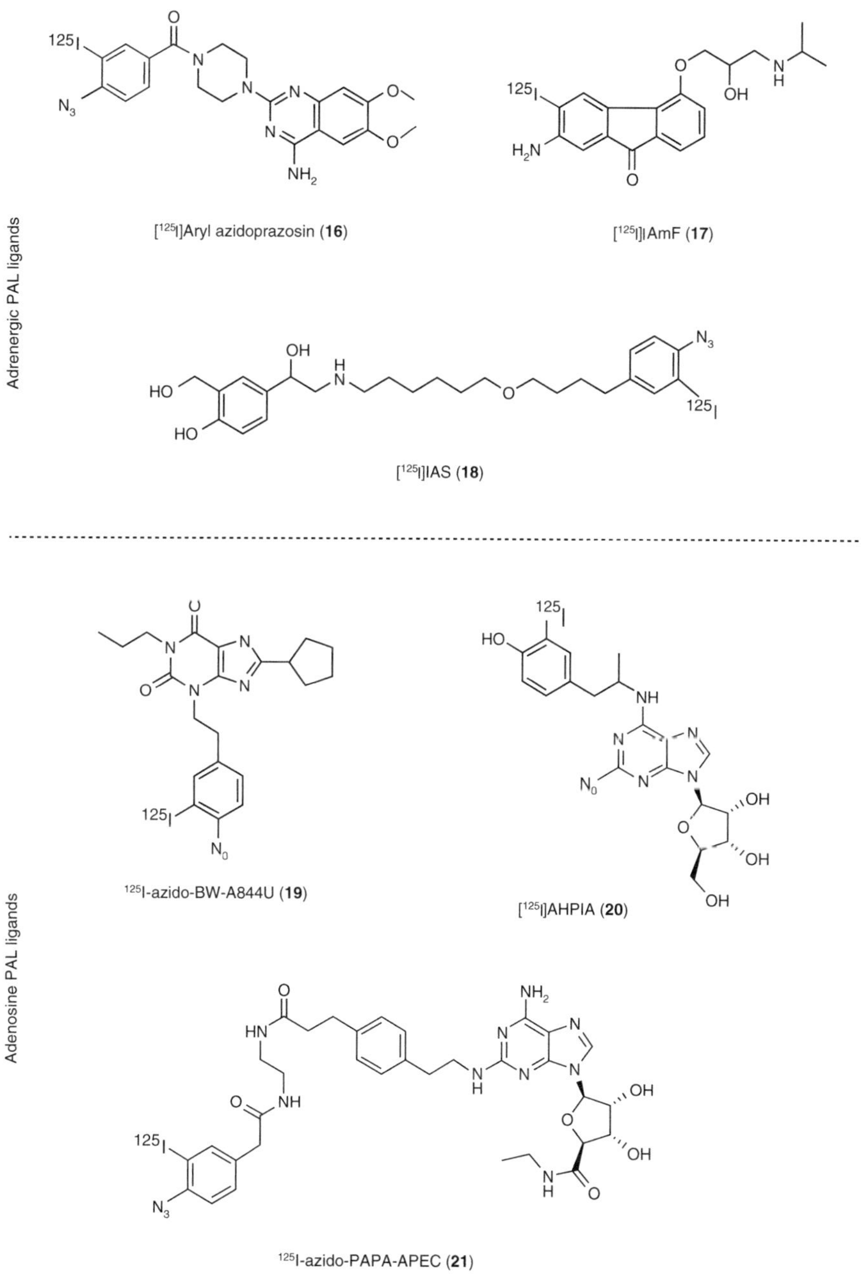

Fig. 12.5 Examples of photoaffinity probes for class A GPCR characterization

12.2.2 Analysis of Crosslink Positions by Mass Spectrometry

Recently, the combined approach of PAL and mass spectrometry (MS) is increasingly becoming popular for analyzing photolabeled proteins and identification of crosslink amino acids (Robinette et al. 2006). Mass spectrometry has enabled measurement of the mass of intact proteins and peptides with extremely high accuracy. Mass spectrometry has also been applied in PAL studies to analyze labeled proteins—mostly soluble proteins—and identify crosslink amino acids by MS/MS sequencing (Fig. 12.6).

Here we present several examples of small molecular photoaffinity ligands with which crosslink amino acids are determined by mass spectrometry (Fig. 12.7). Tomizawa et al. introduced an azide group into nicotinic acetylcholine receptor agonists to synthesize AzEPI (**22**) and AzTHO (**23**), which labeled mollusk Ach-binding protein (AChBP). Crosslink of the ligands was confirmed by fluorography of [^{3}H]-AzEPI and LC-MS analysis of intact AChBP after UV irradiation. The authors also performed LC-MS/MS analysis to reveal two labeled amino acids: Tyr195 and Met116 (Tomizawa et al. 2007, 2009). Trnka et al. performed PAL of bacterial isoform P450$_{Cam}$ with HBP (**24**). The [^{3}H]HBP labeled peptides were tracked by fractionation and counting for radioactivity of digested peptides. MS/MS sequencing of the peptides revealed two methionine residues: Met103 on the B′/C-helix and Met121 on the C-helix (Trnka et al. 2006). Hindi et al. witnessed selective photolabeling of LCK kinase with compound **25**. The *S*-isomer **25** alone labeled LCK and showed dramatic selectivity for five other kinases having seemingly identical binding pockets. MS/MS analysis revealed compound **25** covalently attach to either Gly383 or Leu384 (Hindi et al. 2006). Protein tyrosine phosphatase 1B (PTP1B) is a potential target for treatment of Type II diabetes (LeRiche et al. 2004). LeRiche et al. described the photoaffinity of PTP1B with 3*H*-(diazirin-3-yl) phosphonic acid (**26**). Following PAL, mass spectrum analysis of intact PTP1B showed that approximately 80% of the total protein was modified at 1:1 stoichiometry. Moreover, MS/MS sequencing of the crosslinked PTP1B peptides identified

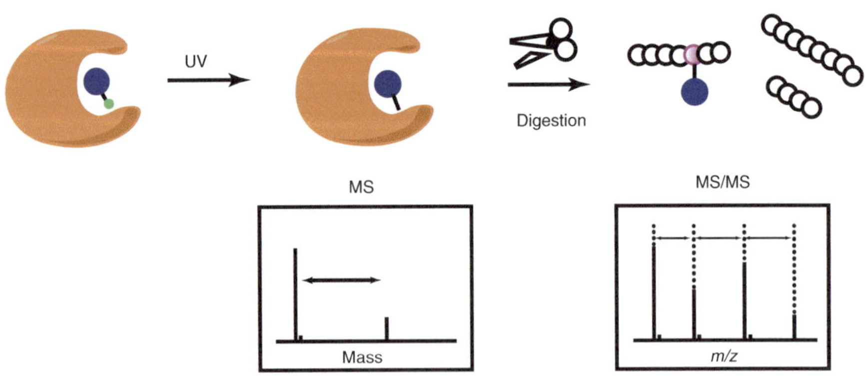

Fig. 12.6 Schematic representation of binding site identification by mass spectrometry

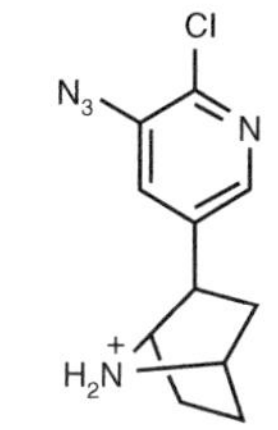

AzEPI (22)

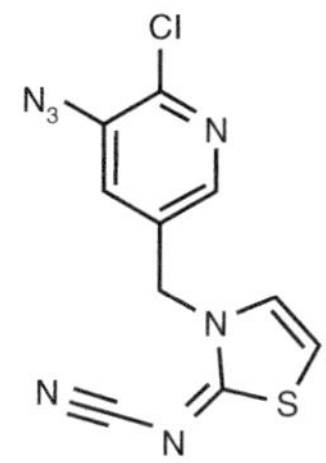

AzTHO (23)

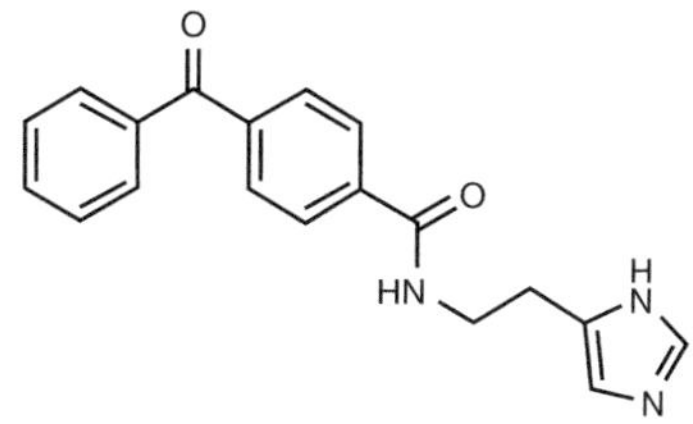

HBP (24)

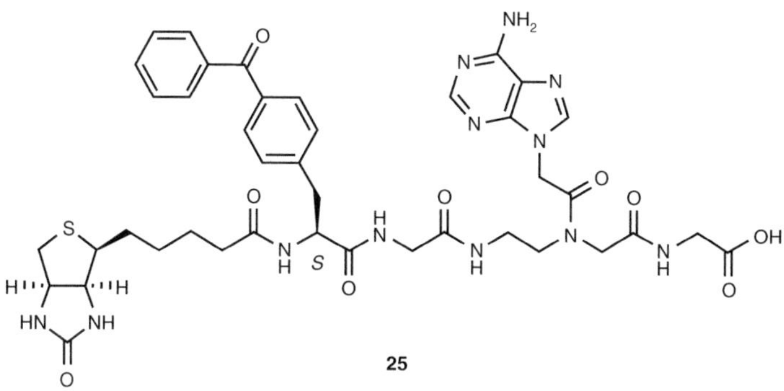

25

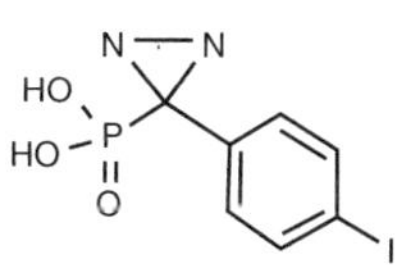

26

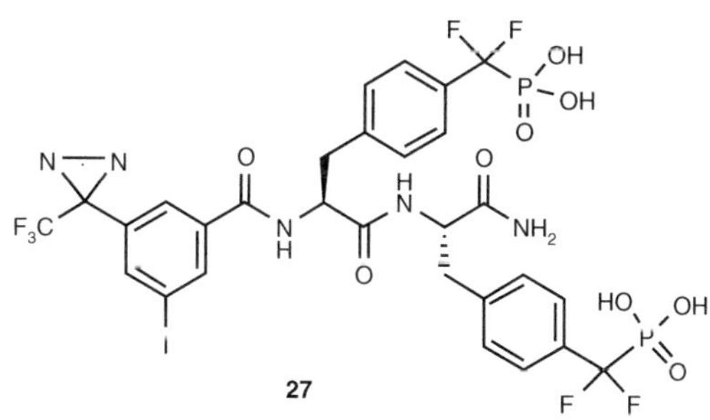

27

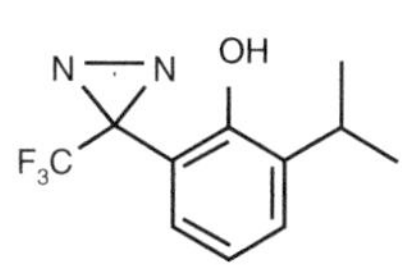

***ortho*-Propofol diazirine (28)**

Fig. 12.7 Representative photoaffinity probes utilized for the identification of crosslink amino acids by mass spectrometry

either Pro180 or Asp181 in the WPD loop to be involved in substrate binding. The authors concluded that Asp181 resided in close proximity to the probe's diazirine ring after determination of the X-ray crystal structure. The same group later reported that labeling of PTP1B with photoprobe **27** exhibited over 50% crosslink at 1:1 stoichiometry by LC-MS analysis based on multiple trials (Skorey et al. 2006). Propofol is the most widely used intravenous anesthetic. Yip et al. described a stable isotope encoding photolabeling approach for $GABA_A$ receptors with protiated and deuterated versions of the propofol analogue **28** having a trifluoromethyl diazirine moiety in the place of one of the two isopropyl side chains (Yip et al. 2013). Photolabeling of $GABA_A$ receptors in intact membranes and protein digest analysis exhibited near-complete sequence coverage, including the TM domain, by MS/MS. Further MS/MS spectra analysis of the protiated and deuterated crosslink peptides identified the site of incorporation as His267.

As exemplified above, mass spectrometry is now being routinely adopted for the identification of crosslink amino acids in PAL. To our knowledge, however, PAL studies determining ligand binding sites in GPCRs by mass spectrometry at the amino acid level are scarce. This may be due to the low expression level of receptors, the difficulty in devising active membrane protein preparations, and the requirement of further analysis steps, such as protease digestion, following amino acid sequence analysis of relatively hydrophobic peptides (Hamouda et al. 2014). Indeed, Rosa et al. reported in 2015 that only six GPCRs had their primary protein sequences completely determined (i.e., over 80% sequence coverage) using proteomic analyses (Rosa et al. 2015), which was less than that of the crystal structures of GPCRs. The precise crosslink site analysis of GPCRs by mass spectrometry remains a challenging task.

12.3 Photoaffinity Labeling of the Human A_{2A} Adenosine Receptor

Fraser described the expression and functional purification of a glycosylation-deficient version of $hA_{2A}AR$ in *Pichia pastoris* having native-like pharmacological properties and demonstrated that the receptor retained its function during the solubilization and purification process (Fraser 2006). Thus, provided that they maintain native-like characteristics, treatment of receptors in a solution state facilitates the monitoring of the photo-crosslink reaction as well as the identification of the cross-link sites by mass spectrometry (Fig. 12.8). More recently, Moss et al. reported the site-specific chemical modification of the human $A_{2A}AR$ with chemically reactive agonists to covalently modify the receptor, although they did not perform PAL (Moss et al. 2014). This has prompted us to conduct PAL using solubilized $hA_{2A}AR$ as a model GPCR for the identification of crosslink amino acids. If successful, this strategy may be applicable for elucidating the ligand–receptor interaction of other GPCRs even if they are allosteric. We will now present our PAL findings on the $hA_{2A}AR$ using a novel diazirine-based photoaffinity probe and cross-link position analysis by mass spectrometry (Muranaka et al. 2017).

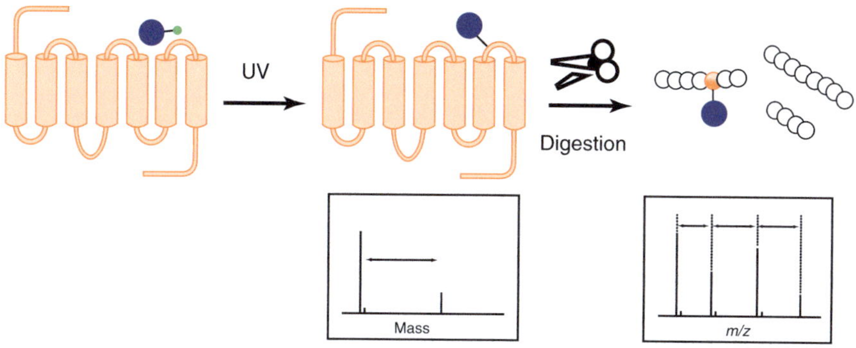

Fig. 12.8 Schematic representation of binding site identification of GPCR PAL by mass spectrometry

12.3.1 Photoaffinity Probe Design

As shown in Fig. 12.3, three major types of photo-reactive groups that show high reactivity under UV light irradiation have been reported (arylazides, benzophenones, and diazirines). Among them, diazirines are the smallest photophor that generate a reactive carbene upon irradiation with light (Brunner 1993). Diazirines are also the preferred phosphors for minimizing structural changes from parent molecules. In our initial $hA_{2A}AR$ labeling trials, we adopted the trifluoromethylphenyl diazirine (TPD) group as a photophor and the pyrazolo[4,3-*e*][1,2,4]triazolo[1,5-*c*] pyrimidine (PTP) nucleus as a template (Redenti et al. 2016). 7FB-PTP (**29**) and 8FB-PTP (**30**) are one of the first reported compounds having a PTP nucleus. 8FB-PTP (**30**) is known as a highly potent antagonist (K_i $rA_{2A}AR$ = 1.2 nM), whereas 7FB-PTP (**29**) showed a decreased affinity (K_i $rA_{2A}AR$ = 12 nM). We designed two novel photoaffinity probes **31** and **32** by substituting the fluorine atom on the phenyl ring at the para position of 7FB-PTP (**29**) and 8FB-PTP (**30**) for a TPD group (Fig. 12.9).

The synthetic route is depicted in Fig. 12.10. The reaction of commercially available benzyl bromide (**33**) with 5-amino-2-(2-furyl)pyrazolo[4,3-*e*][1,2,4] triazolo[1,5-*c*]-pyrimidine (**34**) (Baraldi et al. 1996) in the presence of anhydrous potassium carbonate produced a mixture of N^7 isomers (**31**) and N^8 isomers (**32**) that were separable by column chromatography.

12.3.2 Fluorescent Polarization Experiments

The $hA_{2A}AR$ was expressed in a *Pichia pastoris* system with a 10× His-tag at the C-terminal end. The sedimented membrane fractions were solubilized by overnight incubation at 4 °C in buffer: 16 mM Tris–HCl, pH 7.4, 120 mM NaCl, 2.0% *n*-dodecyl-β-D-maltoside (DDM), and 0.4% cholesteryl hemisuccinate (CHS). The

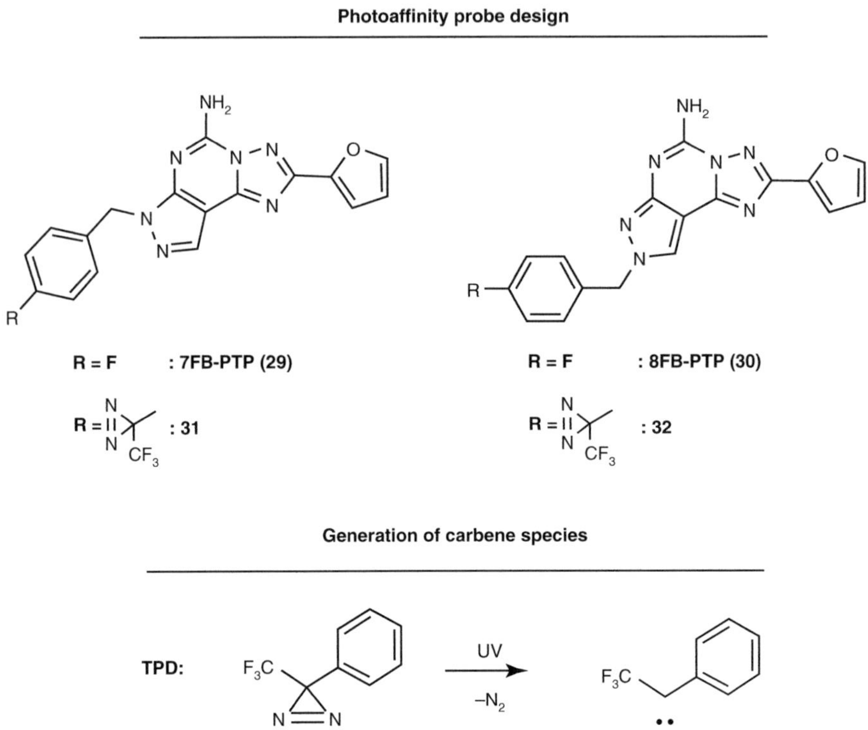

Fig. 12.9 Photoaffinity probe design

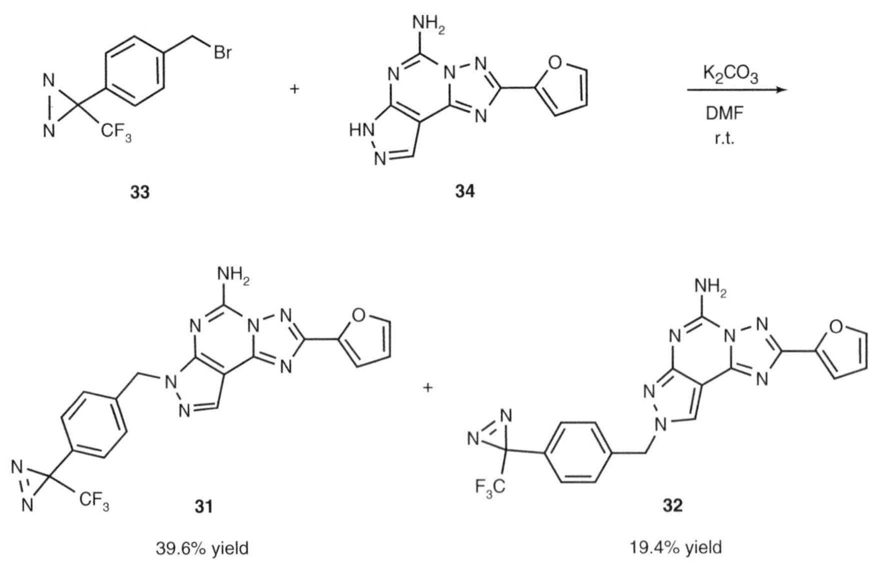

Fig. 12.10 Synthesis of photoaffinity probes

receptor was purified by a cobalt-based affinity resin (TALON resin, Clontech, Palo Alto, CA, USA) and subsequently adopted in LC–MS analyses and PAL studies. The affinity of compounds **31** and **32** for the solubilized $hA_{2A}AR$ was determined by a fluorescence polarization (FP). Saturation binding experiments were performed using the fluorescent tracer MRS5346 (**35**) reported by Kecskés et al. with some modification (Fig. 12.11) (Kecskés et al. 2010; Kumar et al. 2011).

FP signals were measured with an Infinite M1000 plate reader (TECAN, Morrisville, NC, USA) at an excitation wavelength of 470 nm and emission wavelength of 520 nm. Millipolarization (mP) values using four different tracer concentrations of tracer (8, 4, 2, and 1 nM) were recorded. Data analysis was performed to determine the receptor concentrations at which 50% of tracer **35** was bound (Fig. 12.12) (Nikolovska-Coleska et al. 2004). Binding specificity was confirmed in the presence of 10 μM of the antagonist SCH442416 (**36**) (Todde et al. 2000). Apparent K_d values were determined by nonlinear curve fitting using PRISM (version 4.03, GraphPad Software, La Jolla, CA, USA). Based on these, we estimated that less than 2 nM (K_d = 30.8 ± 0.5 nM) of the tracer **35** approached the true K_d value for the solubilized receptor, which was similar to findings reported using membrane preparations from HEK293 cells (K_d = 16.5 ± 4.7 nM) (Kecskés et al. 2010). Taking the S/B ratio into account, 2 nM of the tracer **35** and 25 nM of $hA_{2A}AR$ were used in ensuing competitive binding assays.

FP competitive binding experiments were performed under optimized conditions to determine the affinity of compounds **31** and **32** together with several adenosine ligands, including both antagonists **10**, **11**, and an agonist **12**. The concentration of compounds required to displace 50% of the tracer **35** was elucidated (IC_{50}). As illustrated in Fig. 12.13, the rank order of potency was: ZM241385 (**10**) > compound **32** > or = XAC (**11**) > NECA (**12**) > compound **31**. The K_i values were calculated using the Cheng-Prusoff equation and the known ligands showed similar K_i values to those of the binding experiments with [^{3}H]-ZM241385 using membrane preparations of $hA_{2A}AR$ expressed in *Pichia pastoris* reported by Fraser. The N^8 isomers **32** had a high affinity (K_i = 39.7 ± 0.03 nM) comparable to that of XAC (**11**) (K_i = 50.1 ± 0.03 nM). The N^7 isomers **31**, meanwhile, showed a more than 30-fold

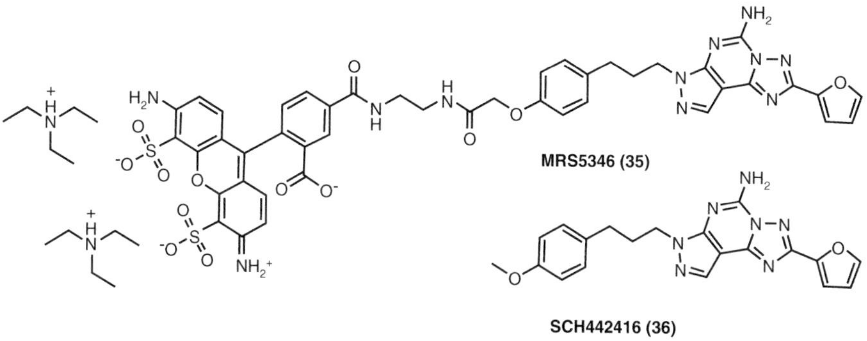

Fig. 12.11 Structures of the fluorescent tracer MRS5346 (**35**) and antagonist SCH442416 (**36**)

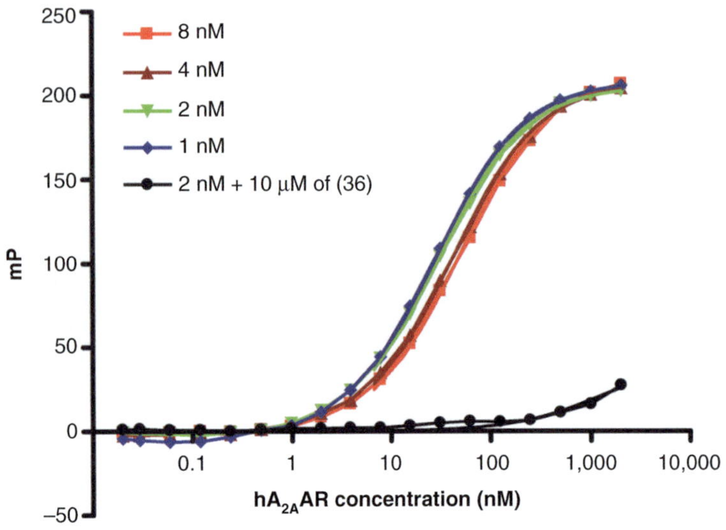

Fig. 12.12 FP saturation binding experiments on solubilized $hA_{2A}AR$ with the tracer **35** at different concentrations: ■, 8 nM; ▲, 4 nM; ▼, 2 nM; ◆, 1 nM; ●, 2 nM in the presence of 10 μM of SCH442416 (**36**) [Transferred from Muranaka H. et al., ACS Med. Chem. Lett., 8, 660–665 (2017).]

Fig. 12.13 Normalized FP competitive binding experiments with 2 nM of the tracer **35** and 25 nM of $hA_{2A}AR$: ■, ZM241385 (**10**); ◆, XAC (**11**); ●, NECA (**12**); ▲, compound **31**; ▼, compound **32**. [Transferred from Muranaka H. et al., ACS Med. Chem. Lett., 8, 660–665 (2017).]

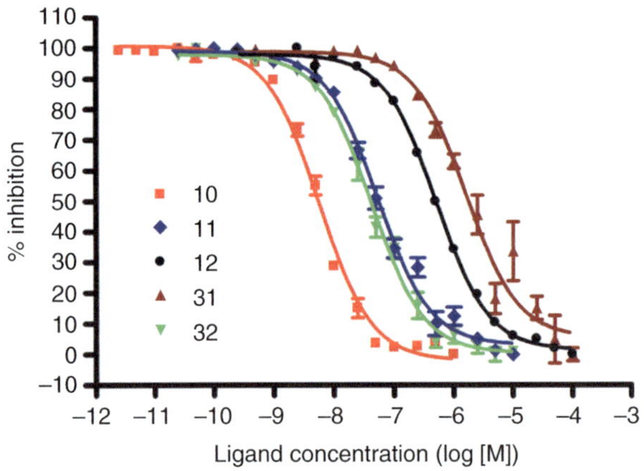

lower affinity (K_i = 1520 ± 0.06 nM) relative to N^8 isomers **32**. The observed structure activity relationships were the same as that of the parent molecules, that is, 8FB-PTP (**30**) was a more potent antagonist than 7FB-PTP (**29**). As a result of the FP competitive binding experiments, we selected compound **32** for further photolabeling experiments.

12.3.3 Photoaffinity Labeling and Mass Spectrometry Analysis

Prior to photolabeling experiments, three reaction mixtures were prepared at 10 μM $hA_{2A}AR$ in the same buffer as FP experiments: a DMSO control, a twofold molar excess of compound **32** relative to $hA_{2A}AR$, and the same mixture containing a fivefold molar excess of ZM241385 (**10**) relative to compound **32** as a competitor. After incubation at room temperature for 30 min, the mixtures were kept on ice and irradiated with a model B-100A UV light (UVP, Upland, CA, USA) for 30 min from a distance of 5 cm. Peak wavelength of the UV lamp was 365 nm and bandwidth was approximately 50 nm. UV light intensity (measured intensity: 5 mW/cm^2) was measured by a UV-37SD digital radiometer equipped with a UVA sensor (CUSTOM, Tokyo, Japan). The progress of the photo-crosslink reaction was monitored by mass spectrometry using a 6520 Accurate-Mass Q-TOF instrument with a 1200 series HPLC (Agilent Technologies, Santa Clara, CA, USA). Chromatographic separation was carried out in a polystyrene divinylbenzene column (PLRP/S, Agilent Technologies) (Whitelegge 2009) with a two-step short-gradient elution. The mobile phase in the first gradient step consisted of 0.1% formic acid water solution and 0.1% formic acid acetonitrile. In the second gradient step, the organic phase was changed to 0.1% formic acid isopropyl alcohol. Excess compound **32** and its phototransformation products together with the detergents and a small proportion of $hA_{2A}AR$ were washed out in the first step, while the relatively hydrophobic $hA_{2A}AR$ retained in the column was eluted in the second gradient step. A mass spectrum was extracted from the main peak of the total ion chromatogram (TIC). The multiply charged ions produced by electrospray ionization were deconvoluted to zero-charge molecular mass values using MassHunter Workstation Software Qualitative Analysis with BioConfirm Software (version B.06.00, Agilent Technologies) by means of the Maximum Entropy deconvolution algorithm. The step-by-step results of a typical LC-MS analysis are shown in Fig. 12.14.

The deconvoluted mass spectrum findings for each mixture are shown in Fig. 12.15. Before UV irradiation, the peak cluster was observed at 37,640 Da, which was in close agreement with the calculated mass of a 10× His-tagged $hA_{2A}AR$ containing four disulfide bonds (calcd for 37,639 Da), indicating there was no detectable post-modification during protein expression (Fig. 12.15a). In the DMSO control, no obvious peak shift was observed after UV irradiation, which suggested that receptor oxidation was almost completely suppressed during UV irradiation and highlighted one of the advantages of the diazirine photophore (Fig. 12.15b). In the presence of compound **32**, a new cluster of peaks was clearly observed at 38,051 Da (Fig. 12.15c). The mass difference of 411 Da between the two major peaks represented the crosslink of a single compound **32** molecule minus N_2 with a negligible difference (calcd for 412 Da). Moreover, the absence of double-labeling peak clusters indicated that specific binding of compound **32** had occurred. In the presence of a fivefold molar excess of ZM241385 (**10**), no peaks were detected around 38,051 Da, implying that compound **32** bound to the same site as did ZM241385 (**10**) (Fig. 12.15d).

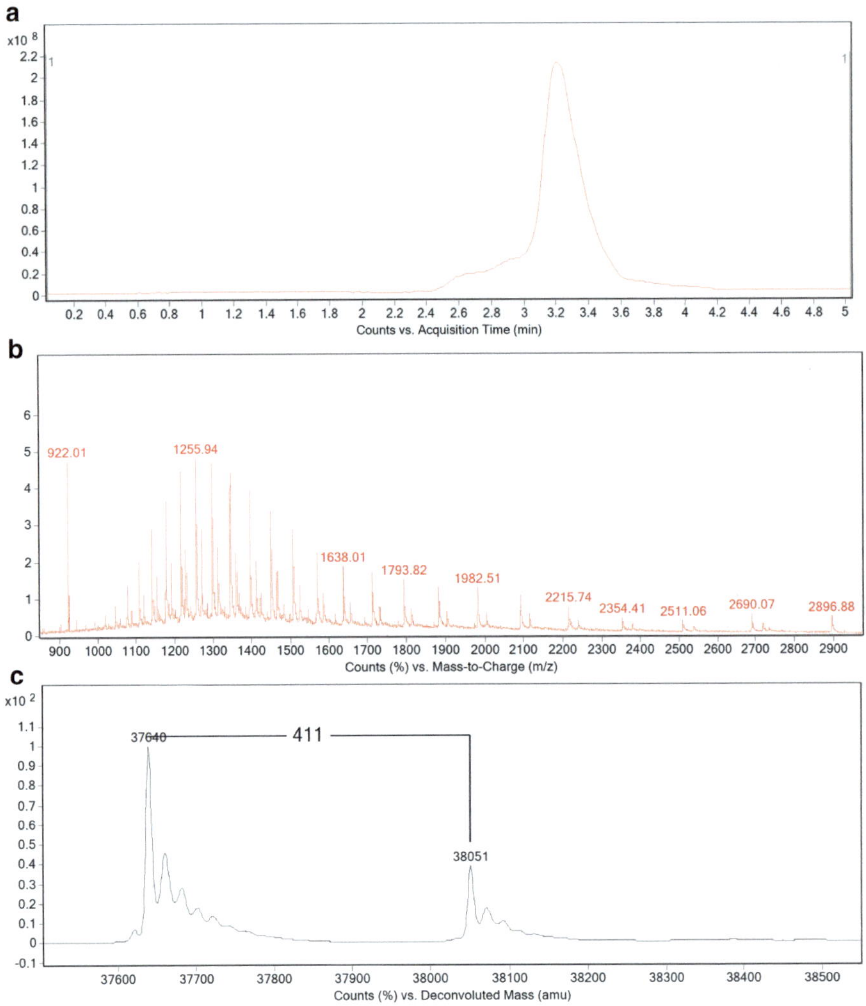

Fig. 12.14 LC–MS results of the $hA_{2A}AR$ after UV irradiation with compound **32**. (**a**) TIC of the second gradient. (**b**) An averaged mass spectrum (3.1–3.5 min). (**c**) A deconvoluted mass spectrum

12.3.4 Determination of Crosslink Amino Acids

Since the crosslink of compound **32** could be ascertained at the protein level, the mixture was ensuingly subjected to protease digestion for identification of precise crosslink positions. The photoreacted mixture was precipitated with trichloroacetic acid and the protein fraction was dissolved in digestion buffer containing a non-ionic surfactant, 0.05% 5-cyclohexyl-1-pentyl-*β*-D-maltoside (CYMAL-5, Anatrace, Maumee, OH, USA) (Zvonok et al. 2007), as well as an ionic surfactant, 0.1% RapiGestSF (Waters,

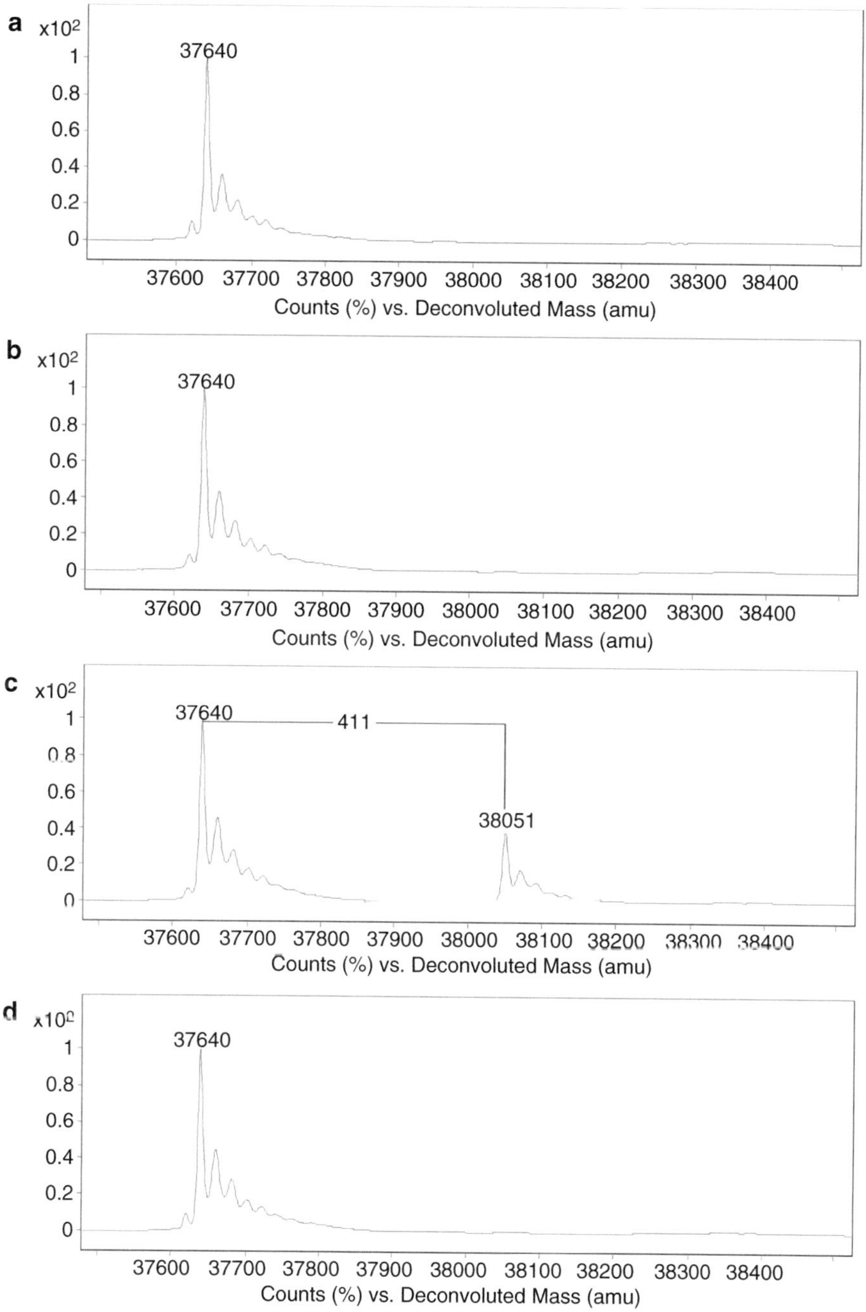

Fig. 12.15 Deconvoluted mass spectrum of the $hA_{2A}AR$. (**a**) Purified $hA_{2A}AR$. The 37,640 Da cluster represents the mass of the $hA_{2A}AR$. (**b**) DMSO control after UV irradiation. (**c**) The photo-labeled $hA_{2A}AR$ with compound **32**. The 38,051 Da cluster corresponds to compound **32** singly crosslinked to the $hA_{2A}AR$ with a loss of N_2. (**d**) Competition experiment in the presence of a fivefold molar excess of ZM241385 (**10**) relative to compound **32**. [Transferred from Muranaka H. et al., ACS Med. Chem. Lett., 8, 660–665 (2017).]

Milford, MA, USA). Disulfides were reduced by tris(2-carboxyethyl)phosphine and alkylated with iodoacetamide. Overnight proteolysis was performed with sequencing-grade chymotrypsin or trypsin (Promega, Madison, WI, USA). The reaction was terminated by the addition of trifluoroacetic acid (final pH < 2) followed by incubation to decompose the ionic surfactant. After centrifugation, the supernatant was subjected to LC-MS/MS analysis, wherein multiply charged ions were automatically selected for further MS/MS sequencing analysis. Several additional ions were manually selected for the second analysis to increase peptide sequence coverage. As a result of peptide mapping by PEAKS Studio (version 7.5, Bioinformatics Solutions Inc., Waterloo, Canada), 89% coverage was obtained to include most of the residues in the extracellular and intracellular loops as well as six of seven TM domains (Fig. 12.16).

To search for crosslink peptides, we compiled a list of all theoretical digested peptide formulae, with modifications for cysteine carbamidomethylation (C_2H_3NO) and one molecule of compound **32**-carbene addition ($C_{19}H_{12}F_3N_7O$), using in-house software. The Agilent MassHunter's Find-by-Formula (FBF) algorithm was employed using the list to search for all possible combinations of the peptide formulae and multi-charged ions at a 10 ppm mass tolerance. Isotopic distribution pattern and the presence of other multi-charged ions for the FBF hit peptides were visually inspected. As a result of the search, quadruply charged peptide ions with m/z = 913.9158 (calcd for m/z = 913.9226, z = 4) together with triply charged

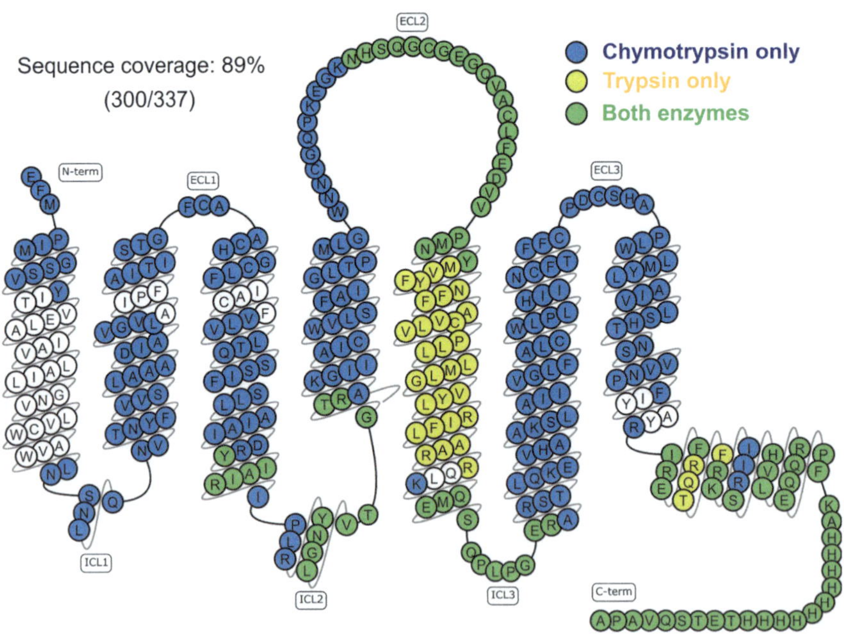

Fig. 12.16 Combined sequence coverage map shown on a snake diagram of the $hA_{2A}AR$ taken from GCPRdb (Isberg et al. 2016). Two extra amino acids (Glu, Phe) at the N-terminus and a 10× His-tag followed by a 1D4 tag at the C-terminal are contained in the sequence. Amino acids are colored to indicate the enzyme used in their identification: *blue*, chymotrypsin; *yellow*, trypsin; *green*, both enzymes. The 37 amino acids in white were not identified. Chymotryptic sequence coverage was 79% (266/337) and tryptic sequence coverage was 35% (119/337) of the entire sequence, respectively. The combined sequence coverage was as high as 89% (300/337)

peptide ions with m/z = 1218.2253 (calcd for m/z = 1218.2278, z = 3) that corresponded to a single peptide (Cys259$^{6.61}$–Phe286$^{7.51}$) were detected (Fig. 12.17).

An additional LC–MS/MS analysis targeted to the quadruply charged crosslink peptide ion (m/z = 913.9226, z = 4) was carried out to determine the crosslink position. Several MS/MS product ion spectra at different collision energies (31, 34, and 37 V) were acquired and the merged fragment ion spectra were analyzed by PEAKS Studio, which assigned sufficient b-series ions to enable sequence identification. The coexistence of b11 (without compound **32** modification) and b13 (with compound **32** modification) ions indicated the crosslinking amino acid as either Met270$^{7.35}$ or Tyr271$^{7.36}$ (Fig. 12.18a). Further to the discovery of a doubly charged ion corresponding to the b12 (with compound **32** modification) ion, we concluded that the most plausible crosslink site was Tyr271$^{7.36}$ in TM7 (Fig. 12.18b).

MS/MS spectra for the unlabeled Cys259$^{6.61}$–Phe286$^{7.51}$ chymotryptic peptide were also acquired. Spectrum comparisons are shown in Fig. 12.19. The same fragment ions (b11 and some smaller b-series ions) were observed in both spectra. On the other hand, similar patterns of fragment ions with a mass shift of compound **32** with a loss of N_2 (b13 and some larger b-series ions) were observed in the labeled peptides. These additional experimental results supported the amino acid residue crosslinked.

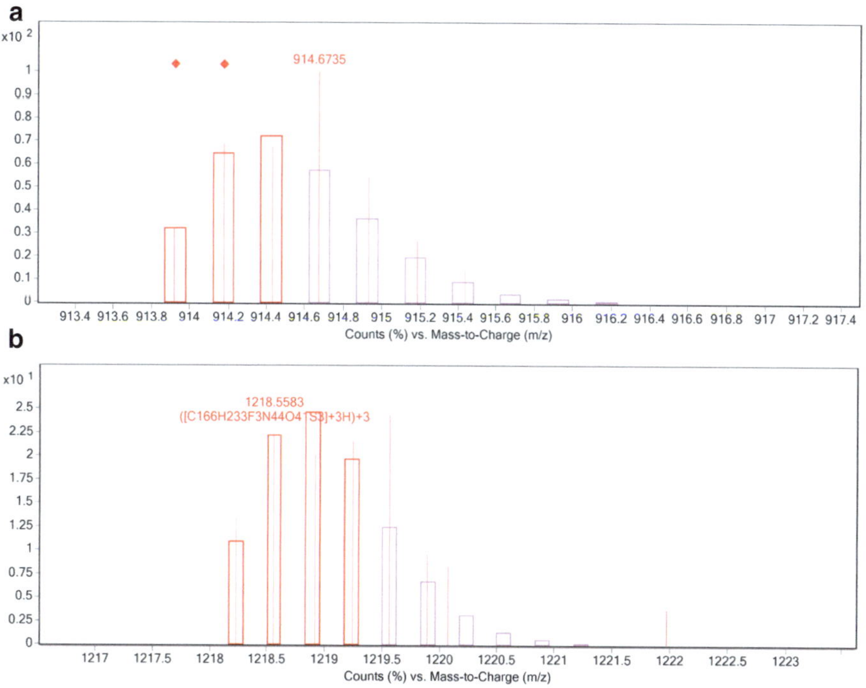

Fig. 12.17 Results of Agilent FBF search. Both ions correspond to the same Cys259$^{6.61}$–Phe286$^{7.51}$ peptide labeled with one molecule of compound **32** with a loss of N_2 ($C_{166}H_{233}F_3N_{44}O_{41}S_3$). (**a**) Quadruply charged peptide ions with m/z = 913.9158 (calcd for m/z = 913.9226, z = 4). (**b**) Triply charged peptide ions with m/z = 1218.2253 (calcd for m/z = 1218.2278, z = 3). The numbering of amino acids is based on the $hA_{2A}AR$

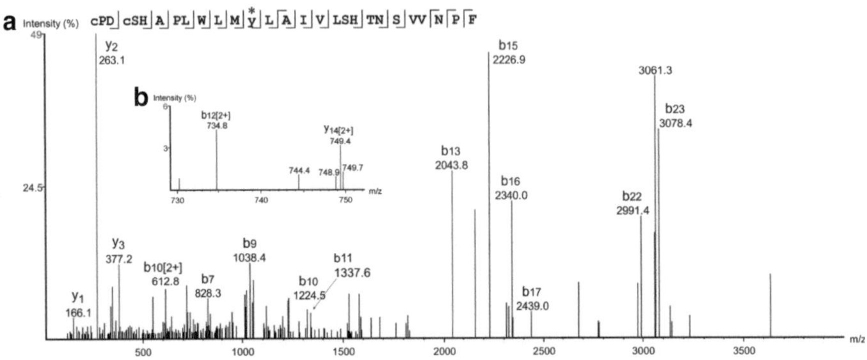

Fig. 12.18 (**a**) MS/MS spectrum for the Cys259$^{6.61}$–Phe286$^{7.51}$ peptide labeled with compound **32** analyzed by PEAKS software. Carbamidomethylcysteines are designated in lower case. The asterisk indicates the crosslink position. The mass difference between C-terminal fragment ions b11 (without compound **32**) and b13 (with compound **32**) identified Met270$^{7.35}$ or Tyr271$^{7.36}$ of the $hA_{2A}AR$ as crosslinked amino acids. (**b**) An enlarged MS/MS spectrum. The weak doubly charged b12 ion indicated the most plausible crosslink site to be Tyr271$^{7.36}$. The numbering of amino acids is based on the $hA_{2A}AR$. [Transferred from Muranaka H. et al., ACS Med. Chem. Lett., 8, 660–665 (2017).]

12.3.5 Docking Study

We lastly investigated the precise location of Tyr271$^{7.36}$ in the crystal structure of the $hA_{2A}AR$. Compound **32** was docked into the ligand binding site of the $hA_{2A}AR$ (PDB code: 3EML) by FRED (version 3.0.1, OpenEye Scientific Software, Santa Fe, NM, USA). One of the docking models showed the TPD moiety of compound **32** to be in close proximity to the Tyr271$^{7.36}$ in TM7 (Fig. 12.20). The tricyclic PTP part of compound **32**, including the 2-furyl ring, overlapped closely with the bicyclic triazolo-triazine component of ZM241385 (**10**). Several mutations of Tyr271(7.36) in $hA_{2A}AR$ have been reported. While one Tyr271(7.36)Ala mutant lost its binding capabilities (Guo et al. 2016; Kim et al. 1995), mutations to other aromatic residues (Phe and His) or Arg did not affect agonists and antagonists binding according to unshown data (Kim et al. 1995). A recent crystal structure of A1AR with a covalently bound xanthine-based antagonist showed that this residue was the precise point for covalent attachment of the antagonist (Glukhova et al. 2017). Overall, Tyr271$^{7.36}$ in TM7 appeared to be the most appropriate crosslink position.

Thus, we successfully conducted PAL of detergent-solubilized $hA_{2A}AR$ using the novel photoaffinity ligand compound **32** having a TPD group and identified the crosslinked amino acid Tyr271$^{7.36}$ in TM7. To our knowledge, this is the first application of PAL on the $hA_{2A}AR$ to elucidate the crosslink position at an amino acid resolution by mass spectrometry.

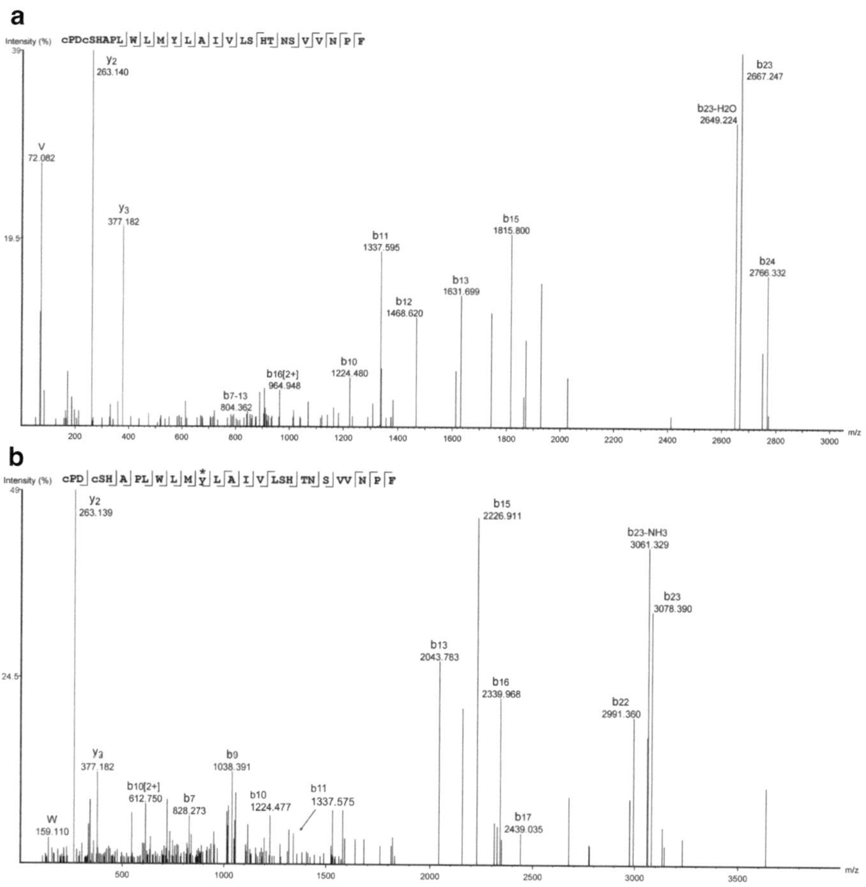

Fig. 12.19 Comparison of MS/MS spectra of the Cys259$^{6.61}$–Phe286$^{7.51}$ chymotryptic peptide. (**a**) Unlabeled and (**b**) labeled with compound **32** with a loss of N_2. Carbamidomethylcysteines are designated in lower case. The asterisk indicates the crosslink position. The numbering of amino acids is based on the $hA_{2A}AR$

12.4 Concluding Remarks

In the drug discovery process, a variety of hit compounds are identified by HTS and/or FBDD. When totally different chemotypes of known ligands are obtained, it is a challenging task to predict the binding site by molecular docking only. Detailed interaction analysis between ligands and receptors is crucial prior to additional medicinal chemistry efforts. Despite the number of GPCR structures growing quite rapidly, few laboratories can successfully generate high resolution crystal structures. PAL is a simple and attractive approach to structural characterization that provides direct biochemical evidence of receptor–ligand interactions.

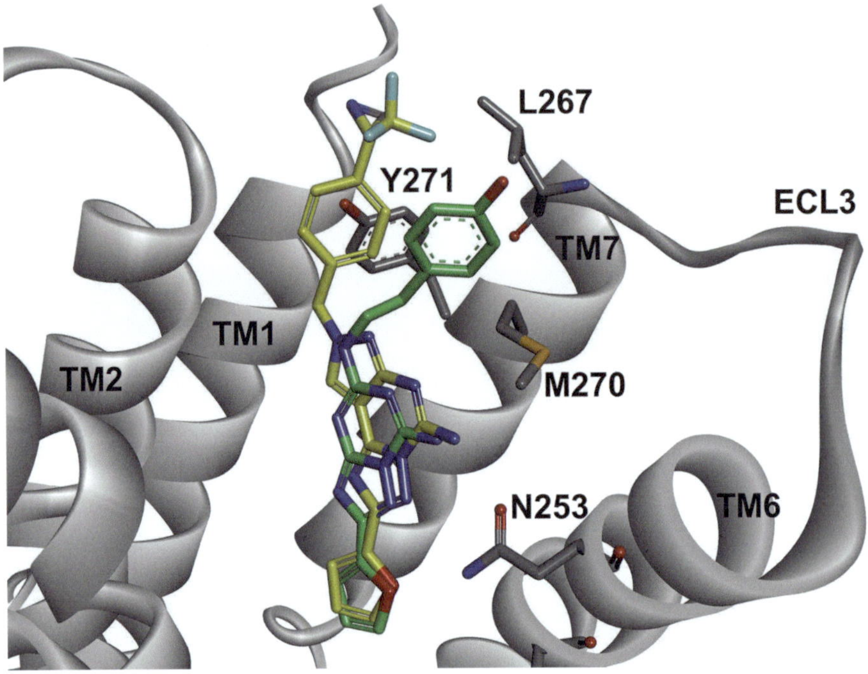

Fig. 12.20 Docking model of compound **32** (*yellow sticks*) binding to the $hA_{2A}AR$ crystal structure (3EML) superimposed onto ZM241385 (**10**, *green sticks*). Key residues are shown as sticks. $Tyr271^{7.36}$ in TM7 was a photolabeled amino acid. [Transferred from Muranaka H. et al., ACS Med. Chem. Lett., 8, 660–665 (2017).]

The identification of photolabeled amino acids for soluble proteins is now routinely available, although the practical analysis conditions for hydrophobic peptides, especially in the TM domain of membrane proteins, still have room for improvement. In this chapter, we introduced PAL of the human adenosine A_{2A} receptor with a novel diazirine-based probe and the successful identification of the most likely crosslink amino acid. Our ligand resided in close proximity to the known orthosteric site; however, PAL may also be applicable for the elucidation of allosteric or hybrid orthosteric/allosteric (bitopic) molecules. In concert with structural information, PAL continues to be an important tool for drug discovery.

Acknowledgments The authors are grateful to Ichio Shimada (Tokyo University, Japan) for providing the cDNA of the $hA_{2A}AR$ and his technical advice on protein production. We thank Yasumaru Hatanaka (Toyama University, Japan) for his advice on PAL experiments. We warmly thank Toshiki Honma for FP data analysis, Shigeru Yonekubo for HPLC analyses, and Yoshinori Nonaka for NMR measurements. We also thank Kosuke Okazaki (Japan Agency for Medical Research and Development, AMED, Japan) for his insights.

References

Al Jaroudi W, Iskandrian AE (2009) Regadenoson: a new myocardial stress agent. J Am Coll Cardiol 54:1123–1130. doi:10.1016/j.jacc.2009.04.089

Baraldi PG, Cacciari B, Spalluto G et al (1996) Pyrazolo[4,3-*e*]-1,2,4-triazolo[1,5-*c*]pyrimidine derivatives: potent and selective A_{2A} adenosine antagonists. J Med Chem 39:1164–1171. doi:10.1021/jm950746l

Brunner J (1993) New photolabeling and crosslinking methods. Annu Rev Biochem 62:483–514. doi:10.1146/annurev.bi.62.070193.002411

Chen J-F, Eltzschig HK, Fredholm BB (2013) Adenosine receptors as drug targets—what are the challenges? Nat Rev Drug Discov 12:265–286. doi:10.1038/nrd3955

Chiu ML, Tsang C, Grihalde N, MacWilliams MP (2008) Over-expression, solubilization, and purification of G protein-coupled receptors for structural biology. Comb Chem High Throughput Screen 11:439–462. doi:10.2174/138620708784911456

Fishman P, Cohen S (2016) The A_3 adenosine receptor (A_3AR): therapeutic target and predictive biological marker in rheumatoid arthritis. Clin Rheumatol 35:2359–2362. doi:10.1007/s10067-016-3202-4

Fraser NJ (2006) Expression and functional purification of a glycosylation deficient version of the human adenosine 2a receptor for structural studies. Protein Expr Purif 49:129–137. doi:10.1016/j.pep.2006.03.006

Fredholm BB, Ijzerman AP, Jacobson KA et al (2001) International Union of Pharmacology. XXV. Nomenclature and classification of adenosine receptors. Pharmacol Rev 53:527

Fredholm BB, Ijzerman AP, Jacobson KA et al (2011) International union of basic and clinical pharmacology. LXXXI. Nomenclature and classification of adenosine receptors-an update. Pharmacol Rev 63:1

Fredriksson R, Lagerström MC, Lundin L-G, Schiöth HB (2003) The G-protein-coupled receptors in the human genome form five main families. Phylogenetic analysis, Paralogon Groups, and fingerprints. Mol Pharmacol 63:1256

Gether U (2000) Uncovering molecular mechanisms involved in activation of G protein-coupled receptors. Endocr Rev 21:90–113. doi:10.1210/edrv.21.1.0390

Glukhova A, Thal DM, Nguyen AT et al (2017) Structure of the Adenosine A1 Receptor Reveals the Basis for Subtype Selectivity. Cell 168:867–877.e13. doi:10.1016/j.cell.2017.01.042

González-Calero G, Cubero A, Klotz K-N (1992) G-protein-coupled A_1 adenosine receptors in coated vesicles of mammalian brain: characterization by radioligand binding and photoaffinity labelling. Cell Signal 4:737–745. doi:10.1016/0898-6568(92)90055-D

Grunbeck A, Sakmar TP (2013) Probing G protein-coupled receptor—ligand interactions with targeted photoactivatable cross-linkers. Biochemistry 52:8625–8632. doi:10.1021/bi401300y

Guo D, Pan AC, Dror RO et al (2016) Molecular basis of ligand dissociation from the adenosine A_{2A} receptor. Mol Pharmacol 89:485–491

Hamouda AK, Jayakar SS, Chiara DC, Cohen JB (2014) Photoaffinity labeling of nicotinic receptors: diversity of drug binding sites! J Mol Neurosci 53:480–486. doi:10.1007/s12031-013-0150-1

Hashimoto M, Hatanaka Y (2008) Recent progress in diazirine-based photoaffinity labeling. Eur J Org Chem 2008(15):2513–2523. doi:10.1002/ejoc.200701069

Hatanaka Y (2015) Development and leading-edge application of innovative photoaffinity labeling. Chem Pharm Bull 63:1–12. doi:10.1248/cpb.c14-00645

Hatanaka Y, Sadakane Y (2002) Photoaffinity labeling in drug discovery and developments: chemical gateway for entering proteomic frontier. Curr Top Med Chem 2:271–288. doi:10.2174/1568026023394182

Hindi S, Deng H, James L, Kawamura A (2006) Selective photolabeling of Lck kinase in complex proteome. Bioorg Med Chem Lett 16:5625–5628. doi:10.1016/j.bmcl.2006.08.023

Isberg V, Mordalski S, Munk C et al (2016) GPCRdb: an information system for G protein-coupled receptors. Nucleic Acids Res 44:D356–D364. doi:10.1093/nar/gkv1178

Jaakola VP, Griffith MT, Hanson MA et al (2008) The 2.6 angstrom crystal structure of a human A_{2A} adenosine receptor bound to an antagonist. Science 322:1211

Jazayeri A, Dias JM, Marshall FH (2015) From G protein-coupled receptor structure resolution to rational drug design. J Biol Chem 290:19489–19495. doi:10.1074/jbc.R115.668251

Kecskés M, Kumar TS, Yoo L et al (2010) Novel Alexa Fluor-488 labeled antagonist of the A_{2A} adenosine receptor: application to a fluorescence polarization-based receptor binding assay. Biochem Pharmacol 80:506–511. doi:10.1016/j.bcp.2010.04.027

Kim J, Wess J, van Rhee AM et al (1995) Site-directed mutagenesis identifies residues involved in ligand recognition in the human A_{2a} adenosine receptor. J Biol Chem 270:13987

Kumar TS, Mishra S, Deflorian F et al (2011) Molecular probes for the A_{2A} adenosine receptor based on a pyrazolo[4,3-*e*][1,2,4]triazolo[1,5-*c*]pyrimidin-5-amine scaffold. Bioorg Med Chem Lett 21:2740–2745. doi:10.1016/j.bmcl.2010.11.082

Lee S-M, Booe JM, Pioszak AA (2015) Structural insights into ligand recognition and selectivity for classes A, B, and C GPCRs. Eur J Pharmacol 763:196–205. doi:10.1016/j.ejphar.2015.05.013

LeRiche T, Skorey K, Roy P et al (2004) Using mass spectrometry to study the photo-affinity labeling of protein tyrosine phosphatase 1B. Int J Mass Spectrom 238:99–106. doi:10.1016/j.ijms.2003.11.023

Moss SM, Jayasekara PS, Paoletta S et al (2014) Structure-based design of reactive nucleosides for site-specific modification of the A_{2A} adenosine receptor. ACS Med Chem Lett 5:1043–1048. doi:10.1021/ml5002486

Muranaka H, Momose T, Handa C et al (2017) Photoaffinity labeling of the human A_{2A} Adenosine Receptor and cross-link Position Analysis by Mass Spectrometry. ACS Med Chem Lett 8:660–665. doi:10.1021/acsmedchemlett.7b00138

Myers JS, Sall KN, DuBiner H et al (2016) A dose-escalation study to evaluate the safety, tolerability, pharmacokinetics, and efficacy of 2 and 4 weeks of twice-daily ocular Trabodenoson in adults with ocular hypertension or primary open-angle glaucoma. J Ocul Pharmacol Ther 32:555–562. doi:10.1089/jop.2015.0148

Nikolovska-Coleska Z, Wang R, Fang X et al (2004) Development and optimization of a binding assay for the XIAP BIR3 domain using fluorescence polarization. Anal Biochem 332:261–273. doi:10.1016/j.ab.2004.05.055

Olah ME, Stiles GL (2000) The role of receptor structure in determining adenosine receptor activity. Pharmacol Ther 85:55–75. doi:10.1016/S0163-7258(99)00051-0

Patel A, Craig RH, Daluge SM, Linden J (1988) ^{125}I-BW-A844U, an antagonist radioligand with high affinity and selectivity for adenosine A_1 receptors, and ^{125}I-azido-BW-A844U, a photoaffinity label. Mol Pharmacol 33:585

Piersen CE, True CD, Wells JN (1994) ^{125}I-2-[4-[2-[2-[(4-azidophenyl)methylcarbonylamino] ethylaminocarbonyl]ethyl]phenyl] ethylamino-5′-N-ethylcarboxamidoadenosine labels transmembrane span V of the A_{2a} adenosine receptor. Mol Pharmacol 45:871–877

Pinna A (2014) Adenosine A_{2A} receptor antagonists in Parkinson's disease: progress in clinical trials from the newly approved istradefylline to drugs in early development and those already discontinued. CNS Drugs 28:455–474. doi:10.1007/s40263-014-0161-7

Rask-Andersen M, Almén MS, Schiöth HB (2011) Trends in the exploitation of novel drug targets. Nat Rev Drug Discov 10:579–590. doi:10.1038/nrd3478. http://www.nature.com/nrd/journal/v10/n8/suppinfo/nrd3478_S1.html

Redenti S, Ciancetta A, Pastorin G et al (2016) Pyrazolo[4,3-*e*][1,2,4]triazolo[1,5-*c*]pyrimidines and structurally simplified analogs. Chemistry and SAR profile as adenosine receptor antagonists. Curr Top Med Chem 16:3224–3257. doi:10.2174/1568026616666160506145831

Robinette D, Neamati N, Tomer KB, Borchers CH (2006) Photoaffinity labeling combined with mass spectrometric approaches as a tool for structural proteomics. Expert Rev Proteomics 3:399–408. doi:10.1586/14789450.3.4.399

Rong Y, Arbabian M, Thiriot DS et al (1999) Probing the salmeterol binding site on the β_2-adrenergic receptor using a novel photoaffinity ligand, [^{125}I] iodoazidosalmeterol. Biochemistry 38:11278–11286

Rosa M, Bech-Serra JJ, Canals F et al (2015) Optimized proteomic mass spectrometry characterization of recombinant human μ-opioid receptor functionally expressed in *Pichia pastoris* cell lines. J Proteome Res 14:3162–3173. doi:10.1021/acs.jproteome.5b00104

Shoichet BK, Kobilka BK (2012) Structure-based drug screening for G-protein-coupled receptors. Trends Pharmacol Sci 33:268–272. doi:10.1016/j.tips.2012.03.007

Shonberg J, Kling RC, Gmeiner P, Löber S (2015) GPCR crystal structures: medicinal chemistry in the pocket. Bioorg Med Chem 23:3880–3906. doi:10.1016/j.bmc.2014.12.034

Singh A, Thornton ER, Westheimer FH (1962) The photolysis of diazoacetylchymotrypsin. J Biol Chem 237:PC3006–PC3008

Skorey K, Waddleton D, Therien M, Leriche T (2006) Enzyme occupancy measurement of intracellular protein tyrosine phosphatase 1B using photoaffinity probes. Anal Biochem 349:49–61. doi:10.1016/j.ab.2005.11.018

Stemmer SM, Benjaminov O, Medalia G et al (2013) CF102 for the treatment of hepatocellular carcinoma: a phase I/II, open-label, dose-escalation study. Oncologist 18:25–26. doi:10.1634/theoncologist.2012-0211

Terman BI, Insel PA (1986) Photoaffinity labeling of α_1-adrenergic receptors of rat heart. J Biol Chem 261:5603–5609

Todde S, Moresco RM, Simonelli P et al (2000) Design, radiosynthesis, and biodistribution of a new potent and selective ligand for in vivo imaging of the adenosine A_{2A} receptor system using positron emission tomography. J Med Chem 43:4359–4362. doi:10.1021/jm0009843

Tomizawa M, Maltby D, Medzihradszky KF et al (2007) Defining nicotinic agonist binding surfaces through photoaffinity labeling. Biochemistry 46:8798–8806. doi:10.1021/bi700667v

Tomizawa M, Talley TT, Park JF et al (2009) Nicotinic agonist binding site mapped by methionine- and tyrosine-scanning coupled with azidochloropyridinyl photoaffinity labeling. J Med Chem 52:3735–3741. doi:10.1021/jm900153c

Trnka MJ, Doneanu CE, Trager WF (2006) Photoaffinity labeling of $P450_{Cam}$ by an imidazole-tethered benzophenone probe. Arch Biochem Biophys 445:95–107. doi:10.1016/j.abb.2005.10.014

Whitelegge JP (2009) HPLC and mass spectrometry of integral membrane proteins. In: Walker JM (ed) Protein protocals handbook. Humana Press, Totowa, NJ, pp 1149–1166

Wu Z, Ruoho AE (2000) A high-affinity fluorenone-based β_2-adrenergic receptor antagonist with a photoactivatable pharmacophore. Biochemistry 39:13044–13052. doi:10.1021/bi001342k

Wu Z, Thiriot DS, Ruoho AE (2001) Tyr199 in transmembrane domain 5 of the β_2-adrenergic receptor interacts directly with the pharmacophore of a unique fluorenone based antagonist. Biochemistry 354:485–491

Yip GMS, Chen Z-W, Edge CJ et al (2013) A propofol binding site on mammalian $GABA_A$ receptors identified by photolabeling. Nat Chem Biol 9:715–720. doi:10.1038/nchembio.1340. http://www.nature.com/nchembio/journal/v9/n11/abs/nchembio.1340.html#supplementary-information

Yuan G, Gedeon NG, Jankins TC, Jones GB (2015) Novel approaches for targeting the adenosine A_{2A} receptor. Expert Opin Drug Discovery 10:63–80. doi:10.1517/17460441.2015.971006

Zvonok N, Yaddanapudi S, Williams J et al (2007) Comprehensive proteomic mass spectrometric characterization of human cannabinoid CB2 receptor. J Proteome Res 6:2068–2079. doi:10.1021/pr060671h

MIX
Papier aus verantwortungsvollen Quellen
Paper from responsible sources
FSC® C105338

If you have any concerns about our products,
you can contact us on
ProductSafety@springernature.com

In case Publisher is established outside the EU,
the EU authorized representative is:
Springer Nature Customer Service Center GmbH
Europaplatz 3, 69115 Heidelberg, Germany

Printed by Libri Plureos GmbH
in Hamburg, Germany